Multifunctional Materials for Tribological Applications

Multifunctional Materials for Tribological Applications

edited by

Robert J. K. Wood

Pan Stanford Publishing

Published by

Pan Stanford Publishing Pte. Ltd.
Penthouse Level, Suntec Tower 3
8 Temasek Boulevard
Singapore 038988

Email: editorial@panstanford.com
Web: www.panstanford.com

British Library Cataloguing-in-Publication Data
A catalogue record for this book is available from the British Library.

Multifunctional Materials for Tribological Applications

ISBN 978-981-4463-90-4 (Hardcover)
ISBN 978-981-4463-91-1 (eBook)

Cover image courtesy: Dr. Liam R. Goodes, nCATS,
University of Southampton, UK

Printed in the USA

Contents

Preface

This book contains chapters from multiple authors covering the increasingly important aspect for engineering surfaces to be multifunctional. As designs of machine components require cheaper, lighter, smarter, longer wearing and more environmentally friendly surfaces that see applications that are hotter, faster, highly pressurized and exposed to other increasingly hostile environments, there is a need for engineering surfaces to be multifunctional and usually these are typically now coated systems. This is a challenging area as usually there is antagonism between obtaining low friction and low wear as well as between high corrosion resistance and low wear.

This book captures the state of the art regarding the emerging needs for multifunctional surface design for controlling wear, friction, corrosion as well as having decorative, self-healing and/or self-sensing capabilities. There are ten chapters written by experts in this field covering a range of multifunctional coatings and materials that relate to a very wide range of engineering applications.

The editor is very grateful to all the contributing authors for making this book project happen and for providing such a wide range of research output that truly explores the future potential for surfaces to be multifunctional by design and the value these surfaces will generate for companies designing next generation machines.

Chapter 1 looks at chemical vapour deposition (CVD) diamond and is written by Dave Wheeler from the UK Atomic Weapons Establishment. Chapter 2 looks at the onerous task of thermal barrier coatings for gas turbine blade applications and is written by Richard Wellman, formally of Cranfield University but now at Rolls-Royce PLC. Chapter 3 looks at the ever-growing demand to use glass as an engineering material and the smart coatings that can transform glass to make self-cleaning windows. Such coatings require to perform tribologically not necessarily in use but in transit. Chapter 4 looks at functionally graded and multilayered carbon-based coatings with a view to link the microstructure and tribological performance together and is written by Prof. Wang from Lanzhou Institute of Chemical Physics, China. The micro-tribological performance of high-temperature coatings for the extreme tribological applications found in cutting tools and high-throughput manufacturing are reviewed and discussed by Ben Beake from MicroMaterials in Chapter 5. The need for multifunctional surfaces that control tribocorrosion and lower friction in various engineering and bio-engineering applications is discussed by the editor of this book in Chapter 6. The potential to develop truly multifunctional surfaces using wet deposition techniques to generate nanostructured metallic coatings is discussed by John Low and Frank Walsh in Chapter 7. Chapter 8 looks at low-friction materials and coatings as well as the development of wear and ultra-low friction coatings based on diamond-like carbon and is written by Ali Erdemir from the Argonne National Laboratory in USA. Chapter 9 looks at other thin physical vapour deposited (PVD) coating systems useful for biomedical applications, automotive piston rings as well as cutting tools and is written by Peter Dearnley (formerly of University of Leeds, UK, but now a visiting Professor at nCATS, University of Southampton, UK). The final chapter looks at the role of viscoelastic and plastic deformations in the creep behaviour of UHMWPE used in replacement knee joints and is written by Federico Quinci in collaboration with Georges Limbert of nCATS, University of Southampton, UK, and other co-workers.

Robert J. K. Wood
Spring 2015

Chapter 1

CVD Diamond: A Multifunctional Tribological Material

David W. Wheeler
AWE, Aldermaston, Reading, Berkshire RG7 4PR, UK
david.wheeler@awe.co.uk

The mechanical and tribological behaviour of diamond coatings produced by chemical vapour deposition (CVD) is reviewed. Its high hardness and strength makes diamond a potentially attractive material in components that may be subject to friction and/or wear. In addition to its mechanical properties, the high thermal conductivity, optical properties, corrosion resistance, biocompatibility and anti-bacterial properties of CVD diamond make it a truly multifunctional material. After considering the mechanical properties of CVD diamond the behaviour of diamond coatings when subjected to abrasive, erosive and adhesive wear is discussed. The chapter ends with a discussion of some examples of where diamond has already found application.

1.1 Introduction

Diamond possesses high hardness and strength, both of which are desirable attributes in components that may be subject to abrasive,

Multifunctional Materials for Tribological Applications
Edited by Robert J. K. Wood

ISBN 978-981-4463-90-4 (Hardcover), 978-981-4463-91-1 (eBook)
www.panstanford.com

erosive or adhesive wear. Its hardness is significantly greater than many common abrasives such as silica and alumina, which results in a high resistance to abrasion and erosion. It also results in a correspondingly high resistance in many environments to adhesive (sliding) wear, in which the wear rate is inversely proportional to hardness. Moreover, diamond is a very inert material and is resistant to most forms of chemical attack, which makes it ideal for multifunctional tribological applications where the environment is also highly corrosive (e.g., erosion–corrosion).

Since the emergence of chemical vapour deposition (CVD) of diamond, which has enabled it to be produced in the form of a coating, the number of potential applications of this material has broadened significantly. Although efforts to deposit diamond from the vapour phase began as far back as the 1950s, it was not until the early 1980s [1–3] that it was possible to deposit diamond, first on diamond, and later on non-diamond substrates. Over the intervening three decades, CVD diamond has made the transition from the research laboratory to the production plant and diamond coatings are now commercially available from a number of suppliers. A 2003 study revealed more than 30 active commercial producers of CVD diamond materials [4] with approximately equal numbers in Europe, Asia and North America. In addition to coatings, CVD can also be used to produce diamond in bulk monolithic form. Indeed, high-power microwave systems are now capable of producing CVD diamond more than 1 mm in thickness and up to 200 mm in diameter [5], while hot filament CVD (HFCVD) can deposit diamond coatings on wafers that are up to 300 mm in diameter [6]. In any discussion of diamond coatings it is important to distinguish between CVD diamond and diamond-like carbon (DLC), the latter being a class of amorphous carbon coatings, some of which exhibit hardness values approaching those of diamond. However, DLC is outside the scope of the present chapter and a more detailed description of this material can be found elsewhere in this book.

Diamond films are usually deposited at temperatures of between 800 and 1000°C and at pressures of 10–50 Torr [7]. The reactant gases consist of a dilute mixture of a hydrocarbon—usually methane (CH_4)—and hydrogen (H_2). In order to deposit high quality diamond the methane content generally does not exceed 1% of the total volume of gas. The choice of hydrocarbon gas is not limited to methane: other gases that have been used include acetylene, methanol and acetone. There are a number of variants of the CVD process that have been used to deposit high-quality diamond films including HFCVD, microwave plasma (MPCVD), direct current (DC) arcjet and many others [5]. The main difference between these methods lies in the energy source used to dissociate the reactant gases. The resultant radicals form on the substrate surface as a mixture of amorphous carbon, graphite and diamond. The function of the hydrogen gas is to etch the non-diamond deposits to leave a purely diamond film, although some residual graphite may remain.

Growth rates of the diamond coating vary widely between the various deposition methods. For HFCVD they are generally between 1 and 10 $\mu m\,hr^{-1}$ [5], while recent improvements in microwave technology have resulted in the advent of 50 kW reactors that can deposit MPCVD diamond at rates of 100–200 $\mu m\,hr^{-1}$. Plasma jet and combustion flame methods can deposit coatings at rates of several hundred micrometres per hour; however, they are limited by the small areas and highly non-uniform microstructures and consequently have fallen from favour in recent years.

Diamond films have been successfully deposited on a wide range of substrates including silicon (Si), silicon carbide (SiC), silicon nitride (Si_3N_4), tungsten (W), molybdenum (Mo) and tungsten carbide (WC). Figure 1.1 shows the growth surface of an 87 μm diamond coating on tungsten carbide. The nature of the diamond coatings can be verified using Raman spectroscopy. Figure 1.2 shows a Raman spectrum taken from a 60 μm diamond film deposited on tungsten. The characteristic Raman shift of diamond occurs at 1332 cm^{-1} while a less prominent broader feature centred around 1500 cm^{-1} indicates the presence of graphite in the coating.

As can be seen from Fig. 1.1, individual diamond grains of up to 50 μm in size can be seen in the coating. As a consequence, the surface roughness (R_a) of that coating is more than 5 μm, making it unsuitable for applications involving sliding contact. The rough nature of an as-deposited diamond surface can be seen in Fig. 1.3, which shows an atomic force microscope (AFM) image of an as-deposited 28 μm microcrystalline diamond coating on tungsten. When in sliding contact with a softer counterface such a surface may

Figure 1.1 SEM micrograph showing the as-grown surface of a 87 μm CVD diamond coating on WC-6% Co substrate. The dimension in the scale bar is 100 μm.

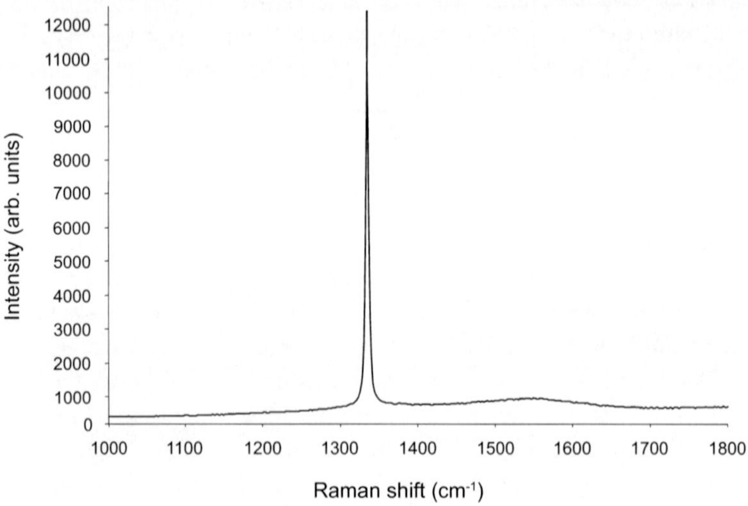

Figure 1.2 Typical Raman spectrum of a CVD diamond coating on a tungsten substrate. The diamond peak is at 1333 cm^{-1}, while the broader hump located at between 1500 and 1600 cm^{-1} denotes the presence of non-diamond carbon in the coating.

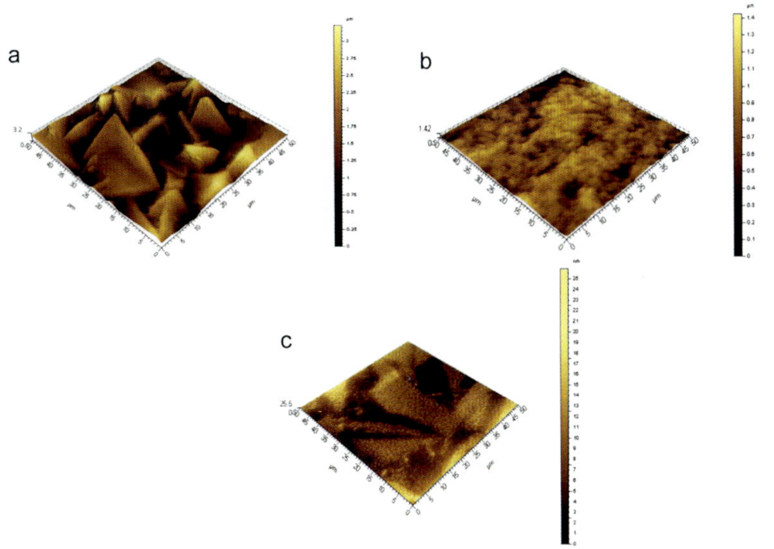

Figure 1.3 Atomic force microscope images of: (a) the deposited surface of a 28 μm CVD diamond coating on tungsten; (b) a 120 μm CVD diamond coating on tungsten which was lapped following deposition; (c) the polished surface of free-standing 500 μm CVD diamond.

be extremely abrasive, leading to high friction and wear. This can be overcome by lapping and/or polishing the diamond film (using either diamond abrasive or chemo-mechanical processes). Figure 1.3 also shows AFM images of lapped and polished CVD diamond, the R_a values of which were 0.2 μm and 8 nm respectively. However, although the lapped and polished surfaces would be suitable for many tribological applications the processes used to achieve such surfaces can be extremely time consuming and are not practicable for non-planar surfaces.

The need for smooth diamond surfaces that do not require post-deposition polishing has resulted in the development of nanocrystalline diamond coatings, which generally have grain sizes of between 10 and 500 nm [8]. Nanocrystalline diamond coatings have been deposited with root mean square (RMS) roughness (R_q) values as low as 10 nm, a desirable attribute for use in micro-electromechanical systems (MEMS), which require thin,

Figure 1.4 SEM micrograph of a nanocrystalline diamond film (left). A microcrystalline diamond film (right) is also included for comparison. Pictures reproduced with the permission of Prof. P. W. May (University of Bristol, England).

continuous, conformal coatings with exceptionally low surface roughness [9].

There are several ways to deposit nanocrystalline diamond. One way is to increase the CH_4 content in the CH_4/H_2 gas mixture from 1% up to approximately 5% [10]. This causes the micrometre-sized diamond grains to decrease by as much as two orders of magnitude. The increased numbers of grain boundaries results in the nanocrystalline diamond films having higher graphite content than the microcrystalline diamond films. Nevertheless, they are still overwhelmingly diamond in both nature and properties. Nanocrystalline diamond films have small grain sizes and the grains are non-facetted with regions of graphite at the grain boundaries. Figure 1.4 shows a scanning electron micrograph of a nanocrystalline diamond film (for comparison a microcrystalline diamond coating taken at the same magnification is also shown). The surface roughness of nanocrystalline diamond can be as low as 50–100 nm [9]. Density measurements have shown that nanocrystalline diamond films exhibit densities close to that of natural diamond. Using surface acoustic wave dispersion measurements, Philip et al. [11] obtained a density of 3.47 ± 0.15 g cm^{-3} for nanocrystalline diamond on silicon. This compares with 3.52 g cm^{-3}, the value for single crystal diamond.

An alternative approach to that outlined above is to use different gas mixtures such as argon (Ar) together with either CH_4 or fullerene (C_{60}). Erdemir et al. [12] used a MPCVD system with these gas mixtures. A typical gas composition was 97% Ar, 2% H_2 and 1% C_{60} or CH_4 at a pressure of 1.33×10^4 Pa; the substrate temperature was between 700 and 950°C. The resultant films had grains of between 10 and 30 nm in size with an RMS surface roughness of between 20 and 40 nm. However, even these impressive figures have been surpassed by ultrananocrystalline diamond films, which can have grain sizes of 2–5 nm and surface roughnesses of 20–40 nm [9].

As with any coating system an important consideration is that of residual stress, which can influence both the life of the coating as well as its mode of deformation. The total film stress is a combination of the intrinsic and thermally induced components [13]. Intrinsic stresses arise from the presence of impurities and structural defects in the coating and the structural mismatch between coating and substrate. Thermal stresses are caused by the differences in thermal expansion coefficient between the coating and the substrate. For ceramic coatings on metal substrates the residual stresses are typically between 1 and 3 GPa [14] and can be either tensile or compressive depending on the nature of the coating and substrate.

In the case of CVD diamond, although variations are seen through the coating thickness [15], for coatings deposited on most substrates the overall residual stress is generally compressive, the most prominent exception being Si_3N_4 substrates [16]. One simple method of assessing the residual stress in diamond coatings is by Raman spectroscopy. The overall residual stress can cause a shift in the position of the characteristic Raman peak to higher or lower wave numbers depending on whether the stress is tensile or compressive. Tensile stresses cause a downward shift in the wave number while an upward shift indicates the presence of a compressive residual stress [17]. This line varies linearly with pressure with a shift of 2.87 cm^{-1}/GPa [18]. Using Raman spectroscopy of CVD diamond coatings deposited on tungsten substrates, the present author found the overall residual stress to be compressive, the magnitude of which was estimated to be between 0.84 and 1.3 GPa [19]. Although excessive residual stresses can result in buckling or delamination of the coating, studies have shown that moderate compressive stresses can be beneficial, for example to increase the tensile stress required to form surface cracks. Roberts [20] investigated the effect of compressive residual stress on the fracture load in a Hertz indentation test indenting glass with a 5 mm diameter stainless steel sphere. He found that the minimum load for fracture of chemically strengthened glass (which had a surface compressive residual stress of 450 MPa) was approaching an order of magnitude higher than that for annealed glass.

1.2 Mechanical Properties of Diamond

Diamond has the highest resistance to plastic deformation of any material, having an elastic modulus (1050 GPa) approximately five times that of stainless steel. It is also the hardest material known to man, with a hardness that can be as high as 100 GPa. In contrast, the most common abrasives such as silica (SiO_2) and alumina (Al_2O_3) exhibit hardness values of approximately 13 GPa and 26 GPa, respectively [21]. As a result, the high hardness and strength of diamond make it an attractive option for components requiring high resistance to wear.

The mechanical properties of CVD diamond are, in many cases, comparable to those of its single-crystal counterpart, although it cannot match it in all respects. Properties such as hardness, elastic modulus and fracture toughness can match, or even exceed, those of natural diamond. The presence of grain boundaries can be beneficial as they can in some instances arrest crack growth. However, the strength of CVD diamond is inferior to natural diamond owing to the presence of microstructural flaws and residual stresses.

One advantage that CVD diamond does possess is that its polycrystalline nature renders it less sensitive to crystallographic orientation. This is in contrast to single-crystal diamond, the hardness of which is dependent on the crystallographic plane that is indented. Brookes [22] found that the values achieved varied from 56 GPa on the (111) to 108 GPa on the (110), the plane with the greatest hardness. This was thought to be due to differences in critical resolved shear stress in different crystallographic directions.

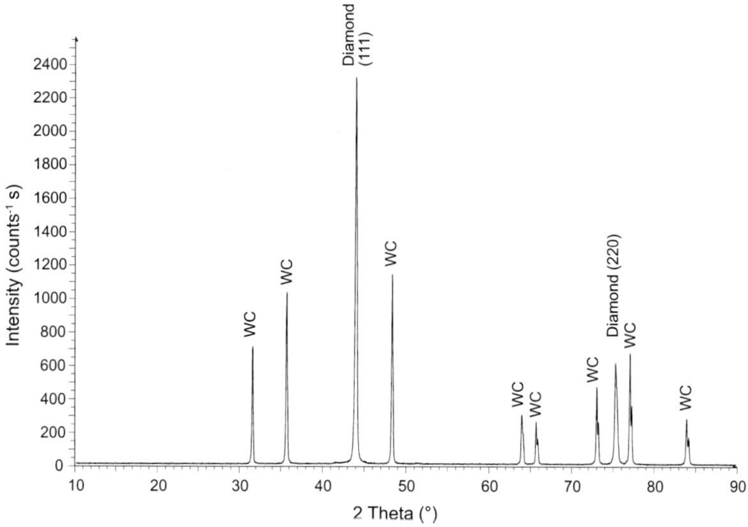

Figure 1.5 X-ray diffraction pattern of a 30 μm CVD diamond coating on tungsten carbide.

Furthermore, the fracture behaviour in single-crystal diamond is dominated by cleavage on the (111) plane, although cleavage can also be induced on other planes [23]. Figure 1.5 shows an X-ray diffraction pattern of a 60 μm CVD diamond coating on tungsten. It can be seen that the grains are orientated predominantly in the (111) and (220) directions. Therefore, this dependence of hardness on orientation is less important than for single-crystal diamond.

Nanoindentation measurements of microcrystalline CVD diamond have recorded hardness values and elastic modulus values as high as 105 GPa and 1147 GPa respectively [24]. Figure 1.6 shows a typical load–depth curve [25] from a nanoindentation test of CVD diamond. These values compare with those of 90 GPa and 1050 GPa for natural diamond [23]. Similarly, a superior fracture toughness of 6.0 MPa$\sqrt{m}$ for CVD diamond has been recorded [26], which compares with values of between 3.4 and 5.0 MPa$\sqrt{m}$ for natural single crystal diamond [23].

The mechanical properties of nanocrystalline diamond coatings have also been measured. Wiora et al. [27] used nanoindentation to measure the hardness and elastic modulus of nanocrystalline

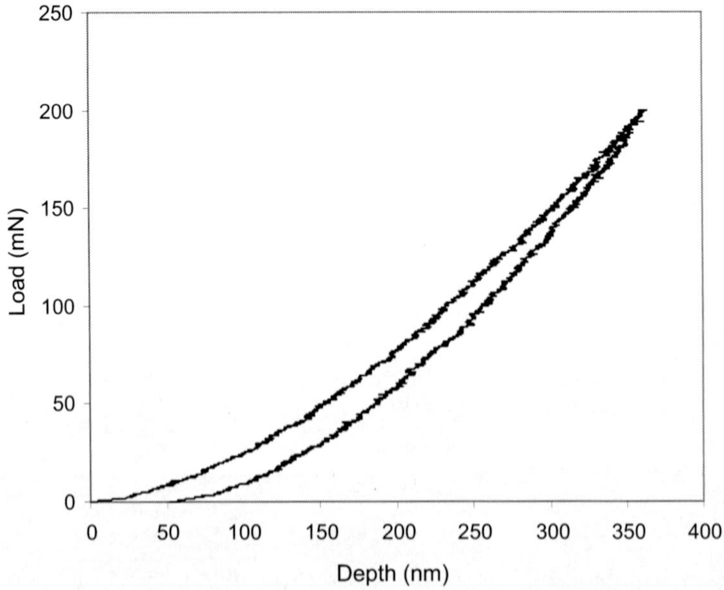

Figure 1.6 Load-depth curve from the nanoindentation test of a 600 μm CVD diamond brazed to tungsten carbide and polished to a surface roughness (R_a) of 8 nm. The hardness was 77.6 GPa and the reduced modulus (E_r) was 578 GPa.

diamond coatings deposited on silicon substrates using a HFCVD process. The coatings were 6 μm in thickness and the grain sizes ranged from 9 to 60 nm. The hardness ranged from 68 to 86 GPa, while the elastic modulus varied from 597 GPa to 857 GPa. Both parameters were observed to decrease with decreasing grain size, a finding attributed to the role of grain boundaries and the existence of grain boundary related defects such as hydrogen impurities and sp^2 carbon. The elastic modulus was also measured using a laser surface acoustic wave technique: the largest grain size specimens exhibited values close to that of natural diamond (1050 GPa), decreasing to about 700 GPa at smaller grain sizes.

Williams et al. [28] used a bulge test to measure the elastic modulus of free-standing nanocrystalline diamond films thinner than 140 nm. They found that the modulus was strongly dependent on the methane content used in the diamond growth process.

The highest moduli (950–1100 GPa) were measured on the lowest methane content (0.5%), dropping to 400–700 GPa at 20% methane. Philip et al. [11] used surface acoustic wave dispersion measurements to measure the elastic modulus of nanocrystalline diamond films deposited on silicon substrates. The films were between 0.6 μm and 4.1 μm in thickness and were grown using methane and hydrogen in a MPCVD reactor. They found wide variations in elastic modulus, from 517 GP to 1120 GPa. The elastic modulus was found to be strongly dependent on nucleation density with the highest values recorded on specimens where the nucleation density was $\geq 10^{12}$ cm^{-2}; however, for films grown with nucleation densities of 10^{10} cm^{-2} or less the modulus was less than half this figure.

In contrast to hardness, elastic modulus and fracture toughness, CVD diamond cannot match the strength of single crystal diamond. The main reason for this is the microstructure of CVD diamond, which contains significantly higher levels of defects (in the form of microfractures) as well as those defects being larger in size (similar size to individual grains). As a result, whereas strength values as high as 3.75 GPa have been recorded for natural diamond [23], the strengths of CVD diamond specimens are less than half this figure. Sussmann et al. [26] conducted pressure burst tests of CVD diamond discs 14 mm and 24 mm in diameter and 179–300 μm in thickness. The results showed that the fracture stress values were in the range 746–1138 MPa with a general trend of thinner samples requiring higher burst pressures.

It should be noted that when quoting the strength of CVD Diamond, it is important also to state the test method by which the data were obtained [29, 30]. This is because a much broader range of flaw sizes can be activated when large volumes are stressed, as is the case with the burst test. As a consequence, lower strengths are often recorded when using this test method compared to tests that stress smaller volumes, such as the indentation test.

1.3 Abrasive Wear

Abrasion is the removal of solid material from a surface by the unidirectional sliding action of discrete particles of another material

[31]. Industrial surveys have revealed abrasion to be the most common cause of wear-related component failures, accounting for more than half the total number [32].

Abrasion occurs between two surfaces in relative motion and can be split into two main categories: (i) two body and (ii) three body. Two-body (or 'grooving') abrasion is caused by hard protuberances on the counterface, while in three-body (or 'rolling') abrasion hard particles are free to roll and slide between two sliding surfaces [33]. Of the two, the increased contact pressures associated with two-body abrasion renders it the more severe form: as a result, wear coefficients are approximately one order of magnitude greater than those for three-body abrasion [34].

The excellent abrasion resistance of diamond coatings has been demonstrated in a number of laboratory studies. Most of these studies have used the micro-abrasion test [35–37] in which a ball is loaded against the coating and rotated in the presence of an abrasive slurry. This test is generally regarded as more suitable for thin coatings than the more traditional abrasion tests such as the dry sand rubber wheel test as it enables the abrasive wear performance of thin coatings to be more easily differentiated. In concert with the abrasive slurry, the rotating sphere causes the formation of a ball-shaped crater in the coating. Together with the sliding distance and the applied load, both of which are known, the crater diameter can be used to give a measure of the specific wear rate (SWR) of the coating and, for thin coatings that are penetrated during the test, of the substrate too.

In micro-abrasion tests CVD diamond coatings have been found to exhibit a wear rate of more than an order of magnitude lower than other CVD coatings such as boron carbide ($B_{13}C_2$) and titanium nitride (TiN) [36]. In a comparison with CrN and TiN coatings Gahlin et al. [35] found that the margin of superiority was even greater at more than three orders of magnitude. They also found that the presence of a compressive residual stress in the coating was not always undesirable as the wear rates of the highly stressed diamond coatings on cemented tungsten carbide substrates were between 5% and 20% of those recorded in tests of the stress-free coatings. They concluded that the compressive stress realigns cracks in a direction parallel to the coating–substrate interface. This prevents cleavage

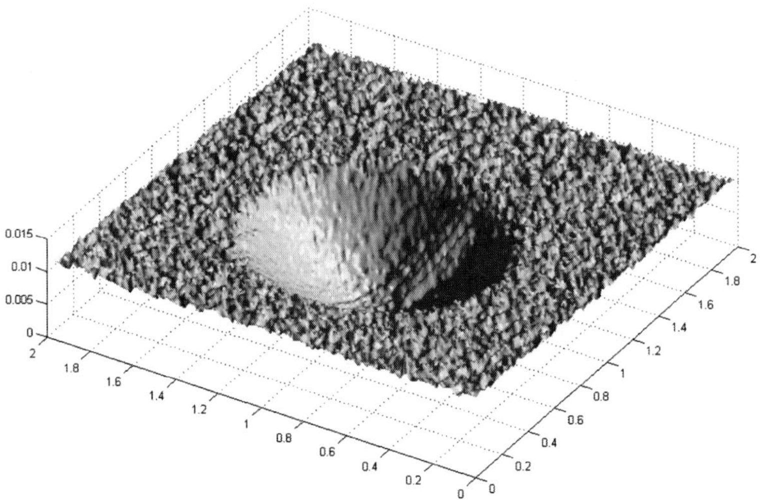

Figure 1.7 Three-dimensional surface profile of a micro-abrasion crater generated in a 120 μm lapped CVD diamond on tungsten. Area of figure: 2 mm × 2 mm (dimensions in all three axes are in mm).

along the (111) planes and hence reduces the wear rates. As the test continues, the surface roughness reduces and this is accompanied by a reduction in wear rate as the contact pressure reduces.

Figure 1.7 shows a three-dimensional surface profile of a crater generated in a micro-abrasion test of a lapped (surface roughness, R_a, of 0.2 μm) diamond coating 120 μm in thickness on a tungsten substrate. The test was conducted using 6 μm diamond abrasive and a 25 mm diameter tungsten carbide ball, which was rotated for 10,000 revolutions. The load was 0.3 N and the rotation speed of the ball was 100 rpm, giving a sliding speed between 1 m s^{-1} and 5 m s^{-1}. It can be seen from Fig. 1.7 that the scar is largely symmetrical with a maximum depth of approximately 13 μm. Also visible is a ridge running through the centre of the scar where the coating has not worn as much as on the adjacent regions. An electron micrograph of the crater can be seen in Fig. 1.8. It can be seen that there are no obvious indications of grooving or scoring, which suggests that the abrasive particles were not embedded in the surface of the ball. Instead, the coating appears to be worn by a three-body abrasive wear process, which results in micro-

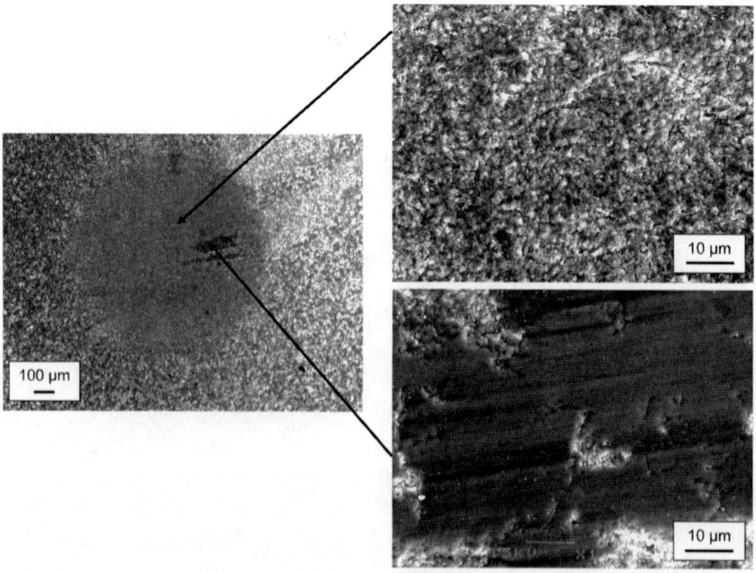

Figure 1.8 SEM micrograph of the wear crater shown in Figure 1.7.

chipping of the diamond grains. The ridges have also been seen in micro-abrasion studies of other materials; Shipway and Hodge [38] have reported similar observations in soda-lime glass. Their formation was attributed to high loads and low slurry viscosity, which prevented effective entrainment of the abrasive particles into the contact resulting in wear that was caused only by a sliding action against the ball.

The high abrasion resistance of CVD diamond has led to it being used in dressing tools for grinding wheels containing abrasives such as Al_2O_3 or SiC. CVD diamond dressing tools have been shown to outlast natural diamond dressers and grinding wheels dressed by CVD diamond tools require less frequent dressing operations [39]. Another application is as abrasive pencils used in the manufacture of micro-mechanical components [40]. The high abrasion resistance of CVD diamond also makes it a suitable coating for tools used to machine materials with an abrasive character such as aluminium–silicon (Al–Si) alloys and metal matrix composites.

1.4 Erosive Wear

1.4.1 *Solid Particle Erosion*

Solid particle erosion occurs in materials subjected to impacts of particles such as sand at velocities of up to several hundred metres per second. It is often a problem in fluidised bed combustion systems, helicopter blades (especially in desert environments) and pipelines and valves used for the transport of liquids and gases in which particulate matter is also present. Over the last few decades the need for more erosion-resistant components has prompted extensive investigations in which the erosion resistance of various hard coatings have been evaluated.

The high hardness and strength of diamond coatings confer a high resistance to erosion from solid particles such as sand. Indeed, the hardness and elastic modulus of CVD diamond can be up to ten times greater than those of silica sand. As a result, the steady state erosion rates of diamond coatings have been shown to be up to 30 times lower than those of cemented tungsten carbide [41–43] while the damage to the impacting particles often exceeds that of the coating. High-speed photography conducted by the present author of silica sand particles impacting on diamond coatings has shown individual particles to fracture into as many as six pieces on impact, even at moderate velocities ($<50\,\text{m s}^{-1}$). At higher velocities, the mean particle diameter has been seen to reduce by more than 60%. As a consequence, the stresses generated in the target by the impacts are insufficient to initiate the erosion damage commonly seen in brittle materials such as radial and lateral cracks and/or Hertzian ring cracks. Erosion damage is most common in the form of micro-chipping of grains. An example of this can be seen in Fig. 1.9, which shows a 30 μm diamond coating on tungsten carbide after 5 hr of erosion by silica sand at a mean impact velocity of $268\,\text{m s}^{-1}$; the same specimen prior to erosion testing is also shown. Other workers have observed different damage features. Feng et al. [44] showed that lower velocity ($34\,\text{m s}^{-1}$ and $59\,\text{m s}^{-1}$) impacts from silica sand particles (300–600 μm in size) on diamond coatings caused the formation of circular ring cracks, the diameters of which could be explained by the Hertz theory of dynamic impact [45].

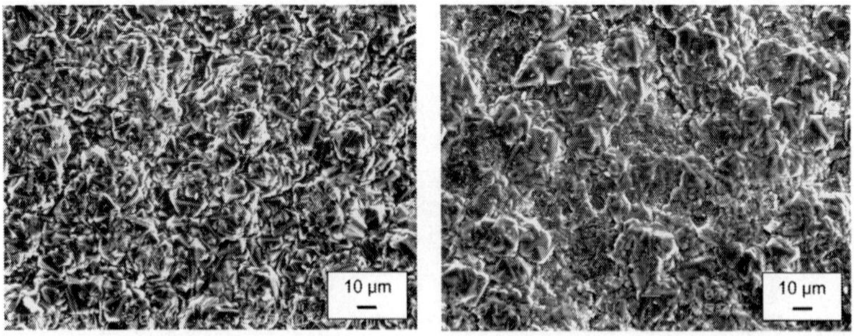

Figure 1.9 SEM micrographs showing a 30 μm CVD Diamond coating on tungsten carbide in the untested condition (left) and after 5 hours of sand erosion at a mean particle impact velocity of 268 m s^{-1} (right).

Although the surface damage to the coating can appear to be modest, the impacting particles also generate sub-surface shear stresses, which also can be explained by the Hertz impact theory [46]. These shear stresses can cause coating delamination via the propagation of cracks at the coating–substrate interface. As the particle impacts continue, this debonding, which can only be observed using scanning acoustic microscopy [47], can later lead to catastrophic failure of the coating.

An important factor governing the life of the coating in erosive conditions is the normalised depth of maximum shear stress z_τ/C_T, where z_τ is the depth of maximum shear stress and C_T is the coating thickness. A study of 30 μm diamond coatings on tungsten subjected to erosion from silica sand at 250 m s^{-1} by Wheeler and Wood [46] found that when the maximum shear stress occurred close to the coating–substrate interface (i.e., $z_\tau/C_T = 1$) the coating failed rapidly and catastrophically. As z_τ/C_T was reduced (i.e., the shear stress acts within the coating) an exponential increase in the number of impacts necessary for coating failure was observed. Moreover, the results showed that there appeared to be a critical normalised depth of maximum shear stress (z_τ/C_T), below which the life of the coating was limited only by the low rate of micro-chipping by the impacting sand particles. For optimum coating performance $z_\tau/C_T < 0.3$ and the stress field contained within the coating. Such findings may be applied to assist the design of coated components intended for use in erosive environments.

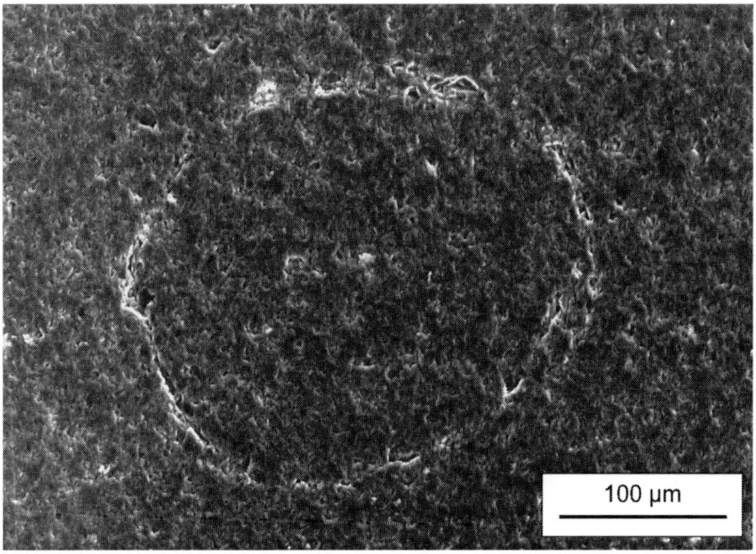

Figure 1.10 SEM micrograph of a typical circumferential crack in the surface of an eroded diamond coating subjected to 63–600 μm silica sand at a mean impact velocity of 268 m s^{-1}.

In addition to causing coating failure, the debonded regions of the eroded coating can also result in the formation of circumferential cracks on the coating surface, an example of which can be seen in Fig. 1.10. These features have been investigated extensively by the present author [42, 47, 48]. In contrast to the findings of Feng et al. [44], the diameters of these circumferential cracks did not agree with those predicted by the theory of dynamic Hertzian contact, being more than double the predicted diameters [42]. Wheeler and Wood [48] demonstrated that the circumferential cracks were formed by the reflection and reinforcement of stress waves at locally debonded regions of the coatings. Comparison of the measured circumferential crack diameters with the theory of stress wave reflection and reinforcement showed good agreement. After the formation of the circumferential cracks, the initiation of damage in the central region within the cracks is often seen, resulting in pin holes in the coating. As the particle impacts continue, growth of this central damage region occurs both vertically (towards the coating–substrate interface) and laterally (towards the circumferential crack). These pin holes therefore appear to provide a visible indication that coating delamination is taking place at the interface, which may be followed by catastrophic failure of the coating.

The high erosion resistance of diamond coatings makes them potentially very attractive for use in components such as choke valves, which are used in the offshore oil industry and which suffer from sand erosion. In a comparison of choke valve trims containing CVD diamond with standard uncoated tungsten carbide valve trims in a high pressure valve loop test facility, valves containing CVD diamond were shown to offer a 50–60% improvement in performance [49]. However, it should be noted that the diamond used in that study was free-standing CVD diamond (0.6 mm in thickness) that was bonded to the WC trims using a high temperature adhesive. The deposition of diamond films onto some non-planar substrates remains a formidable challenge.

1.4.2 *Liquid Droplet Erosion*

Liquid droplet erosion, in which the impact of fast-moving droplets onto a solid surface can result in damage to the target material, is a particular problem in high-speed aircraft, missiles, steam turbine blades and wet gas wells [32]. Impacts from high speed liquid droplets can result in high pressures being generated in the target material. For an impact of a 1 mm radius drop at 330 m s^{-1} (Mach 1) onto a rigid surface, the central pressure reaches a value of 0.6 GPa and lasts for approximately 0.11 μs [50]. Such impacts on brittle materials typically result in the formation of damage in the form of rings of discontinuous cracks surrounding the area of impact. The cracks are caused by the propagation of stress waves (longitudinal, transverse and Rayleigh waves) that are released into the target material a short time after impact. The Rayleigh surface waves, which are tensile in nature, account for the bulk (67%) of the energy associated with impact with the remainder made up by transverse waves (26%) and longitudinal waves (7%) [51]. Their interaction with pre-existing surface flaws larger than a certain critical size results in the formation of circumferential cracks in the target material. In thin targets ($<$1 mm in thickness), the stresses associated with the Rayleigh surface waves can be augmented by reflected longitudinal and transverse waves [52]. The nucleation of these cracks in the target material leads to strength degradation and, where cracks intersect, material removal. Tests by van der Zwaag and Field [53] on soda lime glass subjected to impacts from water droplets have found that the post-impact fracture stress was considerably lower than that of unimpacted specimens.

The damage that can result in components subjected to high-speed liquid droplet impacts has led to extensive research to develop materials and coatings that can offer enhanced resistance to damage of this nature. The high erosion resistance of CVD diamond makes it an attractive material for use in infrared transparent windows and domes, which are used to shield the sensor components of airborne electro-optic systems [54]. Such components can be subjected to impacts from rain droplets travelling at several hundred metres per second. The materials used in these components may be exposed to rain, sand, hail and dust at velocities of up to Mach 2 when carried on an aircraft and at Mach 4 and beyond when mounted on a launched missile [55]. Two of the most commonly used infrared materials zinc sulphide (ZnS) and germanium (Ge), despite their excellent infrared transmission properties, suffer from poor mechanical properties such as strength and fracture toughness. This has led to the development of materials such as CVD diamond as an alternative to ZnS and Ge to reduce the incidence of rain erosion suffered by these components.

In addition to its mechanical properties, CVD diamond also exhibits other properties that render it an attractive window and dome material. It has a high index of refraction and the ability to transmit light from the far infrared to the ultraviolet range. Moreover, its high thermal conductivity (five times that of copper) and its elastic modulus also result in a high thermal shock resistance. The thermal shock figure of merit for diamond is approximately 100 times greater than those of ZnS or sapphire, meaning that diamond can withstand $\sim$100 times the heat flux of the other materials without shattering [56].

A number of investigations have been carried out assessing the liquid impact behaviour of infrared-transmitting materials, including diamond. The parameter used for quantifying the performance of samples subjected to liquid impact is the damage threshold velocity (DTV). The DTV is defined as 'the velocity below which the material remains undamaged regardless of the number of impacts' [57].

Although CVD diamond can deliver significant improvements in DTV compared to ZnS and Ge it cannot match that of natural diamond. Tests on type IIa natural diamond, using 0.8 mm water droplets, were conducted over a range of velocities, beginning at 260 m s^{-1} [58]. After 10 impacts, and following microscopic examination, the velocity was increased in steps of 25 m s^{-1} up to a maximum of 530 m s^{-1}. The sample was still intact at this velocity after 10 impacts but shattered after a further 160. This figure compares favourably with the value of 525 ± 25 m s^{-1} obtained on a synthetic high-temperature high-pressure type Ia diamond [57]. However, this figure may be artificially low due to the small specimen size, which would result in stress wave reflections at the edges and reinforcing the damage induced by the impacts.

Liquid impact studies of CVD diamond have shown a steady improvement in performance over the last few years but the DTV of CVD diamond has yet to reach that of natural diamond. Seward et al. have reported a threshold velocity of 475 ± 25 m s^{-1} for a free-standing disc 500 μm in thickness [50]. However, the highest figure obtained is that recorded by Coad et al. [55] who quoted a DTV for CVD diamond on sapphire of 565 ± 15 m s^{-1}.

Diamond has been deposited on infrared window materials such as ZnS and Ge. Hudson et al. [59] coated Ge windows used in the forward looking infrared imaging system on the Harrier GR7 and AV8B aircraft with diamond. The window was 73.4 mm diameter and 5 mm thick, while the diamond coating was between 5 and 10 μm in thickness. Free-standing diamond domes have also been produced. Wort et al. [60] reported the synthesis of fully polished, crack-free hemispherical diamond domes of 70 mm diameter (free-standing CVD diamond). The dome was approximately 700 μm in thickness after shaping and polishing, which was about half the desired thickness for an in-service component. Nevertheless, its combination of erosion resistance, optical properties and thermal shock resistance make CVD diamond a truly multifunctional material for infrared windows and domes.

1.4.3 *Cavitation Erosion*

Cavitation is the name given to the repeated growth and collapse of bubbles (or cavities) in a fluid due to local pressure fluctuations. Generally, cavities form in low pressure regions where the local pressure is equal to the saturated vapour pressure. When a cavity collapses on, or close to, a solid surface it does so asymmetrically, resulting in the formation of a microjet of liquid. These microjets with a diameter of approximately 1 μm damage the surface by generating high-impact pressure (water hammer), and repeated bubble collapses on the surface lead to fatigue failure and material ejection; the associated pressures can be up to 10^5 bar [61]. When material is removed from the target surface, this process is known as cavitation erosion. Cavitation erosion is a particular problem for liquid circulation equipment (e.g., pumps, valves and pipes) as well as bearings and cylinder linings of high-speed reciprocating engines [32].

The erosion of materials subjected to cavitation tests typically proceeds through four stages, although researchers have disagreed as to the nature of these stages. According to one school of thought, the first stage is the 'incubation' period, during which the extent of erosion is negligible. This is followed by the 'accumulation' period, in which the erosion rate increases to a maximum level. The third stage is that of 'attenuation,' in which the rate of material loss decreases to the final stage, namely the 'steady state' region. An alternative description has divided the process into the following four stages: (i) 'incubation'; (ii) 'transition,' during which erosion begins locally and the observed erosion rate accelerates as it expands over the whole test area; (iii) 'constant rate' stage; and (iv) 'decreasing rate' stage, during which the pits formed during the 'constant rate' stage deepen and the cavitation collapse is attenuated by trapped air or water. The various stages of cavitation erosion are discussed in detail by Preece [62].

According to the studies in the literature, the cavitation erosion of diamond coatings appears to follow the second of the above modes of behaviour, although the final 'decreasing rate' stage is not seen. Indeed, far from the erosion rate decreasing, an acceleration may be observed if the coating is penetrated, thereby exposing a much less erosion-resistant substrate. During exposure to cavitating conditions, after the incubation period, erosion damage in diamond coatings is initiated at defects and rougher regions of the coating. The erosion damage features then increase in size before later coalescing with other similar regions. This is followed by penetration of the coating.

Deuerler et al. [63] investigated the cavitation erosion behaviour of 3 μm microcrystalline and nanocrystalline diamond coatings on tungsten carbide substrates. The frequency was 20 kHz and the amplitude was 40 μm, while the 5 mm diameter sonotrode (the vibrations of which created the cavitation bubble field in front of the specimen) was situated 0.5 mm from the coating surface. The nanocrystalline diamond film offered better erosion resistance than the microcrystalline coating of the same thickness, which was attributed to its lower surface roughness. Indeed, even after a cavitation test of the nanocrystalline coating lasting nearly 14 hr, the surface damage, which was measured by quantitative image analysis of the eroded surfaces, was still below 1%.

Bregliozzi et al. [64] used a similar cavitation test to that of Deuerler et al. [63] (under the same test conditions) to evaluate the adhesion of nanocrystalline diamond coatings (approximately 1 μm in thickness) on silicon substrates that had been subjected to three different surface treatments prior to coating deposition. The Si substrates were single-crystal *p*-type <100>wafers that were in the rough as-cut condition, lapped or chemically mechanically planarised (CMP). Following deposition, the mean surface roughness (R_a) values of the diamond coatings on each substrate were 260, 186 and 57 nm, respectively. They found that all three coatings behaved in the same manner with an initial incubation period when the extent of the surface damage was negligible. However, the length of this incubation period was longer for the coatings deposited on the smoother substrates. Cavitation erosion resistance was inversely proportional to the roughness of the diamond coatings. The erosion damage was in the form of micrometre-sized pits that form first at rough areas of the coating before propagating rapidly across the coating becoming large pores. The diamond deposited on CMP silicon was the most erosion-resistant coating. Its superior performance was attributed to the low roughness of the substrate which resulted in a more uniform coating with better adhesion.

1.5 Adhesive Wear

Adhesive wear occurs when the contacting asperities of two surfaces sliding against each other form junctions, transmit tangential motion and subsequently fracture. On detachment, the material emerges as wear debris [65] or is transferred between the sliding surfaces. Adhesive wear accounts for the second largest number of wear-related component failures after abrasion.

1.5.1 *Friction of Diamond*

Natural diamond exhibits a low coefficient of friction (μ) sliding against itself. With the exception of high vacuum, where it can exceed 1.0, values of μ of between 0.05 and 0.15 have been recorded in most environments [30]. However, the friction behaviour of diamond on diamond is dependent on a number of factors including atmosphere, load, sliding contact geometry, crystallographic orientation of the diamond, direction of sliding, surface preparation and repeated sliding over the same path [66, 67].

One explanation for the low friction of diamond has been the termination of dangling bonds at the surface by hydrogen atoms. When this occurs on both sliding surfaces, the two sets of hydrogen atoms shear past each other, chemical bonding is prevented and a low coefficient of friction ensues. Work by Martin et al. [68] using a pin-on-flat test in an ultra-high vacuum (UHV) chamber equipped with surface analytical instrumentation has shown that the adsorption of oxygen-containing species onto the diamond surface results in the same behaviour. When friction tests of diamond sliding against diamond are performed in the presence of water the coefficient of friction was greatly reduced. Field and Pickles [30] used a reciprocating pin-on-flat and found that the coefficient of friction stabilised at approximately 0.05: when the test was performed in air, it was more than double this figure (0.12).

CVD diamond coatings have been shown to exhibit similar frictional behaviour to natural diamond. In friction force measurements in ambient air (relative humidity ~40%) using AFM, the behaviour of ultrananocrystalline diamond was found to be indistinguishable from single-crystal diamond at the nanometer scale [69]. However, owing to the polycrystalline nature of CVD diamond films, the effect of crystallographic orientation on friction coefficient is less evident. Moreover, the coefficient of friction is heavily dependent on the surface roughness of the coating. Hayward et al. [70] measured the friction of a reciprocating diamond stylus sliding (at a speed of $0.3\ \text{mm s}^{-1}$) against diamond coatings on Si of various roughnesses. The initial radius of the sphere was 200 μm and the nominal force was 2 N. They found that rougher coatings generated higher friction coefficients. One test of a coating having an initial R_a of 210 nm recorded an initial friction coefficient of approximately 0.4. By the end of the test, the friction coefficient had reduced by half its initial value and the R_a had decreased to 56 nm. Another test of a smoother coating (initial R_a 7.1 nm) recorded an initial friction coefficient of 0.1, which reduced to approximately 0.05 (final R_a 1.4 nm). These values compare with typical friction coefficients of between 0.04 and 0.08 for diamond styluses rubbing against single crystal (100) diamond. Similar observations were reported by Bull et al. [71], who found that the coefficient of friction for an as-grown coating stabilised at around 0.2; when the coating was polished, it stabilised at approximately 0.05. The as-grown coatings also took longer to stabilise. Another factor that can affect the friction coefficient is the purity of the diamond coating. Perry et al. [72] found that diamond films with a higher amorphous carbon content exhibited reduced friction under UHV (1×10^9 Torr) when sliding against a hemispherical diamond tip. The coefficient of friction was in the region of 0.6 on coatings with substantial levels of amorphous carbon, rising to approximately 0.85 on coatings with greater levels of sp^3-bonded diamond. However, in tests conducted in ambient atmosphere, the opposite, albeit less marked, trend was observed: the coefficient of friction was observed to decrease from ~0.19 to ~0.16 with decreasing amorphous carbon content.

In tests involving as-grown CVD diamond sliding against itself, an initially high value of friction coefficient is often seen. This is particularly apparent in micro-crystalline diamond films, where the initial surface roughness (R_a) can be more than 1 μm and has been attributed to the mechanical interlocking of surface asperities [73, 74]. However, as the sliding continues, the protruding diamond asperities become increasingly truncated, exposed edges blunted and the surface roughness decreases. The presence of any large debris particles can cause grooving of the worn coating surface, and self-polishing at a micro-scale level by two- and three-body abrasion also takes place [75]. These processes are often accompanied by a progressive reduction in the coefficient of friction until the steady-state sliding regime is attained. An example of this can be seen in the work of Bull et al. [71] who conducted tests of as-grown CVD diamond on Si substrates. They found that the initial coefficient of friction was in the region of 0.35, before declining and stabilising at between 0.1 and 0.2. Once steady-state conditions are achieved, the coefficient of friction remains approximately constant, although if coating failure occurs it may increase once more.

Many of the experiments described above were conducted under low loads and low sliding velocities. However, while they allow the friction behaviour to be elucidated, the test conditions are hardly representative of in-service conditions experienced by many components, for example mechanical seals. A more relevant study for such components was that undertaken by Schade et al. [74], who used a ring-on-ring test to study the friction and wear behaviour of self-mated diamond on SiC. The coatings were between 2.4 and 13 μm in thickness and were of four different surface roughnesses, ranging from 0.32 μm (micro-crystalline diamond) to 0.09 μm (nanocrystalline diamond). A force of 70 N (equivalent to a load of 0.4 MPa) and a sliding speed of 2.0 m s^{-1} for a total sliding distance of 10,000 m were the test parameters. Although they found that friction declined from initially high levels (approximately 0.55) to lower steady-state values (~0.3) over the course of the tests, large fluctuations in the coefficient of friction were often seen prior to a steady-state value being reached. Moreover, the temperature near the sliding contact (measured by a thermocouple placed at the outer shell face of the static ring) area rose rapidly soon after the start

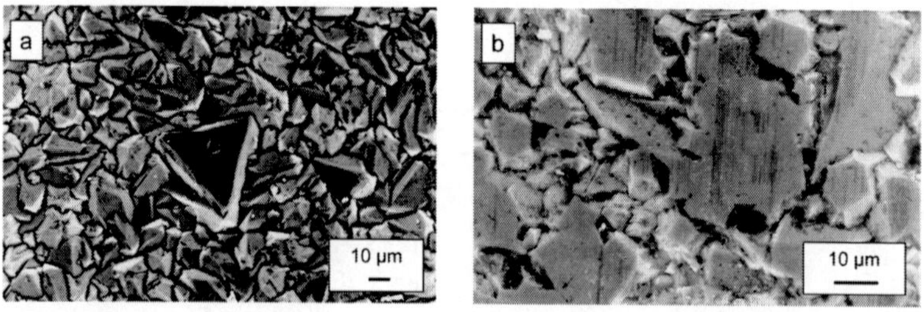

Figure 1.11 SEM micrograph of CVD diamond in (a) the untested condition; and (b) after sliding against a CVD diamond counterface at 2.5 m s^{-1} for 1 hour.

of the test, in some cases reaching over 200°C before declining to between 150°C and 180°C. The authors plotted mean steady-state coefficient of friction against SWR to rank the performance of the various coatings. The best performing coating (i.e., lowest friction and lowest wear rate) was the fine-grained coating with small faceted crystallites. The mean friction coefficients were between 0.27 and 0.31 and the wear rates were between 8.3×10^{-17} and 2.4×10^{-16} m^3 N^{-1} m^{-1}. The worst performer was the nanocrystalline coating: this was attributed to adhesion of the two sliding surfaces, delamination of fragments of the coating during the tests and the coating thickness, which was less than the other specimens.

Figure 1.11 shows a scanning electron micrograph of a CVD diamond surface before and after a pin-on-disc test sliding against a diamond coated disc also coated with CVD diamond. This (previously unpublished) work undertaken by the present author [76] investigated the wear performance of CVD diamond sliding against itself at a velocity of 2.5 m s^{-1} and a load of 50 N (contact pressure 5 MPa). This test was carried out without lubrication on coatings that were in the as-grown condition (the R_a being between 1 μm and 2 μm). The figure shows that the diamond coatings underwent progressive flattening of their asperities during the test, resulting in a polished appearance by the end of the test. The initial surface roughness (R_a) of the pin was 1.05 μm, which had reduced to 0.006 μm; a similar reduction in the R_a of the disc was seen from 1.52 μm to 0.90 μm.

1.5.2 *Effect of Lubrication on Friction of Diamond*

The presence of lubricants can also affect the friction behaviour of diamond coatings in sliding contacts. Feng and Field [77] used a reciprocating stylus test in which a natural diamond stylus was traversed across the surface of both CVD diamond coatings as well as natural diamond at a velocity of 0.2 mm s^{-1} and the load varied between 0.05 N and 2 N. They found that water had a large effect on the coefficient of friction, reducing it from 0.1 (the value in air) to 0.05. In contrast, the addition of liquid paraffin had the opposite effect, and in tests where it was present, the coefficient of friction actually increased to approximately 0.14. The coefficient of friction of diamond sliding on CVD diamond coatings was found to be two to three times higher than that of diamond on polished natural diamond; this was attributed to the higher surface roughness of the former. The effects of water and lubricating oils on the friction of natural diamond were similar to CVD diamond.

In a similar study, Grillo and Field [78] examined the effect of the environment, load and sliding velocity on the tribological properties of natural diamond sliding against natural and CVD diamond. The CVD diamond used in these tests was a 20 μm coating deposited on silicon nitride and polished to a surface roughness of 10 nm. They found that the friction coefficient of both natural and CVD diamond was reduced by the presence of water at the interface of contact. Moreover, the pH of the water also influenced the coefficient of friction with the lowest friction observed at a pH of approximately 7. This effect was most evident at low loads and sliding speeds. Evidence from electron energy loss spectroscopy of wear debris suggested some transformation of diamond to sp^2 hydridised carbon during sliding. It was proposed that the latter acts as a lubricant, thus reducing the coefficient of friction.

1.5.3 *Friction of Nanocrystalline Diamond*

The considerable influence of surface roughness on the friction behaviour of CVD diamond, together with the difficult and time-consuming nature of polishing as-grown diamond coatings, has led many researchers to focus on nanocrystalline diamond as the most promising way of procuring low friction diamond coatings for sliding contacts. Abreu et al. [73] investigated the friction and wear of nanocrystalline diamond coatings on silicon nitride substrates under unlubricated conditions (50–60% relative humidity). They used a reciprocating ball on flat test, in which both ball and flat specimens were of the same coating–substrate combination. The loads were between 10 N and 40 N, which corresponded to mean Hertzian contact stresses of between 4.4 GPa and 7.1 GPa. The effect of surface roughness was compared as the Si_3N_4 substrates had received three different treatments prior to coating deposition, which resulted in the diamond coatings having RMS surface roughness (R_q) values of between 41 nm and 206 nm. They observed that all three surface conditions exhibited the same behaviour, namely an initially high coefficient of friction (0.5–0.7) before reducing to steady-state figures of between 0.02 and 0.03. The diamond deposited on the smoothest Si_3N_4 substrate exhibited poorer adhesion than the other specimens, which resulted in premature detachment of the coating. However, following coating failure, the friction coefficient of the exposed Si_3N_4 substrate sliding against the smooth diamond film on the ball remained low.

Abreu et al. [79] also investigated the behaviour of nanocrystalline diamond sliding against itself in lubricated conditions. The coatings were deposited onto silicon nitride substrates and tested in a reciprocating ball on flat geometry with water lubrication. The loads were between 10 and 90 N, giving mean Hertzian contact pressures of between 1.9 GPa and 4.0 GPa. The mean sliding velocity was 9 mm s^{-1} and the total sliding distance was approximately 690 m. The coatings were seen to behave in a similar manner to those tested in dry conditions. However, in the presence of water, the critical load required to cause film delamination increased from 60 N (dry) to 85 N (water lubricated). For the tests below the critical delamination load, the friction coefficient was initially about 0.33 before decreasing until steady-state conditions were reached, by which time it had reduced to between 0.02 and 0.05. The wear coefficient values were between 10^{-18} m^3 N^{-1} m^{-1} and 10^{-17} m^3 N^{-1} m^{-1}, which is within the mild wear regime, with the coefficients of the balls being one order of magnitude lower than those of the flat specimens.

1.5.4 *Friction of Diamond Sliding against Non-Diamond Counterfaces*

In addition to its low friction behaviour, diamond also exhibits significantly lower wear rates when sliding against other materials. Kohzaki et al. [80] employed a ball-on-disc configuration to compare the frictional behaviour of diamond films with that of SiC. The coefficient of friction of the diamond coatings (0.1–0.2) was much lower than uncoated SiC (0.6). The wear rate for the as-grown diamond disc (8×10^{-17} m^3 N^{-1} m^{-1}) was more than two orders of magnitude lower than SiC. Further improvements were seen in the polished diamond film, the value of surface roughness (20 nm) caused the wear rate to drop to 10^{-18} m^3 N^{-1} m^{-1} and the coefficient of friction to 0.09.

In dry sliding tests against silicon nitride (Si_3N_4) balls, Erdemir et al. [12] found that the friction coefficient of nanocrystalline diamond declined from its initial value of approximately 0.4 to 0.1 upon the establishment of steady-state conditions. In dry nitrogen (N_2) the steady-state friction coefficient was even lower, at approximately 0.05. In contrast, the friction coefficient of the Si_3N_4 ball sliding against the rougher microcrystalline diamond remained high and unsteady in both environments. Moreover, the microcrystalline diamond surfaces proved to be very damaging to the counterface as demonstrated by the wear rates of Si_3N_4 balls slid against the microcrystalline diamond ($\sim 2 \times 10^{-13}$ m^3 N^{-1} m^{-1}), which were more than two orders of magnitude higher than those slid against the nanocrystalline diamond coatings.

When sliding against non-diamond counterfaces, it might be expected that the hardness of diamond would result in the counterface incurring a disproportionate amount of wear, However, while this is true in most cases, it is not universal. Hollman et al. [81] examined the wear behaviour of diamond sliding against Al, Al-17Si and steel. Diamond was found to have a very low wear rate against Al and Al-17Si, one reason for its use in machining Al alloys. However, against steel, the diamond exhibited a high wear rate. Raman spectroscopy of the diamond surface indicated that some transformation to amorphous carbon had occurred. It is for this reason that cubic boron nitride is preferred to diamond for the machining of ferrous metals.

The low friction behaviour of CVD diamond makes it an attractive material for use in pumps, bearings and mechanical seals. In addition to its low friction behaviour, the high thermal conductivity of diamond ensures the rapid dissipation of heat from the contact region. CVD diamond has also been used as a coating for wire drawing dies: researchers at the Fraunhofer Institute in Germany have succeeded in depositing diamond onto the internal surfaces of silicon nitride dies with an inside diameter of 2 mm [82]. The coating varied both in thickness (from 6 μm to 10 μm) and in morphology (from microcrystalline to nanocrystalline) depending on the location within the die. The dies were subsequently used successfully for the drawing of aluminium wires, with the finished product being of acceptable quality both in respect to surface quality and drawing forces.

1.6 Applications

CVD diamond is being exploited in an ever-expanding variety of applications, ranging from machine tools to medical components. This section discusses a few of the components to which diamond coatings have been successfully applied. They range from the macroscopic, such as mechanical seals, to the microscopic, such as components used in MEMS.

1.6.1 *Machine Tools*

One of the earliest, and most common, applications of CVD diamond is in machine tools, in which the coating is deposited on tungsten carbide substrates. However, difficulties can be encountered in depositing diamond on cemented WC owing to the presence of certain binders, most notably cobalt (Co), which can catalyse the formation of graphite and results in poor adhesion of the coating. This can be overcome by etching the WC surface to remove Co from the surface layers, depositing an interlayer or boronising the surface. The last two options act as a diffusion barrier to prevent reaction of Co with the gases used in the CVD process to ensure that the deposited film is overwhelmingly diamond in nature.

CVD diamond is used for the machining of a wide range of materials including graphite, composites, aluminium alloys, plastics and wood; of these, graphite machining alone accounted for an estimated 70% of CVD diamond coated tool sales in 2003 [4]. Diamond-coated tools have also been used to machine ceramics and metal matrix composites, where they have exhibited lower wear rates and produced better surface finishes than the sintered polycrystalline diamond (PCD) [83]. Moreover, they can also enable machining to be carried out at higher speeds than, for example, tungsten carbide tools.

Another advantage of CVD diamond over PCD tooling is the wider range of tooling geometries that can be coated. They range from coated WC inserts to drill bits, reamers and countersinks. One such example of more complex tools is a diamond-coated tungsten carbide end mill, which features a fully coated flute length of more than 100 mm [4]. On a smaller scale, diamond has been used to coat drills of only 1 mm in diameter, an example of which being the tungsten carbide printed circuit board drill coated with CVD diamond in Fig. 1.12 [84]. An alternative to coating is to braze free-standing pieces of CVD diamond, which can be up to 1 mm in thickness, to the tool [85]. However, this does not allow the bonding of CVD diamond to anything other than the simplest of geometries.

One group of materials for which CVD diamond tools have been used to machine is aluminium–silicon (Al–Si) alloys; the presence of Si can confer an abrasive character to the alloy, which can result in rapid wear to the tools used to machine the alloy [86]. An example of this is the $AlSi_7Mg$ rear axle supports used in the BMW 5-series [87]. For the manufacture of these components it was deemed necessary for the machining to be carried out without any lubricant, which prevented the use of PCD tooling owing to the complexity of the tool geometry required to facilitate chip removal. The diamond-coated tooling, which consisted of a 4 μm thick nanocrystalline layer with a grain size of between 20 nm and 200 nm, delivered a 10-fold increase in tool life over the uncoated WC-Co tools.

Other materials for which CVD diamond has been used to machine include wood-based composites such as medium density

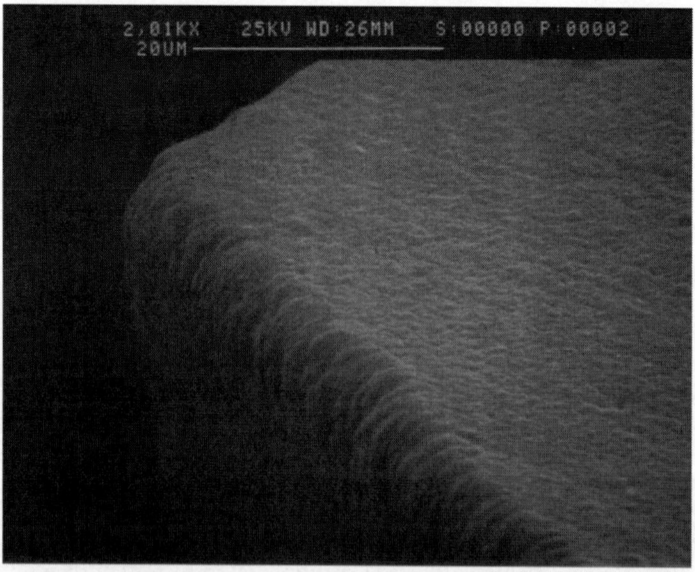

Figure 1.12 Micrograph of a 1-mm-diameter diamond-coated printed circuit board drill [84].

fibreboard [88], polymer composites [89] and sintered WC-Co used in forging dies [90]. For the machining of carbon fibre reinforced polyetheretherketone (PEEK) composites [89] diamond coatings, 2 μm in thickness, were applied to WC tools and used to machine the composites and were compared with uncoated WC and PCD. The use of CVD diamond tools resulted in lower cutting forces and power consumption, which improves the energy efficiency of the machining process. In the drilling of carbon fibre-reinforced composites, diamond-coated drills have been shown to produce better quality holes with less delamination and fibre breakout around the entrances and exits of the holes [91].

Belmonte et al. [90] examined the performance of CVD diamond coated tools in the dry turning of WC-27 wt% Co, which is used in forging dies. The diamond tools were in the form of 300 μm diamond inserts brazed to the WC-Co tool. During turning, the diamond inserts were seen to wear by cratering of the rake face; no obvious signs of flank wear were observed. The authors recorded a 10-fold reduction in the machining time compared with the conventional diamond wheel grinding method.

CVD diamond has also found application as inserts in chain saws used in stone cutting. Turchetta et al. [92] compared CVD diamond inserts with uncoated tungsten carbide in the cutting of marble. It was found that significantly lower cutting forces were measured using the diamond inserts, together with an associated reduction in energy consumption of more than 15%. This also enabled the CVD diamond inserts to be used at higher cutting speeds, thereby reducing the machining time.

Dai et al. [93] tested a diamond-coated WC-Co cutting tool in machining aluminium bronze and Al-22 wt.% Si. They found that in cutting the aluminium bronze the life of the coated tool was 27 times that of the uncoated tool; for the Al-22 wt.% Si alloy the margin of superiority between the coated and uncoated tools was 55 times. They also found that the coated tools gave a better surface finish of the machined materials and reduced the temperature of the work piece.

Another example of tools incorporating diamond is that of diamond-coated blades that are made by the German manufacturer GFD under the brand name 'Diamaze PSD' (plasma sharpened diamond). These blades, which are used for the precision cutting of paper, cardboard, fabric and plastic, are made from tungsten carbide that is coated with diamond. After the deposition process the coating is subjected to a plasma sharpening process, which can reduce the radius of curvature from between 10 μm and 15 μm down to less than 0.2 μm [94].

1.6.2 *Medical Applications*

CVD diamond coatings have also shown promise in a number of medical applications. In addition to its high wear resistance, the biocompatibility of CVD diamond has been found to be comparable to that of titanium and stainless steel, both of which are used frequently in implantable devices [95]. This has led to a number of applications including bioinert coatings for microchips used in artificial retinas, valves used in artificial hearts and wear-resistant orthopaedic implants [96]. In addition to its biocompatibility, CVD diamond has also been found to exhibit anti-microbial properties. Jakubowski et al. [97] compared the susceptibility of nanocrystalline diamond, AISI Type 316L stainless steel and a titanium alloy (Ti-6Al-4V) to *Escherichia coli* (*E. coli*) colonisation. Diamond was found to offer significantly greater resistance to *E. coli* than either Ti-6Al-4V or stainless steel.

In addition to its biocompatibility and anti-microbial properties, diamond coatings can also be subjected to plasma treatments that can modify the surface chemistry and therefore the surface energy, which can affect the osteoblast (bone-forming cell) responses. One beneficiary of the ability to tailor the surface chemistry could be diamond-coated orthopaedic implants. Yang et al. [98] subjected nanocrystalline diamond coatings on Si substrates to three different plasma treatments to produce either hydrogen-, oxygen- or amine-terminated diamond films. AFM images of the treated films revealed no significant topographical differences between the three treated coatings. However, the treatments produced significant differences in the surface energies and, hence, the hydrophilicity of the surfaces. CVD diamond surfaces are usually hydrogen terminated, which results in a hydrophobic surface with water contact angles of between 80° and 90°. In contrast, the oxygen-terminated diamond surface was more hydrophilic (contact angle 66°). However, both the hydrogen- and oxygen-terminated surfaces were surpassed by the NH_3-treated surface, which exhibited the lowest water contact angle and highest surface energy. Moreover, significantly increased osteoblast attachment was observed on coatings with higher surface energies.

One application where the low friction behaviour and high wear resistance of diamond has been exploited is that of surgical blades. Single-crystal diamond blades have been used for many years in ophthalmology, cardio-thoracic surgery and neurosurgery. They offer a number of advantages over the commonly used stainless steel blades, including sharper surfaces, which, coupled with low friction against living tissue, leads to lower cutting forces and greater control of the cut by the surgeon. As a consequence, there is less tearing of the tissue, which diminishes the amount of scar tissue produced, resulting in faster healing times. Moreover, the high wear resistance of diamond offers enhanced blade life. An example of this is in hand surgery where diamond blades have been used in the excision of Dupuytens tissue, the hardness of which can lead to rapid blunting of steel blades. In contrast, a single diamond blade can be used in many operations. The use of diamond blades has also resulted in significantly reduced operating theatre time, to the benefit of patient, surgeon and hospital [99]. In recent years, diamond-coated surgical blades with an approximate surface roughness of 2–40 nm have been developed in Germany by GFD [100]. Like the single-crystal blades, the coated blades have a much smaller radius of curvature than those of stainless steel: this is achieved by employing a post-deposition plasma sharpening process (described in Section 1.6.1) to reduce the radius of curvature.

Another component for which the wear resistance and bioinertness of CVD diamond have been combined to offer superior performance is in dental cutting tools such as burs and drills [101]. Although diamond has been used previously in dental tools, it has been in the form of diamond particles embedded in a metal matrix, typically nickel. Such tools suffer from heterogeneity of the diamond grain shapes, poor adhesion and the tendency of the metallic binder to contaminate oral tissue [102, 103]. Sein et al. [104] used HFCVD to deposit a coating of 15–17 μm in thickness onto a tungsten carbide dental burr, which resulted in a highly uniform coating with a grain size of 8–10 μm. Machining tests of borosilicate glass and acrylic/porcelain teeth found that the diamond coating had reduced the wear rate of the tool by a factor of three compared with the uncoated WC tool. CVD diamond has also been deposited onto molybdenum, which is used as a base material for fabricating diamond dental burrs [105]. The performance of diamond-coated burrs was compared with conventional burrs by using them in a dental laboratory operation on borosilicate glass at 30,000 rpm. In contrast to the conventional burrs, which failed after 30–60 operations, the diamond-coated burr exhibited little deterioration after 1000 operations.

Diamond coatings have also been investigated for use as a coating for the Temporomandibular (jaw) joint. This joint consists of an upper temporal bone, which is part of the skull, and the lower mandible (jaw bone). The mating surfaces of the joint are known as the condyle and the fossa. In temporomandibular prostheses wear debris has been blamed for a range of adverse bodily reactions including osteolysis, inability to chew and blindness [106]. This has prompted efforts to identify more wear-resistant materials for these implants. Papo et al. [106] deposited diamond coatings onto Ti-6Al-4V substrates designed to imitate the condyle and fossa; the thickness of the coatings on each component was approximately 3 μm and 2.25 μm respectively. Two variants of coating were deposited, one wholly nanocrystalline diamond coating and the other a nanocrystalline/microcrystalline/nanocrystalline multilayer. Both coatings were tested in a mandibular movement simulator, the motion of which simulated the cyclic chewing action of a human jaw, and compared with the uncoated Ti-6Al-4V substrates tested under the same conditions. The tests were performed for 2.25×10^5 loaded cycles, which corresponds to approximately two years of clinical use. The diamond coatings remained intact at the end of the tests; in contrast, the uncoated Ti-6Al-4V substrates exhibited extensive wear.

1.6.3 *Mechanical Seals*

Mechanical seals are devices for sealing a rotating shaft consisting of a sealing interface located between a pair of radial faces, one rotating and the other stationary. They are used in a wide variety of applications including the process, automotive and aerospace industries. While the operating conditions vary depending on the specific application, in general the net specific loads are usually between 0.1 and 5 MPa and the sliding speeds are within the range of 1–20 m s^{-1}. The temperatures of the contact zone between the two counterfaces can be from 20 to 300°C [107].

Mechanical seals often work in mixed and boundary lubricated conditions in which the working fluid acts as the lubricant [108]. When operating correctly, the sealing faces are separated by a very thin fluid film of between 0.25 μm and 8 μm thick [109], making it comparable to the residual waviness of the seal, which is about 0.1–1 μm peak-to-peak [107]. Although mechanical seals do not normally operate under dry sliding, such conditions are often encountered intermittently, particularly during the start-up phase of operation.

In selecting a material for service in a mechanical seal, the desirable properties required for such a component include high elastic modulus, high tensile strength, low coefficient of friction, high thermal conductivity, low coefficient of thermal expansion and good corrosion resistance [110]. The current leading seal face combination is a high-duty mechanical carbon (e.g., carbon-graphite) running against a reaction-bonded or sintered SiC [111]. Carbon-graphite is particularly suitable on account of its chemical inertness, low cost and self-lubricating properties [112]. Other common materials used in mechanical seals include Al_2O_3 and WC.

Although carbon–graphite has demonstrated its suitability as a seal face material in many applications, it is susceptible to wear in the presence of abrasive slurries, which are encountered in pumps used in the mineral and ore processing industries. In such environments a SiC/SiC face seal combination offers enhanced wear resistance; however, in the event of inadequate lubrication in the sealing gap high friction and rapid wear can result.

The high wear resistance (resulting from its high hardness) of diamond, together with its low friction behaviour, high elastic modulus, high thermal conductivity and resistance to chemical attack, gives it a number of advantages over carbon-graphite. Diamond therefore has the potential to deliver both reduced power consumption as well as reduced wear, and hence prolong the operating life. Furthermore, its high thermal conductivity reduces the heat generation which can cause thermal distortion of the seal faces that can result in unacceptable leakage rates.

Diamond began to receive attention from researchers in the 1990s as a potential seal material. One of the first studies of diamond-coated seals was by Hollman et al. [108], who compared diamond-coated (3.5 μm thickness) WC face seals with a range of other commercially available seal rings, including WC, SiC, Al_2O_3 and sintered carbon impregnated with antimony. The tests were performed dry and the contact pressure and sliding velocity were 2.5 MPa and 3.5 m s^{-1}, respectively. A steady-state coefficient of friction of 0.2 was recorded for CVD diamond sliding against itself, while for diamond sliding against SiC, it was even lower at 0.1. In contrast, self-mated WC and SiC both exhibited steady-state coefficients of friction of approximately 0.7.

Kelly et al. [113] used a thrust washer test rig to evaluate the friction and wear of diamond-coated discs in sliding contact with rotating washers made from WC, SiC or diamond-coated SiC. The diamond coatings were between 2 μm and 4 μm in thickness depending on the deposition conditions and the tests were conducted under conditions of marginal water lubrication at a sliding speed of 10 mm s^{-1} and a pressure of approximately 0.5 MPa. In tests where both the disc and counterface were diamond coated the coefficient of friction varied between 0.05 and 0.07. At the end of the test, the coating remained intact with only a lightly polished track visible on both disc and washer. However, when the diamond-coated disc was run against a SiC washer, rapid failure of the coating was observed, even though the coefficient of friction had stabilised at around 0.05. In another test of a diamond-coated disc against a WC washer, the reverse was observed: the coefficient of friction was significantly higher (average 0.4), although the coating on the disc remained intact. The measured wear volume in the WC was also approximately 50% higher than that recorded on the SiC.

Jones [114] conducted a similar investigation. A diamond-coated SiC pad was loaded against an uncoated SiC cylinder at a velocity of 0.5 m s^{-1}. The coating was 5–6 μm in thickness and its as-deposited surface roughness (R_a) was 0.25 μm; no post-deposition lapping or polished was carried out. During the tests, which were performed under dry conditions, the load was increased incrementally up to the point of contact instability. The diamond film offered an approximately 25-fold improvement in pressure-velocity capability.

Tome et al. [115] coated Si_3N_4 rings with diamond coatings of between 8 μm and 25 μm in thickness; the grain sizes were between 2 μm and 3.5 μm, while the initial surface roughness values were between 0.15 μm and 0.26 μm. Although the asperities of the as-grown diamond films hindered full sealing in the early stages of the ring-on-ring tests, once they had been worn down by self-polishing, this full sealing was attained. The coefficient of friction stabilised at approximately 0.05 and remained at this level throughout the remainder of the tests, the longest of which were for sliding distances of up to 5700 km with no measurable wear.

Nanocrystalline diamond coatings have also been evaluated for use in mechanical seals Mubarok et al. [116] used a ring-on-ring test apparatus to evaluate Si_3N_4 seals, each coated with 2 μm of nanocrystalline diamond. The surface roughness (R_a) of the deposited coating was 62 nm, which rendered any running-in procedures unnecessary. In the tests, which were conducted in a water environment, the loads were between 300 N and 925 N and sliding speeds between 2 m s^{-1} and 4 m s^{-1}. The mean steady-state friction coefficients were between 0.01 and 0.02, while the wear, which was monitored by weight loss, was below measurable limit. The one exception to this was the test conducted at the highest load/sliding speed combination (925 N/4 m s^{-1}) where seizure occurred. Examination of the sliding surfaces revealed that the diamond coating had been delaminated from the substrate. This was attributed to excessive heating at the contact, resulting in water vaporisation that eliminated the boundary lubrication layer (and cooling from water layer) and removal of the coating.

Kovalchenko et al. [117] evaluated the performance of SiC seals coated with a 1 μm film of ultrananocrystalline diamond. Prior to deposition, the seals were polished with three different grades of diamond powder (6 μm, 12 μm and 30 μm). They were then tested in a dynamic seal testing machine, the diamond-coated seals sliding against either an uncoated SiC or carbon counterface. After a running-in procedure, the tests were performed at a water pressure of 0.7 MPa for 50 hr and a rotation speed of 3600 rpm. Post-test examination of the diamond-coated seals revealed that the substrates that had been polished with the 12 and 30 μm diamond powder prior to deposition did not exhibit any measurable wear. Indeed, when studied using optical microscopy, the wear tracks did not differ in appearance from that of the unworn surface. In contrast, regions of the coating deposited on the substrate polished with 6 μm diamond powder had delaminated during the running-in period. Surface profilometry and Raman spectroscopy indicated that failure had occurred at the coating–substrate interface.

In 2007, Burgmann Industries GmbH & Co. introduced a diamond-coated mechanical seal for use in multiphase pumps, slurry applications, the chemical and pharmaceutical industries and domestic hot water applications. The diamond coating, which is up to 8 μm in thickness, is deposited using a HFCVD process. The seal has proved to be successful on an abrasive multiphase pump

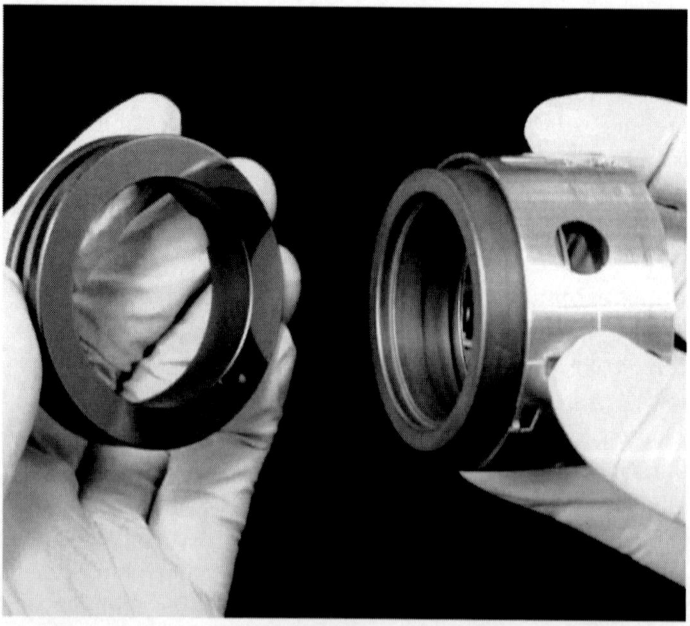

Figure 1.13 A 1.375 in (34.9 mm) diamond-coated mechanical pump seal. Picture reproduced with the permission of C.F. West (Advanced Diamond Technologies Inc.).

application in the Canadian oil sands fields [118]. In the following year, Advanced Diamond Technologies Inc. (ADT) introduced a range of mechanical seals (an example of which is shown in Fig. 1.13) that incorporated ultrananocrystalline diamond films and that can be run directly against blister-resistant carbon or silicon carbide counterfaces [119]. A further addition to the range in 2010 was specifically optimised to withstand fluid loss or dry running in mechanical seals for fluid pumps. When run against SiC, the energy dissipation due to friction was seen to reduce by 75% compared with running SiC against itself [120].

1.6.4 *Micro-Electromechanical Systems*

Another area where diamond coatings have shown promise is for microelectromechanical systems (MEMS). The term 'MEMS' covers a wide range of components including sensors, actuators, resonators, accelerometers and micromotors. The dimensions of these devices are typically smaller than 100 μm [121]. Since the emergence of the first MEMS prototypes in the early 1980s, the ensuing three decades have witnessed a rapid growth in the technology to the extent that it is beginning to reach commercial exploitation, although, at present, commercial MEMS devices are limited to those that have either no motion or limited motion without sliding contacts such as linear sliding, hinge motions or rotations [122]. Nevertheless, the total market for MEMS devices is approximately US$ 10 billion per year and according to one prediction could reach US$ 100 billion per year by 2025 [123].

Until now, the most common MEMS material has been silicon, which can be either in single-crystal or polycrystalline form [124]. Complex shapes, such as levers, gears and pinwheels have been produced using standard lithographic patterning and etch methods, while fully assembled devices such as micromotors and gear trains have been produced using multiple masks and multilayer deposition and etching techniques [125]. Owing to the high surface area-to-volume ratio of MEMS devices, tribological effects are likely to be important factors in determining performance [126]. Although friction and wear are minimal in most MEMS sensors, for MEMS actuators failures are often caused by friction and wear of the rubbing surfaces [127]. Such failures can be exacerbated by the poor mechanical properties and tribological behaviour of Si, notably its high coefficient of friction and rapid wear. Furthermore, in devices such as micromotors the small clearances between rotor and hub are of the order of 2 μm, making them vulnerable to seizure in the event of wear debris being generated [124]. Therefore, low-friction and near-zero wear rates are required for the reliable operation of these machines.

The low friction and high wear resistance of diamond make it potentially very suitable for use in MEMS components. The brittle fracture strength of diamond is 23 times that of Si and the projected wear life of diamond MEMS moving mechanical assemblies (MMA) is 10,000 times greater than that of Si MMAs [9]. It has also been shown that the diamond films can be doped with elements such as boron in order to vary their electrical resistivity. This enables the films to be insulating or semiconducting in character [128].

Another problem encountered with Si is stiction caused by adhesion of contacting surfaces. Stiction, which combines adhesion and static friction [129], can be defined as a phenomenon that causes two surfaces in close proximity to yield unusually high adhesive forces in the normal and tangential directions due to such effects as the presence of a thin liquid layer at the contact interface or a potential difference across the interface [124]. This propensity for stiction, which can be aggravated by the hydrophilic nature of silicon, may be reduced by the use of diamond, the hydrogen-terminated surface of which is hydrophobic. In addition, the tendency for stiction may also be diminished by the greater stiffness of diamond that can also help to overcome the adhesive forces.

A number of prototype diamond MEMS components have been reported in the literature. The most suitable form of diamond is likely to be nanocrystalline diamond films as their smooth surfaces will deliver a low coefficient of friction without the need for post-deposition polishing [130]. Hunn et al. [131] described the fabrication of a free-standing single-crystal diamond gear 400 μm in diameter and 13 μm in thickness. Krauss et al. [9] have also described the production of free-standing 'propellers' 1 μm in thickness made from ultrananocrystalline diamond, grown by selective deposition followed by etching of the Si substrate. Other components include mechanical sensors [132], thermal actuators [132], micro-switches [133] and micro-grippers driven by electrostatic comb actuator with 10 μm teeth width and 5 μm spacing [134]. Wang et al. [135] have also described the production of diamond film micro-motors (shown in Fig. 1.14), the dimensions of which were a rotor diameter of 150 μm and a thickness of 2.0 μm. The gap between rotor and bearing was 1.6 μm, while the gap between the rotor and the stator was 3.0 μm.

Diamond-coated gear wheels have also been used in watches. Silicon gears coated with nanocrystalline diamond are produced by the Swiss company Diamaze Microparts SA and have been incorporated into watch movements made by the Swiss manufacturer Ulysse Nardin [136]. In addition to the high wear resistance and low friction of the diamond coatings, the low thermal expansion coefficient and high strength of diamond mean that dimensional stability is

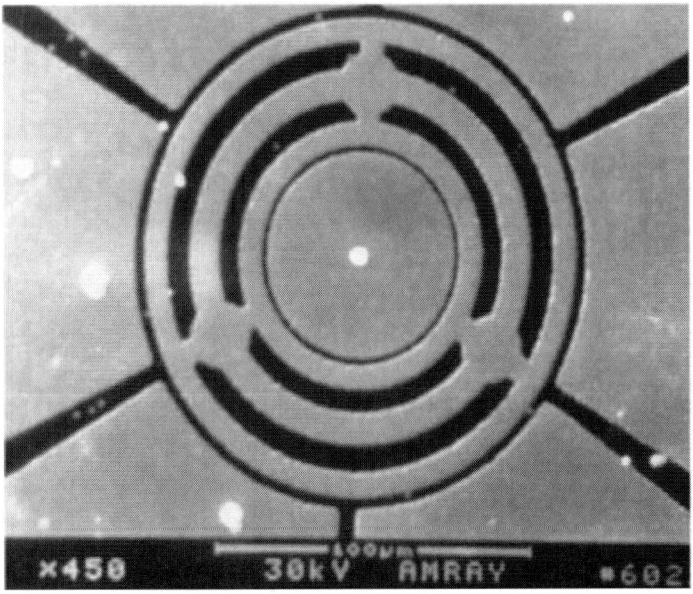

Figure 1.14 SEM micrograph of the structure of a diamond film micromotor fabricated by selective growth using DC bias-enhanced MPCVD [135].

maintained, a consequence of which is improved chronometrical accuracy.

CVD diamond has also been used—both as a coating and in monolithic form—in AFM probes. AFM probes are usually made from silicon or silicon nitride. However, when probe wear occurs, blunting of the tip can lead to inaccurate measurements. Chung and Kim [137] evaluated silicon cantilevers that were coated with 50–200 nm of diamond. An AFM-based wear test was used to study their performance when sliding against silicon nitride and copper; the loads used were between 1 and 5 μN and the sliding velocity was 1 $\mu m\ s^{-1}$. Wear was measured from SEM images of the probes before and after the tests. They found no significant probe wear when sliding against copper. Against silicon nitride the wear rate was determined to be between 3.9 and $5.0 \times 10^{-16}\ m^3\ N^{-1}\ m^{-1}$. Liu et al. [138] have reported the fabrication of monolithic AFM probes made from ultrananocrystalline diamond by a combination of microfabrication and HFCVD. The grain size of the diamond

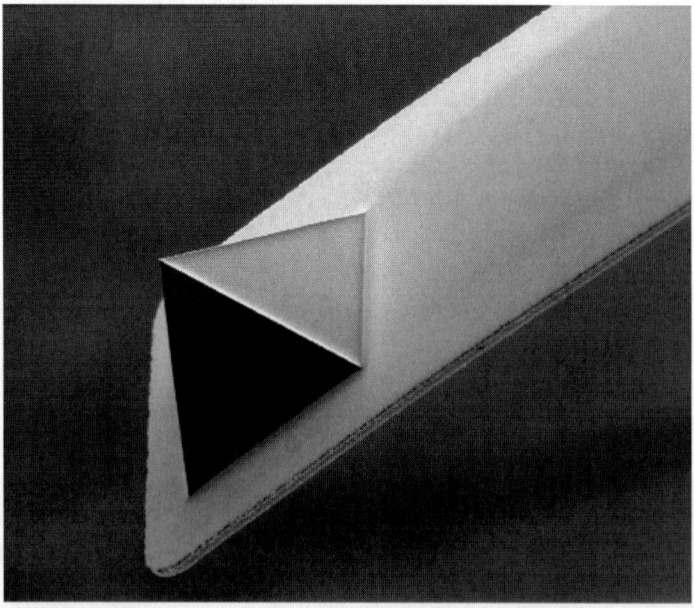

Figure 1.15 Monolithic diamond atomic force microscope probe. Picture reproduced with the permission of C.F. West (Advanced Diamond Technologies Inc.).

was less than 5 nm and the tip radii were between 30 nm and 40 nm. Under contact mode scanning conditions such probes exhibited significantly enhanced wear resistance compared with the more commonly used silicon nitride probes. Figure 1.15 shows a monolithic ultrananocrystalline diamond AFM probe.

1.7 Economics of CVD Diamond

The previous sections of this chapter have shown the capabilities of diamond coatings to improve the operating lives of components used in tribological applications. However, it is the financial benefits resulting from the application of diamond components that will determine their uptake in commercial products. Such benefits, both in components and in manufacturing operations, will now be discussed. In addition, the increasing prominence of environmental considerations means that this aspect must also be taken into account. For this reason, the potential environmental benefits from using diamond-coated components in tribological applications are also discussed in this section.

1.7.1 *Costs of Diamond Coating*

In any new technology, the associated costs decrease with time making it more affordable. CVD diamond is no exception to this and over the last two decades the costs have fallen as deposition processes have improved and production rates have increased. CVD diamond can be grown for about $5 per carat [139] (1 carat = 0.2 g).

One way by which costs can be reduced is to increase the numbers of components that can be coated at a time. Advances in deposition technology have increased the number of components that can be coated in a single deposition run. The latest HFCVD reactors can accommodate over 100 round tools or several hundred cutting inserts [123]. Further cost reductions can be achieved by increasing the numbers of reactors. In one study of the deposition of diamond burr tips [140], the projected cost per piece dropped from $7 to $1 as the number of reactors increased from 1 to 100.

1.7.2 *Machine Tools*

Although the costs of diamond-coated tools are several times those of their uncoated counterparts, the extra costs are more than outweighed by the increased life of the coated tools. As an example, CVD diamond coated tools are typically four to six times the cost of uncoated tungsten carbide tooling; however, this increased cost can be accompanied by an increase in lifetime of between 10 and 20 times. This translates into net tooling cost reductions of between 40% and 80% [91]. Similar savings have been achieved following the adoption of diamond-coated tooling in the manufacture of automotive crankshafts and cylinder heads, which are increasingly being made from aluminium–silicon (Al–Si) alloys. The increased life of diamond-coated tools has resulted in an 80% reduction in tooling costs and a concomitant increase in output arising from fewer tool changes (and associated downtime) and the ability to employ higher feed rates using diamond-coated tools than are possible with uncoated tools [141]. Indeed, during flat milling of cylinder heads, the use of tools coated with a CVD diamond multilayer coating resulted in a productivity increase of 15% [142].

In the aviation industry, a striking illustration of the economic benefits of CVD diamond tooling is the experience of Lockheed Martin Aeronautics Co., which manufactures components for the F-35 Joint Strike Fighter. For the post-mould machining of carbon-fibre reinforced wingskins, the adoption of tools coated with CVD diamond led to a reduction in the number of tools needed for one complete wingskin from 24 to 2. The associated cost savings per aircraft are in the region of $80,000: if the full number of aircraft currently planned (2,783) are manufactured a total saving of $222.6 million has been estimated [143].

1.7.3 *Mechanical Seals*

Mechanical seal failures are responsible for over 70% of pump problems [144]. One beneficiary of improvements to the operating lives of mechanical seals will be in pumping systems. According to the US Department of Energy, over 50% of pump life cycle costs result from energy and maintenance expenses, and energy savings of 20% or more are possible with systems optimisation [119].

Two ways by which such costs may be reduced are: (1) extending the seal operating lives; and (2) using components with low friction surfaces. Both may be achieved by using CVD diamond. Firstly, the increased operating lives of diamond-coated seals render component replacements less frequent. Indeed, analyses have shown that the return on investment due to the significantly increased service life achieved by a diamond-coated mechanical seal can be obtained after just 18 months [145]. Secondly, the low friction behaviour of diamond coatings can reduce the frictional losses and therefore energy consumption. In 1998 Hollman et al. [108] estimated that if an energy loss of 1 kW could be reduced to 300 W the energy cost would be reduced by US$ 600 per year (US$ 0.1 per kWh running 24 hours a day). A more recent illustration of the savings that could result from using diamond-coated seals is given by Eagle Burgmann GmbH, which in 2012 applied a diamond coating to its Cartex cartridge seal. This type of seal can be used in the chemical, pharmaceutical, paper, food and waste water industries. According to the company, an uncoated SiC Cartex seal with 8000 operating hours per year converts 3200 kWh into friction; however, when coated with diamond this figure is reduced by approximately 2500 kWh over the same period, which equates to approximately € 325 in industrial electricity prices [146].

1.7.4 *Environmental Benefits*

In addition to the cost savings, there are potential environmental benefits associated with the adoption of diamond coatings in tribological applications. In 2010, Nosonovsky and Bhushan [129] outlined the principles of green tribology. Among the twelve principles formulated were: (1) minimisation of heat and energy dissipation; (2) minimisation of wear; and (3) reduction or complete elimination of lubrication. As was discussed in earlier sections of this chapter, some of the properties of diamond coatings offer the means by which these aims may be achieved. The low friction behaviour and high thermal conductivity can help to minimise heat and energy dissipation. Friction is the primary source of energy dissipation: according to some estimates, about one-third of the energy consumption in the United States is spent to overcome friction. Moreover, the high wear resistance of diamond coatings offers the potential for minimising wear. Furthermore, the ability of diamond coatings to operate in the absence of lubrication (e.g., in machining as well as in mechanical seals) offers the potential for reducing or eliminating the need for lubrication.

1.8 Conclusions

CVD diamond is a truly multifunctional material, combining wear resistance (by virtue of its high hardness and strength) with low friction, high thermal conductivity and chemical inertness. This combination of properties make diamond coatings ideal both for components that require low friction and/or high wear resistance in corrosive environments as well as those requiring biocompatibility or anti-microbial characteristics. Diamond films can also be treated or doped to alter their surface energy or electrical resistivity, thereby enabling the properties to be tailored to specific applications.

Diamond deposition is now a mature technology, which has successfully made the transition from the research laboratory to the production plant, where it is used for applications that require a high resistance to wear. Diamond-coated machine tools are now the preferred option for the machining of Al–Si alloys, composites and ceramics. However, despite these successes, there remain many other components which diamond has the potential to increase the operating life.

The benefits of using CVD diamond in tribological applications also extend to the environmental field. The lower levels of friction in machining and mechanical seals can result in reduced energy consumption. Moreover, the diamond-coated machine tools enable machining to be carried out in the absence of cutting fluids, which often contain constituents that are harmful to the environment.

Over the next few years, the number and range of new applications for diamond-coated components is expected to increase. CVD diamond remains a field where the potential applications are far from being exhausted.

Acknowledgements

The author would like to thank Dr. J. Zekonyte (University of Southampton) for the atomic force microscopy images and nanoindentation measurements. Prof. P. W. May (University of Bristol) for providing two of the electron micrographs and C. F. West (Advanced Diamond Technologies Inc.) for allowing two of the figures to be included in this chapter.

References

1. B.V. Spitsyn, L.L. Bouilov, B.V. Derjaguin, Vapor growth of diamond on diamond and other surfaces, *J. Cryst. Growth*, 52 (1981), 219–226.
2. M. Kamo, Y Sato, S. Matsumoto, N. Setaka, Diamond synthesis from gas phase in microwave plasma, *J. Cryst. Growth*, 62 (1983), 642–645.

3. S. Matsumoto, Y. Sato, M. Tsutsumi, N. Setaka, Growth of diamond particles from methane-hydrogen gas, *J. Mater. Sci.*, 17 (1982), 3106–3112.
4. B.L. Cline, H.S. Cline, Status and trends in the commercialization of polycrystalline CVD diamond materials, *Finer Points* (Summer/Fall 2003), 18–23.
5. P.W. May, Diamond thin films: a 21st century material, *Phil. Trans. R. Soc. Lond. A*, 358 (2000), 473–495.
6. J. Herlinger, sp3's experience using hot filament CVD reactors to grow diamond for an expanding set of applications, *Thin Solid Films*, 501 (2006), 65–69.
7. M.N.R. Ashfold, P.W. May, J.R. Petherbridge, K.N. Rosser, J.A. Smith, Y.A. Mankelevich, N.V. Suetin, Unravelling aspects of the gas phase chemistry involved in diamond chemical vapour deposition, *Phys. Chem. Chem. Phys.*, 3 (2001), 3471–3485.
8. P.W. May, Chemical vapour deposition—a route to microcrystalline, nanocrystalline, ultrananocrystalline and single crystal diamond films, *Carbon Based Nanomaterials* (ed. N. Ali, A. Öchsner and W. Ahmed), Trans Tech, Switzerland (2010), 145–176.
9. A.R. Krauss, O. Auciello, D.M. Gruen, A. Jayatissa, A. Sumant, J. Tucek, D.C. Mancini, N. Moldovan, A. Erdemir, D. Ersoy, M.N. Gardos, H.G. Busmann, E.M. Meyer, M.Q. Ding, Ultrananocrystalline diamond thin films for MEMS and moving mechanical assembly devices, *Diamond Relat. Mater.*, 10 (2001), 1952–1961.
10. P.W. May, M.N.R. Ashfold, Y.A. Mankelevich, Microcrystalline, nanocrystalline and ultrananocrystalline diamond chemical vapour deposition: experiment and modelling of the factors controlling growth rate, nucleation and crystal size, *J. Appl. Phys.*, 101 (2007), 053115.
11. J. Philip, P. Hess, T. Feygelson, J.E. Butler, S. Chattopadhyay, K.H. Chen, L.C. Chen, Elastic, mechanical and thermal properties of nanocrystalline diamond films, *J. Appl. Phys.*, 93 (2003), 2164–2171.
12. A. Erdemir, G.R. Fenske, A.R. Krauss, D.M. Gruen, T. McCauley, R.T. Csencsits, Tribological properties of nanocrystalline diamond films, *Surf. Coat. Technol.*, 120–121 (1999), 565–572.
13. E.D. Nicholson, J.E. Field, The mechanical and thermal properties of thin films, *J. Hard Mater.*, 5 (1994), 89–132.
14. U. Wiklund, J. Gunners, S. Hogmark, Influence of residual stresses on fracture and delamination of thin hard coatings, *Wear*, 232 (1999), 262–269.

15. D. Roy, Z.H. Barber, T.W. Clyne, Strain gradients along the growth direction in thin diamond film deposited on silicon wafer, *J. Appl. Phys.*, 94 (2003), 136–139.
16. R.P. Martinho, F.J.G. Silva, A.P.M. Baptista, Wear behaviour of uncoated and diamond coated Si_3N_4 tools under severe turning conditions, *Wear*, 263 (2007), 1417–1422.
17. D.S. Knight, W.B. White, Characterisation of diamond films by Raman spectroscopy, *J. Mater. Res.*, 4 (1989), 383–393.
18. H. Boppart, J. van Straaten, I.F. Silvera, Raman spectra of diamond at high pressures, *Phys. Rev. B*, 32 (1985), 1423–1425.
19. D.W. Wheeler, PhD Thesis, University of Southampton (2001).
20. S.G. Roberts, Hertzian testing of ceramics, *Brit. Ceram. Trans.*, 99 (2000), 31–38.
21. P.H. Shipway, I.M. Hutchings, The role of particle properties in the erosion of brittle materials, *Wear*, 193 (1996), 105–113.
22. C.A. Brookes, Plastic deformation and anisotropy in the hardness of diamond, *Nature*, 228 (1970), 660–661.
23. J.E. Field, *The Properties of Natural and Synthetic Diamond*, Academic Press, London (1992).
24. S. Chowdhury, E. de Barra, M.T. Laugier, Study of mechanical properties of CVD diamond on SiC substrates, *Diamond Relat. Mater.*, 13 (2004), 1625–1631.
25. J. Zekonyte, D.W. Wheeler, unpublished data (2010).
26. R.S. Sussmann J.R. Brandon, G.A. Scarsbrook, C.G. Sweeney, T.J. Valentine, A.J. Whitehead, C.J.H. Wort, Properties of bulk polycrystalline CVD diamond, *Diamond Relat. Mater.*, 3 (1994), 303–312.
27. M. Wiora, K. Bruhne, A. Floter, P. Gluche, T.M. Willey, S.O. Kucheyev, A.W. van Buuren, A.V. Hamza, J. Biener, H.J. Fecht, Grain size dependent mechanical properties of nanocrystalline diamond films grown by hot filament CVD, *Diamond Relat. Mater.*, 18 (2009), 927–930.
28. O.A. Williams, A. Kriele, J. Hees, M. Wolfer, W. Muller-Sebert, C.E. Nebel, High Young's modulus in ultra thin nanocrystalline diamond, *Chem. Phys. Lett.*, 495 (2010), 84–89.
29. A.R. Davies, J.E. Field, C.S.J. Pickles, Strength of free-standing chemically vapour-deposited diamond measured by a range of techniques, *Phil. Mag.*, 83 (2003), 4059–4070.
30. J.E. Field, C.S.J. Pickles, Strength, fracture and friction properties of diamond, *Diamond Relat. Mater.*, 5 (1996), 625–634.

31. M.A. Moore, A review of two-body abrasive wear, *Wear*, 27 (1974), 1–17.
32. R.J.K. Wood, Understanding surface wear in engineering materials, *Surface Coatings for Protection against Wear* (ed. B.G. Mellor), Woodhead Publishing Ltd., Cambridge (2006), 1–57.
33. I.M. Hutchings, *Tribology*, Edward Arnold, London (1992).
34. J.A. Williams, *Engineering Tribology*, Oxford University Press, Oxford (1994), p. 181.
35. R. Gahlin, A. Alahelisten, S. Jacobson, The effects of compressive stresses on the abrasion of diamond coatings, *Wear*, 196 (1996), 226–233.
36. K. Bose, R.J.K. Wood, Influence of load and speed on rolling micro-abrasion of CVD diamond and other hard coatings, *Diamond Relat. Mater.*, 12 (2003), 753–756.
37. F.J.G. Silva, A.J.S. Fernandes, F.M. Costa, V. Teixeira, A.P.M. Baptista, E. Pereira, Tribological behaviour of CVD diamond films on steel substrates, *Wear*, 255 (2003), 846–853.
38. P.H. Shipway, C.J.B. Hodge, Microabrasion of glass – the critical role of ridge formation, *Wear*, 237 (2000), 90–97.
39. W. Pricken, Dressing of vitrified bond wheels with CVDRESS and MONODRESS, *Ind. Diamond Rev.*, 59 (1999), 225–231.
40. J. Gabler, L. Schafer, B. Menze, H.W. Hoffmeister, 2003, Micro abrasive pencils with CVD diamond coating, *Diamond Relat. Mater.*, 12 (2003), 707–710.
41. D.W. Wheeler, R.J.K. Wood, Erosive wear behaviour of thick chemical vapour deposited diamond coatings, *Wear*, 225–229 (1999), 523–536.
42. D.W. Wheeler, R.J.K. Wood, Solid particle erosion of CVD diamond coatings, *Wear*, 233–235 (1999), 306–318.
43. D.W. Wheeler, R.J.K. Wood, CVD diamond: erosion-resistant hard material, *Surf. Eng.*, 19 (2003), 466–470.
44. Z. Feng, Y. Tzeng, J.E. Field, Solid particle impact of CVD diamond films, *Thin Solid Films*, 212 (1992), 35–42.
45. S.P. Timoshenko, J.N. Goodier, 1970, *Theory of Elasticity*, McGraw-Hill, New York.
46. D.W. Wheeler, R.J.K. Wood, Erosion damage in diamond coatings by high velocity sand impacts, *Phil. Mag.*, 87 (2007), 5719–5740.
47. D.W. Wheeler, R.J.K. Wood, High velocity sand impact damage on CVD diamond, *Diamond Relat. Mater.*, 10 (2001), 459–462.

48. D.W. Wheeler, R.J.K. Wood, Fracture of diamond coatings by high velocity sand erosion, *Phil. Mag.*, 89 (2009), 285–310.
49. D.W. Wheeler, R.J.K. Wood, D. Harrison, E. Smith, Application of diamond to enhance choke valve life in erosive duties, *Wear*, 261 (2006), 1087–1094.
50. C.R. Seward, J.E. Field, E.J. Coad Liquid impact erosion of bulk diamond, diamond composites and diamond coatings, *J. Hard Mater.*, 5 (1994), 49–62.
51. G.F. Miller, H. Pursey, On the partition of energy between elastic waves in a semi-infinite solid, *Proc. R Soc. Lond. A*, 233 (1955), 55–69.
52. F.P. Bowden, J.E. Field, The brittle fracture of solids by liquid impact, by solid impact, and by shock, *Proc. R. Soc. Lond. A*, 282 (1964), 331–352.
53. S. van der Zwaag, J.E. Field, Rain erosion damage in brittle materials, *Eng. Fract. Mech.*, 17 (1983), 367–379.
54. R.H. Hopkins, W.E. Kramer, G.B. Brandt, J.S. Schruben, R.A. Hoffman, K.B. Steinbruegge, T.L. Peterson, Fabrication and evaluation of erosion-resistant multispectral optical windows, *J. Appl. Phys.*, 49 (1978), 3133–3139.
55. E.J. Coad, C.S.J. Pickles, C.R. Seward, G.H. Jilbert, J.E. Field, The erosion resistance of infrared transparent materials, *Proc. R. Soc. Lond. A*, 454 (1998), 213–238.
56. D.C. Harris, Review of Navy Program to develop optical quality diamond windows and domes, Naval Systems Air Command Report (2002), http://www.dtic.mil/cgi-bin/GetTRDoc?AD=ADA421202. Accessed 29 Feb 2012.
57. G.H. Jilbert, C.S.J. Pickles, E.J. Coad, J.E. Field, Diamond: an erosion resistant aerospace material, *Applications of Diamond Films and Related Materials: 3rd Int. Conf.* (1995), pp. 561–564.
58. C.R. Seward, C.S.J. Pickles, R. Marrah, J.E. Field, Rain erosion data on window and dome materials, *Proc. SPIE*, 1760 (1992), 280–290.
59. M.D. Hudson, C.J. Brierley, A.J. Miller, A.E.J. Wilson, Fabrication and testing of diamond coatings on infrared windows for the Harrier GR7 and AV8-B systems, *Proc. SPIE*, 3060 (1997), 196–202.
60. C.J.H. Wort, C.S.J. Pickles, A.C. Beale, C.G. Sweeney, M.R. McClymont, R.J. Saunders, R.S. Sussmann, K.L. Lewis, Recent advances in the quality of CVD diamond optical components, *Proc. SPIE*, 3705 (1999), 119–128.
61. F. Deuerler, H. Gruner, M. Pohl, L. Tikana, Wear mechanisms of diamond-coated tools, *Surf. Coat. Technol.*, 142–144 (2001), 674–680.

62. C.M. Preece, Cavitation erosion, (ed. C.M. Preece) *Treatise on Materials Science and Technology Vol. 16*, Academic Press, New York (1979), 249–306.

63. F. Deuerler, M. Pohl, C. Heßing, Micro- and nanostructured diamond films under wear charging in cavitation test, *Surf. Coat. Technol.*, 174–175 (2003), 732–737.

64. G. Bregliozzi, W. Haenni, H. Haefke, Cavitation erosion behavior of nanocrystalline diamond thin films on silicon substrates, *J. Mater. Sci.*, 39 (2004), 6603–6606.

65. S.K. Biswas, Sliding wear of materials, (ed. I.M. Hutchings) *New Directions in Tribology*, Mechanical Engineering Publications (1997), London, pp. 159–175.

66. Y. Enomoto, D. Tabor, The frictional anisotropy of diamond, *Proc. R. Soc. Lond.* A, 373 (1981), 405–417.

67. M. Seal, The friction of diamond, *Phil. Mag. A*, 43 (1981), 587–594.

68. J.M. Martin, T. Le Mogne, C. Donnet, N. Millard-Pinard, K.V. Ravi, M.N. Gardos, The role of dangling bonds on sliding diamond, *Proc. 3rd Int. Symp. Diamond Materials*, Honolulu, Electrochem. Soc. (1993), pp. 947–954.

69. A.V. Sumant, D.S. Grierson, J.E. Gerbi, J. Birrell, U.D. Lanke, O. Auciello, J.A. Carlisle, R.W. Carpick, Toward the ultimate tribological interface: surface chemistry and nanotribology of ultrananocrystalline diamond, *Adv. Mater.*, 17 (2005), 1039–1045.

70. I.P. Hayward, I.L. Singer, L.E. Seitzman, Effect of roughness on the friction of diamond on CVD diamond coatings, *Wear*, 157 (1992), 215–227.

71. S.J. Bull, P.R. Chalker, C. Johnston, V. Moore, The effect of roughness on the friction and wear of diamond thin films, *Surf. Coat. Technol.*, 68–69 (1994), 603–610.

72. S.S. Perry, J.W. Ager, G.A. Somorjai, Combined surface characterization and tribological (friction and wear) studies of CVD diamond films, *J. Mater. Res.*, 8 (1993), 2577–2586.

73. C.S. Abreu, M. Amaral, A.J.S. Fernandes, F.J. Oliviera, R.F. Silva, J.R. Gomes, Friction and wear performance of HFCVD nanocrystalline diamond coated silicon nitride ceramics, *Diamond Relat. Mater.*, 15 (2006), 739–744.

74. A. Schade, S.M. Rosiwal, R.F. Singer, Influence of surface topography of HF-CVD diamond films on self-mated planar sliding contacts in dry environments, *Surf. Coat. Technol.*, 201 (2007), 6197–6205.

75. C.S. Abreu, E. Salgueiredo, F.J. Oliviera, A.J.S. Fernandes, R.F. Silva, J.R. Gomes, CVD diamond water lubricated tribosystems for high load planar sliding, *Wear*, 265 (2008), 1023–1028.
76. D.W. Wheeler, unpublished data (1997).
77. Z. Feng, J.E. Field, Friction of diamond on diamond and chemical vapour deposition diamond coatings, *Surf. Coat. Technol.*, 47 (1991), 631–645.
78. S.E. Grillo, J.E. Field, The friction of natural and CVD diamond, *Wear*, 254 (2003), 945–949.
79. C.S. Abreu, M. Amaral, F.J. Oliviera, J.R. Gomes, R.F. Silva, HFCVD nanocrystalline diamond coatings for tribo-applications in the presence of water, *Diamond Relat. Mater.*, 18 (2009), 271–275.
80. M. Kohzaki, K. Higuchi, S. Noda, Frictional properties of chemical vapor deposited diamond thin films, *Mater. Lett.*, 9 (1990), 80–82.
81. P. Hollman, A. Alahelisten, T Bjork, S. Hogmark, CVD-diamond coatings in sliding contact with Al, Al-17Si and steel, *Wear*, 179 (1994), 11–16.
82. L. Schafer, M. Hofer, R. Kroger, The versatility of hot-filament activated chemical vapor deposition, *Thin Solid Films*, 515 (2006), 1017–1024.
83. C.H. Shen, Performance and applications of CVD thin film diamond coated tools, *Key Eng. Mater.*, 138–140 (1998), 25–55.
84. C. Faure, W. Hanni, C.J. Schmutz, M. Gervanoni, Diamond-coated tools, *Diamond Relat. Mater.*, 8 (1999), 830–833.
85. I.E. Clark, P.K. Sen, Advances in the development of ultrahard cutting tool materials, *Ind. Diamond Rev.*, 58 (1998), 40–44.
86. H. Yoshikawa, A. Nishiyama, CVD diamond coated insert for machining high silicon aluminum alloys, *Diamond Relat. Mater.*, 8 (1999), 1527–1530.
87. J. Halwax, R. Pfaffenberger, CVD diamond tools machining aluminium in the new 5-series BMWs, *Ind. Diamond. Rev.*, 64 (2004), 21–22.
88. J.Y. Sheikh-Ahmad, J.S. Stewart, H. Feld, Failure characteristics of diamond-coated carbides in machining wood-based composites, *Wear*, 255 (2003), 1433–1437.
89. J.P. Davim, F. Mata, Chemical vapour deposition (CVD) diamond coated tools performance in machining of PEEK composites, *Mater. Design* 29 (2008), 1568–1574.
90. M. Belmonte, P. Ferro, A.J.S. Fernandes, F.M. Costa, J. Sacramento, R.F. Silva, Wear-resistant CVD diamond tools for turning of sintered hardmetals, *Diamond Relat. Mater.*, 12 (2003), 738–743.

91. N.C. Engdahl, CVD Diamond-Coated Rotating Tools for Composite Machining, http://www.crystallume.com/html/composite_machining.html (Accessed 29 Aug 2012).

92. S. Turchetta, L. Carrino, W. Polini, CVD diamond insert in stone cutting, *Diamond Relat. Mater.*, 14 (2005), 641–645.

93. M. Dai, K. Zhou, Z. Yuan, Q. Ding, Z. Fu, The cutting performance of diamond and DLC-coated cutting tools, *Diamond Relat. Mater.*, 9 (2000), 1753–1757.

94. http://en.blades.diamaze-gfd.com/press_blades_en.html (Accessed 8 Mar 2012).

95. L. Tang, C. Tsai, W.W. Gerberich, L. Kruckeberg, D.R. Kania, Biocompatibility of chemical-vapour-deposited diamond, *Biomaterials*, 16 (1995), 483–488.

96. O. Auciello, A.V. Sumant, Status review of the science and technology of ultrananocrystalline diamond (UNCD^{TM}) films and application to multifunctional devices, *Diamond Relat. Mater.*, 19 (2010), 699–718.

97. W. Jakubowski, G. Bartosz, P. Niedzielski, W. Szymanski, B. Walkowiak, Nanocrystalline diamond surface is resistant to bacterial colonization, *Diamond Relat. Mater.*, 13 (2004), 1761–1763.

98. L. Yang, Y. Li, B.W. Sheldon, T.J. Webster, Altering surface energy of nanocrystalline diamond to control osteoblast responses, *J. Mater. Chem.*, 22 (2012), 205–214.

99. M. Seal, High technology applications of diamond, (ed. J.E. Field) *The Properties of Natural and Synthetic Diamond*, Academic Press, London (1992), pp. 607–636.

100. http://en.blades.diamaze-gfd.com/ (Accessed 8 Mar 2012).

101. H. Sein, W. Ahmed, C. Rego, Application of diamond coatings onto small dental tools, *Diamond Relat. Mater.*, 11 (2002), 731–735.

102. C.F.M. Borges, P. Magne, E. Pfender, J. Heberlein, Dental diamond burs made with a new technology, *J. Prosthet. Dent.*, 82 (1999), 73–79.

103. W. Ahmed, H. Sein, M. Jackson, R. Polini, Chemical vapour deposition of diamond films onto tungsten carbide dental burs, *Tribol. Int.*, 37 (2004), 957–964.

104. H. Sein, W. Ahmed, C.A. Rego, A.N. Jones, M. Amar, M. Jackson, R. Polini, Chemical vapour deposition diamond coating on tungsten carbide dental cutting tools, *J. Phys.: Condens. Matter*, 15 (2003), S2961–S2967.

105. M. Amar, W. Ahmed, H. Sein, A.N. Jones, C.A. Rego, Chemical vapour deposition of diamond coatings onto molybdenum dental tools, *J. Phys.: Condens. Matter*, 15 (2003), S2977–S2982.
106. M.J. Papo, S.A. Catledge, Y.K. Vohra, C. Machado, Mechanical wear behavior of nanocrystalline and multilayer diamond coatings on temporomandibular joint implants, *J. Mater. Sci.: Mater. Med.*, 15 (2004), 773–777.
107. B.S. Nau, Mechanical seal face materials, *Proc. Inst. Mech. Eng. J.: J. Eng. Tribol.*, 211 (1997), 165–183.
108. P. Hollman, H. Bjorkman, A. Alahelisten, S. Hogmark, Diamond coatings applied to mechanical face seals, *Surf. Coat. Technol.*, 105 (1998), 169–174.
109. M.J. Neale, *The Tribology Handbook* (2nd ed.), Butterworth-Heinemann, Oxford (1995).
110. B.M. Steinetz, Seal technology, (ed. M. Kutz) *Mechanical Engineers' Handbook: Materials and Mechanical Design Vol. 1* (3rd ed.), John Wiley, New Jersey (2006), pp. 1161–1203.
111. G.A. Jones, On the tribological behaviour of mechanical seal face materials in dry line contact Part I. Mechanical carbon, *Wear,* 256 (2004), 415–432.
112. L. Thorvart, The materials key to seal performance, *Mater. World*, 2 (1994), 519–521.
113. P.J. Kelly, R.D. Arnell, M.D. Hudson, A.E.J. Wilson, G. Jones, Enhanced mechanical seal performance through CVD diamond films, *Vacuum*, 61 (2001), 61–74.
114. G.A. Jones, On the tribological behaviour of mechanical seal face materials in dry line contact Part II. Bulk ceramics, diamond and diamond-like carbon films, *Wear* 256 (2004), 433–455.
115. M.A. Tome, A.J.S. Fernandes, F.J. Oliveira, R.F. Silva, J.M. Carrapichano, High performance sealing with CVD diamond self-mated rings *Diamond Relat. Mater.*, 14 (2005), 617–621.
116. F. Mubarok, J.M. Carrapichano, F.A. Almeida, A.J.S. Fernandes, R.F. Silva, Enhanced sealing performance with CVD nanocrystalline diamond films in self-mated mechanical seals, *Diamond Relat. Mater.*, 17 (2008), 1132–1136.
117. A.M. Kovalchenko, J.W. Elam, A. Erdemir, J.A. Carlisle, O. Auciello, J.A. Libera, M.J. Pellin, D.M. Gruen, J.N. Hryn, Development of ultrananocrystalline diamond (UNCD) coatings for multipurpose mechanical pump seals, *Wear*, 270 (2011), 325–331.

118. Diamond faces deliver performance. *Sealing Technology* (August 2007), pp. 2–3.

119. Diamond pump seals launched. *Sealing Technology* (April 2008), p. 1.

120. Diamond wear surfaces optimised for dry running. *Sealing Technology* (April 2010), p. 16.

121. R.S. Muller MEMS: Quo Vadis in Century XXI? *Microelectron Eng.*, 53 (2000), 47–54.

122. S.H. Kim, M.T. Dugger, A. Erdemir, A.L. Barnette, E. Hsiao, M.J. Marino, O. Eryilmaz, MEMS lubrication with alcohol vapour, *Tribology*, 4 (2010), 109–114.

123. V. Singer, A growing global demand for CVD diamond, *Finer Points*, (Summer 2012), 10–13.

124. K. Komvopoulos, Surface engineering and microtribology for micro-electromechanical systems, *Wear*, 200 (1996), 305–327.

125. O. Auciello, J. Birrell, J.A. Carlisle, J.E. Gerbi, X. Xiao, B. Peng, H.D. Espinosa, Materials science and fabrication processes for a new MEMS technology based on ultrananocrystalline diamond thin films, *J. Phys.: Condens. Matter*, 16 (2004), R539–R552.

126. S.M. Spearing, Materials issues in micro-electromechanical systems (MEMS), *Acta Mater.*, 48 (2000), 179–196.

127. W. Wang, Y. Wang, H. Bao, B. Xiong, M. Bao, Friction and wear properties in MEMS, *Sens. Actuators A*, 97–98 (2002), 486–491.

128. Z. Cao, D. Aslam, Fabrication technology for single-material MEMS using polycrystalline diamond, *Diamond Relat. Mater.*, 19 (2010), 1263–1272.

129. M. Nosonovsky, B. Bhushan, Green tribology: principles, research areas and challenges, *Phil. Trans. R. Soc. A*, 368 (2010), 4677–4694.

130. I.S. Forbes, J.I.B. Wilson, Diamond and hard carbon films for micro-electromechanical systems (MEMS)—a nanotribological study, *Thin Solid Films*, 420–421 (2002), 508–514.

131. J.D. Hunn, S.P. Withrow, C.W. White, R.E. Clausing, L. Heatherly, C.P. Christensen, Fabrication of single-crystal diamond microcomponents, *Appl. Phys. Lett.*, 65 (1994), 3072–3074.

132. E. Kohn, P. Gluche, M. Adamschik, Diamond MEMS—a new emerging technology, *Diamond Relat. Mater.*, 8 (1999), 934–940.

133. S. Ertl, M. Adamschik, P. Schmid, P. Gluche, A. Floter, E. Kohn, Surface micromachined diamond microswitch, *Diamond Relat. Mater.*, 9 (2000), 970–974.

134. T. Shibata, Y. Kitamoto, K. Unno, E. Makino, Micromachining of diamond film for MEMS applications, *J. Microelectromech. Syst.*, 9 (2000), 47–51.

135. X.D. Wang, G.D. Hong, J. Zhang, B.L. Lin, H.Q. Gong, W.Y. Wang, Precise patterning of diamond films for MEMS application, *J. Mater. Proc. Technol.*, 127 (2002), 230–233.

136. http://en.microparts.diamaze-gfd.com/economy_linie_en.html (Accessed 25 April 2012).

137. K.H. Chung, D.E. Kim, Wear characteristics of diamond-coated atomic force microscope probe, *Ultramicroscopy*, 108 (2007), 1–10.

138. J. Liu, D.S. Grierson, N. Moldovan, J. Notbohm, S. Li, P. Jaroenapibal, S.D. O'Connor, A.V. Sumant, N. Neelakantan, J.A. Carlisle, K.T. Turner, R.W. Carpick, Preventing nanoscale wear of atomic force microscopy tips through the use of monolithic ultrananocrystalline diamond probes, *Small*, 6 (2010), 1140–1149.

139. D.W. Olson, Diamond, Industrial, *U.S. Geological Survey Minerals Yearbook*, (2003), 22.1–22.9.

140. V.J. Trava-Airoldi, E.J. Corat, N.G. Ferreira, N.F. Leite, CVD-Diamond: an overview of research and development at INPE, *Brazil. J. Phys.*, 27 (1997), 88–92.

141. http://www.cemecon.com/coating_competence/coating_ materials/ccdiasupreg_supfiberspeed/index_eng.html (Accessed 24 Nov 2012).

142. http://www.cemecon.com/coating_competence/coating_materials/ccdiasupreg_supmultispeed/index_eng.html (Accessed 24 Nov 2012).

143. http://www.diamondtc.com/media/hpc/index.html (Accessed 21 Nov 2012).

144. M. Zou, I. Green, Real-time condition monitoring of mechanical face seal, (ed. D. Dowson *et al.*) *Tribology for Energy Conservation*, Elsevier Science B.V. (1998), pp. 423–430.

145. http://eagleburgmannnow.com/ebnow/index.cfm/stories/diamond-coating-increases-service-life-of-mechanical-seals/ (Accessed 5 Dec 2012).

146. Eagle Burgmanns new-generation Cartex cartridge seal benefits from DiamondFace coating. *Sealing Technology* (October 2012), p. 1.

Chapter 2

Thermal Barrier Coating Systems: Multilayer Multifunctional Surface Engineering

Richard G. Wellman
Cranfield University, Bedford MK43 0AL, UK
r.wellman@cranfield.ac.uk

This chapter looks at thermal barrier coatings (TBCs) from the view point of their standing and performance as multifunctional tribological coatings, concentrating on the ceramic topcoat while briefly touching on the performance and attributes of the bond coat. After a brief overview of TBC systems, designing for low thermal conductivity and the concepts behind integrating sensors into the ceramic topcoat are discussed. The rest of the chapter examines the ceramic topcoat of the TBC system from a functional perspective, including the effects of sintering, erosion and CMAS (calcium, magnesium alumino-silicate) attack (corrosion).

While the current industrial standard ceramic topcoat of 7–8 wt.% yttria-stabilised zirconia has been in service for a number of decades, it is fast approaching its operational limits and new systems are being investigated. However, changes to the composition of the ceramic insulating topcoat to incorporate sensing technology or to

Multifunctional Materials for Tribological Applications
Edited by Robert J. K. Wood

ISBN 978-981-4463-90-4 (Hardcover), 978-981-4463-91-1 (eBook)
www.panstanford.com

reduce thermal conductivity often results in a reduction of erosion resistance, while increases in turbine operating temperatures have resulted in the advent of CMAS attack of the ceramic topcoat.

2.1 Introduction

Thermal barrier coatings (TBCs) are a unique class of multilayer coatings that have been specifically designed to protect the metallic components operating in the extreme conditions experienced in gas turbines [1]. Although designed primarily for thermal protection, due to the nature of the environment in which they operate and recent developments in sensing technology, modern TBCs need to be truly multifunctional. The drive to high thermally efficient engines resulted in an increase in operating temperature, specifically the turbine entry temperature. Since the super alloys used in the manufacture of high pressure (HP) turbine blades and the nozzle guide vanes are already operating in temperatures exceeding their melting points, TBCs are now integral in the component surviving the design life and have become prime reliant. Internal cooling and air film boundary layers combined with yttria-stabilised zirconia (YSZ) TBCs allow HP turbine components to operate in the extreme environment of the modern gas turbine engine. However, the task of the TBC system, which includes the bond coat (BC), the thermally grown oxide (TGO) and the ceramic topcoat (typically yttria partially stabilised zirconia—YSZ) involves far more than mere thermal protection.

The latest TBC system is expected to have an extremely low thermal conductivity, less than $1.5\,\mathrm{W\,K^{-1}\,m^{-1}}$, to have a BC that is oxidation and corrosion resistant, to be erosion resistant, to be resistant to sintering at operating temperatures of the ceramic surface in excess of 1200°C, as well as resistant to chemical attack by low melting point CMAS type fluxes. Furthermore, in the future TBCs are more than likely to incorporate some form of phosphorescent sensors for in situ temperature and/or lifing measurements.

There are really two issues that need to be considered in the design and operation of a good TBC system, thermal conductivity and durability. However, as already mentioned, there are a number of ways in which a TBC can degrade and lose durability, issues which have to be considered in the design stage of developing a TBC.

For many years oxidation of the BC and subsequent spallation was accepted as the primary cause of failure, and issues around oxidation and thermal cycling have been extensively researched [2–4]. Erosion of TBCs [5] was always considered a secondary problem; more recently with the use of electron beam (EB) physical vapour-deposited (PVD) TBCs on turbine blade aerofoils, the effects of erosion has become more of an issue and has subsequently received more attention. Early work concentrated on the effect of variables like velocity, impact angle and erodent properties [6–9], while more recent work has examined erosion mechanisms [10–12] and the effects of aging and dopant additions [13].

With the drive for higher operating temperatures and the associated increase in the service operating temperature of TBCs, CMAS (calcium mangnesium alumino-silicatcs, essentially atmospheric dust) attack has become a life-limiting issue. CMAS attack (or thermo-chemical degradation) is the chemical degradation of the ceramic topcoat of a TBC by molten low melting point slags. When the ceramic surface temperature is greater than the CMAS melting point, typically around 1230–1260°C, [14] the CMAS will infiltrate the TBC resulting in extensive degradation of the columnar microstructure and 'locking' the tops of the columns together on cooling, resulting in a loss of strain tolerance and possible spallation on cooling.

The schematic in Fig. 2.1 illustrates the various degradation mechanisms that can occur to TBCs during service in gas turbine engines. Although the diagrams are all of EB PVD TBCs, plasma-sprayed (PS) TBCs will suffer from the same issues, although there will be some differences in the mechanisms. During laboratory testing most of these issues are investigated independently, whereas in reality any number of these mechanisms could occur simultaneously.

In summary, the modern TBC system is in fact a multilayer, multifunctional tribological coating that needs to be chemically inert and erosion/impact resistant with a low thermal conductivity and a thermal expansion coefficient similar to the substrate, which ideally also incorporates sensing capabilities. This chapter aims to look at how TBCs are designed to achieve their multifunctionality.

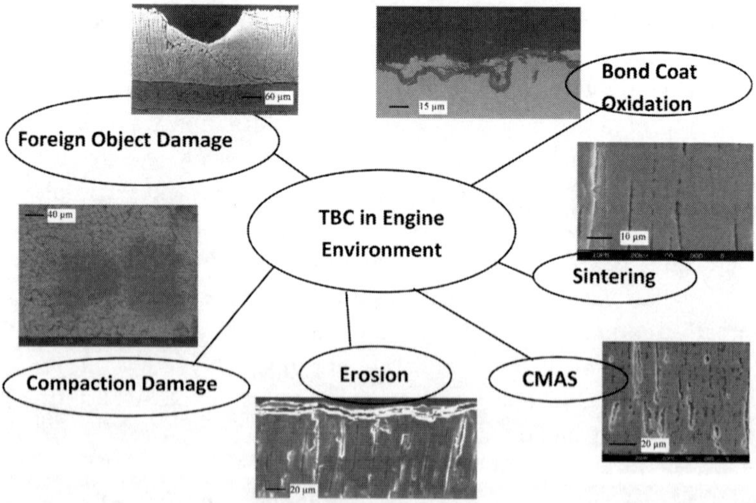

Figure 2.1 Schematic illustrating the various types of life-limiting damage that can occur to EB PVD TBCs in an engine environment, PS TBCs will have similar issues although the mechanisms might differ to a certain extent [29].

2.2 TBCs: The Basics

Current TBC systems can be broadly divided into two main categories, EB PVD and air plasma sprayed (APS) TBCs, depending on the route used for depositing the ceramic topcoat. However, recent developments include PS PVD and sol gel TBCs, PS PVD TBCs are still a relatively new concept while sol gel application is a more established technology; however, neither are currently in use in gas turbine engines. Whatever route is used to apply the ceramic topcoat, the basic TBC system in a gas turbine engine is still very similar consisting of a cooled substrate, which is normally a superalloy (sometimes in the single crystal form) with three subsequent layers:

A BC system: The BC, which is applied to the substrate, has a number of functions that includes imparting oxidation/corrosion resistance to the superalloy while minimising the thermal mismatch between the superalloy substrate and the ceramic topcoat. There are two main families of BCs, the established MCrAlY family of corrosion protection coatings, which are often applied by thermal spraying, and diffusion aluminide coatings. A recent modification of the diffusion coating is the gamma, gamma prime ($\gamma + \gamma'$) platinum aluminide coatings [15–18]. One of the fundamental roles of the BC is to act as an aluminium reservoir to supply aluminium to the TGO.

The TGO: Due to the fact that the ceramic topcoat of a TBC system is deposited onto the TGO, the morphology and the growth rate of the TGO is critical to the durability of the TBC system. Ideally the TGO should be slow growing, defect free, adherent alpha-alumina layer that does not react chemically with the ceramic topcoat. The study of how the TGO grows, degrades, cracks chemically reacts and finally fails (via delamination) due to thermal cyclic exposure is the subject of numerous research papers [2–4, 19]. Although the behaviour and adherence of the TGO to both the BC and the topcoat ultimately determines the life of the TBC system (under oxidation/corrosion conditions), it is the ceramic topcoat that supplies the thermal properties to the TBC system.

The ceramic topcoat: The primary function of the ceramic topcoat is to act as a thermal insulator reducing the transfer of heat to the substrate metal. However, it also needs to be chemically inert, have a good thermal shock and erosion resistance, have a thermal expansion similar to the substrate while having a low thermal conductivity [20]. The current 'industry standard' essentially established in the 1980s [21] is ZrO_2 partially stabilised with 7–8 wt.% Y_2O_3 (YSZ); however with the drive to higher operating temperatures, YSZ is fast reaching its temperature limit, due to phase transformations, sintering and chemical attack [22] and numerous alternative ceramics are currently being investigated. These include zirconia with multiple rare earth stabilisers [23–25] and pyrocholres such as gadolinia zirconate [26]; however, it has been suggested that some of the alternative materials have inferior erosion resistance and cyclic lives compared to the std YSZ [27, 28].

The SEM micrographs in Fig. 2.2 show the differences in the microstructure of an EB PVD TBC and a PS TBC. Although applied using the same material, the PS TBC has a lower conductivity ($0.8\ \mathrm{W\,k^{-1}\,m^{-1}}$) as opposed to $1.8\ \mathrm{W\,k^{-1}\,m^{-1}}$ for the EB PVD version

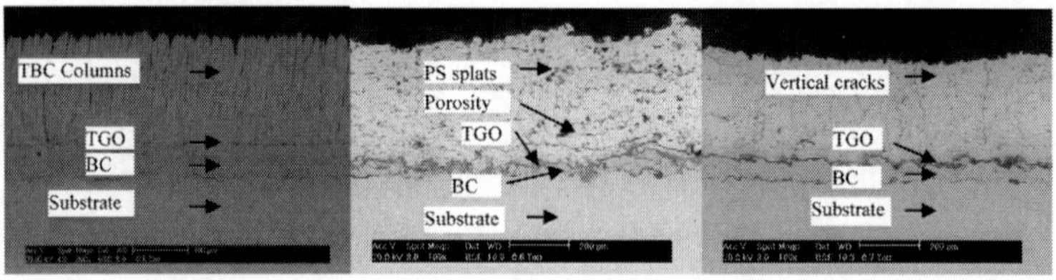

(a) (b) (c)

Figure 2.2 SEM micrographs of EB PVD, PS and vertically cracked PS TBCs [29].

of the YSZ topcoat [25], which is purely due to the differences in the microstructure [25].

2.3 Designing for Low Thermal Conductivity

As observed by Nicholls et al. [25], the thermal conductivity of YSZ TBCs is strongly influenced by the microstructure of the coating, which in turn is a function of the deposition route. These microstructures that are shown in the micrograph in Fig. 2.2 not only affect the thermal properties but also the surface roughness, adhesive strength, Young's modulus and the erosion resistance of the coatings. The orientation and type of microstructural defects, for example, splat boundaries, column boundaries and porosity, all contribute to reducing thermal conductivity. It is the planar scattering centres of the PS TBC (the intersplat boundaries) that contribute to its low thermal conductivity but also to its low strain tolerance and high erosion rate, compared to EB PVD TBCs. The chemical composition of the TBC also affects the thermal conductivity, and it has been shown that as the yttria content, or the content of other rare earth (RE) oxide dopants, increases the thermal conductivity decreases in YSZ systems [25]. Porosity also influences thermal conductivity of EB PVD TBCs, both the size and shape distribution of the pores and to a lesser extent the cross sectional area of the pore perpendicular to the heat flux, will all influence the overall thermal conductivity of the system [30].

Ultimately, there are not very many options available to the design engineer to reduce thermal conductivity in these systems and one is limited to lowering the mean free path of the heat carriers, increasing the porosity or lowering the velocity of the heat carriers [25]. In crystal structures, lattice imperfections, such as vacancies, dislocations, grain boundaries and impurities, all act as phonon scattering centres thus lowering the thermal conductivity. Reduction of thermal conductivity by modification of either photon or phonon transport can be achieved by [25]:

(1) The introduction of lattice defects, through the addition of dopants or by ion bombardment during deposition, thus reducing phonon conduction.
(2) Colouring the TBC (via the addition of RE oxides) to increase the opaqueness to infra-red radiation, reducing photon conduction.
(3) The introduction of structures (layers parallel to the substrate) to scatter both photons and phonons.
(4) The incorporation of nanograined structures to scatter both phonons and photons.
(5) Increasing the porosity of the system.

As illustrated in the graph in Fig. 2.3, the addition of rare earth oxides can have a significant effect on the thermal conductivity of EB PVD TBCs. However, these additions not only affect erosion rate

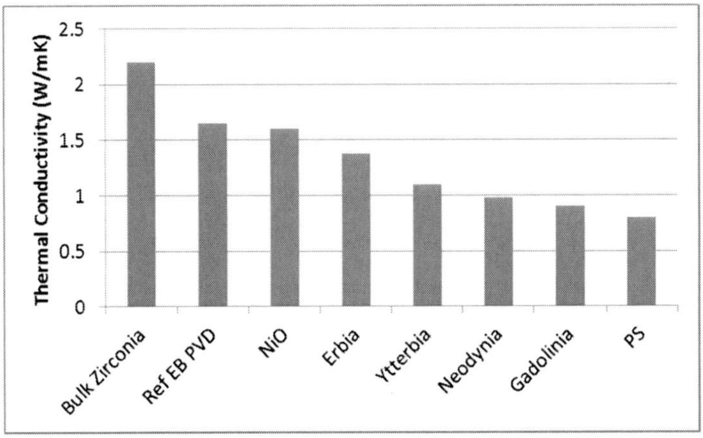

Figure 2.3 Thermal conductivity of doped EB PVD TBCs. Graph replotted from Ref. [23].

but as the level of dopants increases, the thermal cyclic life of the coating system decreases. While recent studies by Wang et al. [31] have shown that the use of nanostructured YSZ powders layered with lanthanum zirconate can also lower the thermal conductivity of sprayed TBC systems. However, as will be discussed later, some of these routes have detrimental effects on other tribological properties of the TBC and thus, as is often the case in surface engineering, there needs to be a trade-off between the various properties required for a given application.

2.4 Sintering of TBCs

Another issue that affects the overall life and performance of a TBC is the degree to which it will sinter during service and how the sintering will affect the mechanical and thermal properties of the coatings. When standard 7YSZ TBCs are exposed to temperatures above 1200°C for extended periods of operation phase transformations (with the associated changes in volume) and sintering can become a significant issue, which results in coating degradation and even spallation [32–34]. Sintering of TBCs not only affects the mechanical properties of TBCs, but also their thermal properties. It has been shown that after just a few hours at 1400°C, the in plane Young's modulus of PS TBCs increases from 20 to in excess of 50 GPa [35], while for EB PVD TBCs an increase in Young's modulus from 30 to in excess of 80 GPa was observed in coatings aged at 1200°C for up to 100 hr [36].

In PS TBCs, sintering results in the reduction of porosity, increased inter-splat bonding and the healing of through thickness micro cracks. However, it is this fine inter-splat porosity and cracks, especially those that are perpendicular to the heat flux, which contribute to the low thermal conductivity of PS TBCs as well as their strain compliance. Sintering of PS TBC has been shown to increase the thermal conductivity of the coatings by 50% [35]. However, Xie et al. also demonstrated that sintering of PS TBCs was highly dependent on the percentage of impurities in the coating particularly of alumina and silica and showed that by reducing the amount these components from 0.2 wt.% to less than 0.1 wt.%, the sintering resistance and phase stability of the TBC was increased significantly.

While EB PVD TBCs have a totally different microstructure to PS TBCs, they still contain a high degree of porosity, both inter-columnar porosity and intra-column porosity. It is this inter-column porosity as a result of the columnar microstructure and weakly connected columns that give EB PVD TBCs their strain compliance; however, this is reduced when the columns start to sinter together at the bridging contact points. While the intra-column porosity associated with the feather-like sub-column microstructure contribute to the low thermal conductivity of EB PVD TBCs. Sintering induces changes in both the morphology and distribution of both the open and the closed porosity increasing the Young's modulus of the coating while also increasing the thermal conductivity [37].

2.5 TBCs Designed as Sensors

A significant amount of research over the last few years has concentrated on lifing prediction of TBCs, since the life of a TBC is strongly dependent on the growth of the TGO, the temperature of the BC is an important factor. Topcoat spallation is related to TGO thickness and hence TGO growth rate that in turn is a function of time at temperature. An increase of 50°C in operating temperature of a TBC system can triple the growth rate of the TGO resulting in premature failure [38]. Considering that temperature measurement of gas turbine components are extremely unreliable, it is not surprising that life prediction models have a high degree of uncertainty. Technology enabling accurate and reliable measurement of temperature of gas turbine components would not only improve life prediction models but would also enable operators to run their engines at maximum efficiency. This need led Choy et al. [39] to introduce the notion of sensor TBCs, whereby the introduction of a dopant, which could act as a thermographic phosphor, would enable temperature measurement of TBCs.

Luminescence is the emission of photons that occurs when electrons return to the ground state from an excited state. The excited state can be caused by a number of factors including the absorption of light, in which case it is referred to as photoluminescence. Materials can be made luminescent by introducing optically active centres. RE ions, which absorb and emit visible spectrum, can be used as luminescent activators in TBC systems, furthermore, some of these also lower the thermal conductivity of the TBC [25].

Initially, YSZ doped with europia were used to demonstrate the feasibility of the concept [40], by using a layered system it was proven to be possible to measure temperature through 50 μm of std YSZ TBC. Research in the field of phosphorescent dopants for TBC systems has progressed rapidly with a wide range of phosphors being tested as possible dopants for TBC temperature sensing.

The major challenge that needed to be overcome was the temperature capability of the phosphors, early systems only had limited functional temperature ranges some as low as 400°C. Since the surface temperature of a TBC can be as high as 1300°C, the challenge was to identify systems that could achieve such functionality that were compatible with the TBC structure and chemistry and did not degrade with time at temperature. As the research in this field progressed, it became apparent that the phosphorescencent materials could also act as a means of detecting aging of TBCs. This is related to aging affecting the local atomic surroundings of the optically active ion [39, 41, 42]. Luminescence has also been used to study delamination and erosion of TBC systems [43–45].

Steenbakker et al. [46] designed and produced a multilayer phosphorous heat flux gauge for EB PVD TBC, comprising of an outer layer of YAG:Tm and an inner layer of YSZ:Dy, as illustrated in Fig. 2.4. During testing, it was demonstrated that temperature measurements could be measured simultaneously at two different depths in the TBC using a single excitation source. Furthermore, aging tests at 1200°C for 1000 hr showed that the YAG phase had a good thermal stability.

As will be discussed later in this chapter, erosion of TBCs can be a significant issue resulting in loss of thickness and consequently loss of thermal protection, which can result in premature failure of the entire coating system. Since small losses/thinning of the TBC cannot be detected by current inspection routines (visual inspections using boroscopes) significant safety margins are designed into the system

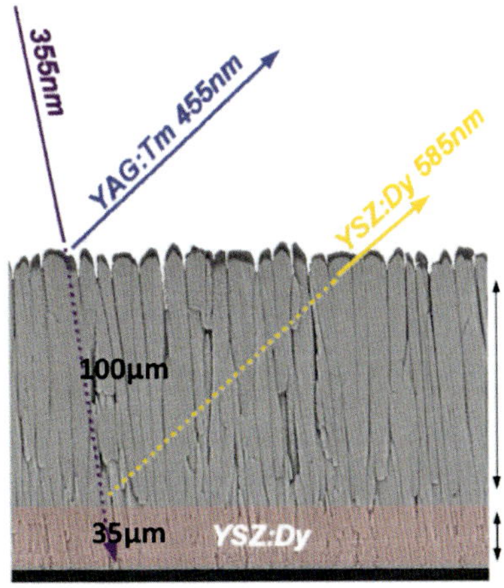

Figure 2.4 SEM micrograph of the multilayer sensing EB PVD TBC with the excitation and emission wavelengths [46].

thus limiting the overall efficiency. Quantitative non-destructive in situ evaluation of TBC degradation would enable condition based maintenance. Thus enabling TBCs to operate at their potential and reducing downtime and maintenance costs. Phosphor TBCs used in a multilayer system, as illustrated in Fig. 2.5, could be used to detect not only temperature but also depth of erosion damage. This system has subsequently been proven in an engine test.

Pilgrim et al. [44] have demonstrated the concept, which has been proposed previously [47–49], that through careful design of instrumentation and coating structure, it is possible to measure the depth of erosion in a TBC with a resolution equal to the thickness of the layers used.

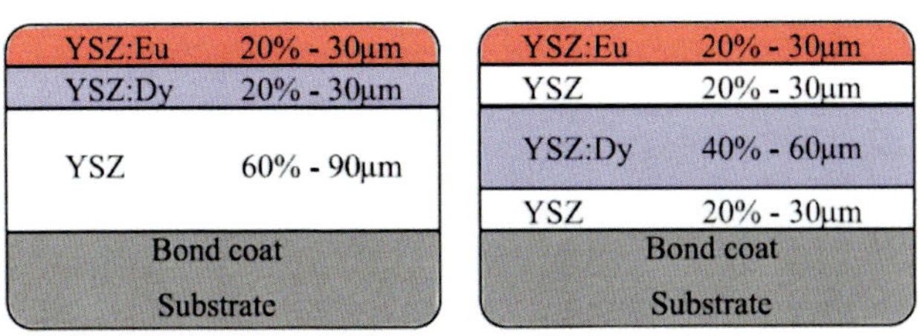

Figure 2.5 Schematic of the coating design [44].

2.6 Erosion of TBCs

So the question then becomes how do all of these factors affect the erosion performance of TBC systems? Erosion resistance is not something that is designed into TBC systems so much as a consequence of the resultant microstructure and composition resulting from the design of its thermal, sensing or sintering properties. However, before looking at how various different aspects such as sintering, composition, CMAS infiltration, microstructure and aging affect erosion of TBCs, it is worthwhile to consider the various different erosion (impact damage) mechanisms that can occur in TBC systems. One of the most important issues to remember is that TBCs are not erosion resistant and although EB PVD TBCs generally have a significantly lower erosion rate compared to PS TBCs [7], this is still considerably higher than both bulk YSZ and the metallic substrate. The graph in Fig. 2.6 is a plot of the erosion rates of EB PVD and PS TBCs and a bulk zirconia at two different temperatures; note that the erosion rate is plotted as a log scale. The erosion tests were conducted using 100 μm alumina particles at 140 m s^{-1} for the room temperature tests and 230 m s^{-1} for the high temperature erosion tests. The reason for these differences in erosion rates is a combination in the differences in hardness and

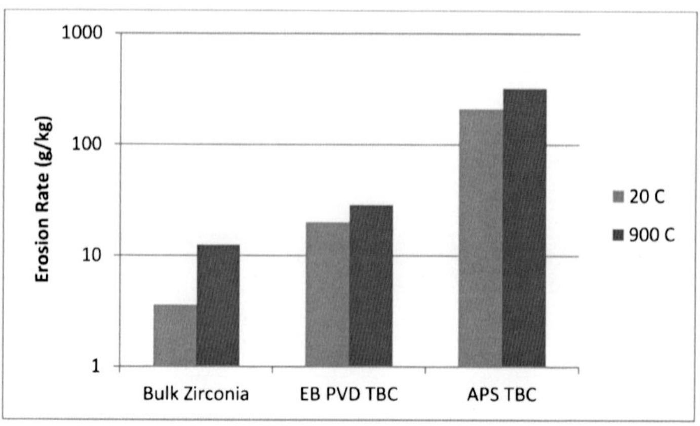

Figure 2.6 Erosion rates of bulk zirconia, EB PVD TBC and PS TBC. Graph replotted from Ref. [6].

microstructure, which together lead to differences in the erosion mechanisms that are occurring.

The bulk polycrystalline zirconia has the highest hardness and erodes via a classical brittle erosion mechanism involving the initiation and propagation of median and lateral cracks, as described by Lawn et al. [50–53]. Material is lost when the lateral cracks intersect with either the surface or other lateral cracks; this mechanism requires that a certain degree of damage be maintained within the eroding material for mass loss to occur.

Plasma sprayed coatings on the other hand erode via the initiation and propagation of cracks at the inter-splat boundaries. This results in a lower impact energy required to initiate cracks along, what are essentially, pre-existing flaws in the coating. Furthermore the erosion rate is thus also a function of splat size and more material is lost per event than for an EB PVD TBC.

However, EB PVD TBCs can exhibit a number of different responses to impact depending on a number of factors such as particle size, velocity density and column size, hardness and density, which can be neatly encompassed in the contact footprint diameter, D, to column diameter, d, ratio (D/d). The contact footprint diameter, D, will be a function of the impact conditions and the coating, and particle mechanical properties so encompasses all the impact and system variables. As a function of the column diameter, this will determine the number of columns with which the impacting particle interacts and ultimately this will determine the erosion mechanism and hence erosion rate [54]. This concept was an extension of the observation that the hardness of an EB PVD TBC, tested using a nanoindenter, was a function of the applied load (from 0.1 N to 2 N), due to the fact that as the load increased the number of columns interacting with the indenter increased and hence the compliance of the system increased [55]. So at low loads, provided the indenter was aligned to the centre of a column, the hardness of a single column was measured. At high loads, loads greater than 2N, the hardness of the coating 'as a whole' was measured, this is illustrated in the graph in Fig. 2.7.

Impact damage mechanisms of EB PVD TBCs were first categorised, from information available in the literature and experimental observations, into three different modes, erosion, compaction

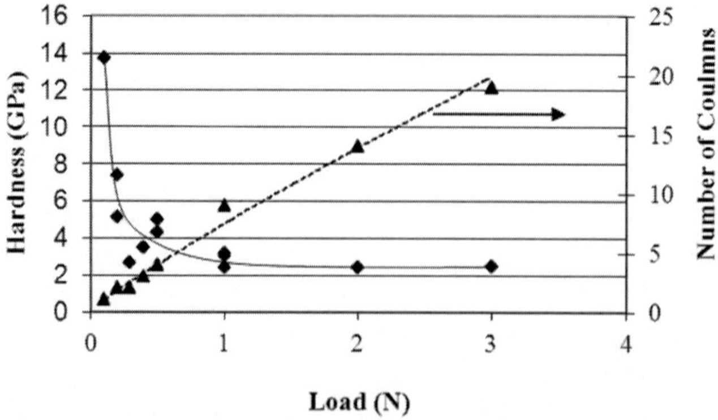

Figure 2.7 Effect of increasing load on the hardness of EB PVD TBCs [54, 55].

and foreign object damage (FOD) in 2003 [56]. Shortly afterwards, this was modified to include a second type of FOD (Type 2), which involved plastic deformation of the columnar microstructure [57] and finally refined in 2005 [11]. The different modes of impact damage can be summarised as:

Erosion (near surface lateral cracking—Mode I): This is the gradual chipping away of the coating, with impact damage typically confined to the top 30 μm of the coating. This mechanism operates during impact by small particles travelling at high velocities and cracks are initiated in individual columns and generally do not propagate into neighbouring columns; the D/d ratio is typically less than 2. A number of neighbouring columns need to be cracked before material is lost from the coating. This mechanism is independent of temperature and will occur at both low and high temperatures; however, the erosion rates will be different as factors such as hardness, fracture toughness and Young's Modulus, which affect erosion rate, will be affected by temperature [29].

Compaction damage (Mode II): This damage mechanism involves the densification of the top of the coating without either cracking or gross deformation of the coating. This is an intermediate transition mechanism between erosion and FOD and occurs when the contact footprint covers a number of different columns, the D/d ratio is typically between 3 and 10. Compaction damage is seldom observed

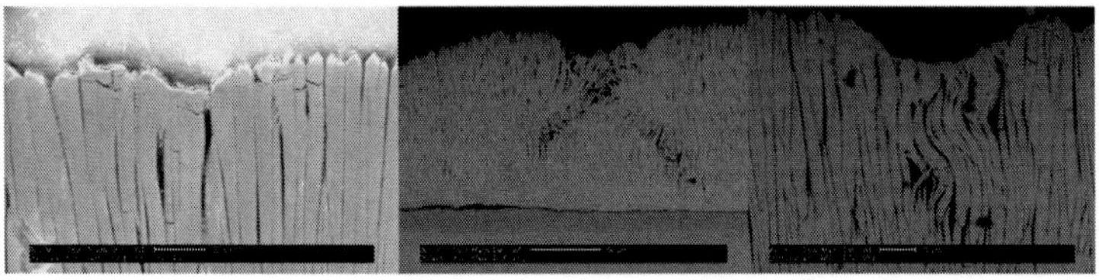

(a) (b) (c)

Figure 2.8 SEM micrographs of EB PVD TBCs showing the damage caused by impacting particles: (a) erosion, (b) FOD Type I and (c) FOD Type II [54].

on its own and can be observed in the presence of both Mode I and Mode III [10–12, 29].

Foreign object damage (Mode III): Is defined as gross deformation of the coating, which can penetrate to the BC, with associated regions of densification, cracking and kink bands, Type I FOD. Under FOD conditions, the contact area is significantly larger than the average column diameter and occurs over a significant number of columns, the D/d ratio is typically greater than 12. Under high temperature conditions, it is possible for the EB PVD columns to deform plastically in the form of 'S' bends without the associated cracking and kink band formation, Type II FOD, as a means of adsorbing the impact energy.

A fundamental point to realise is that these damage mechanisms are not mutually exclusive and it is perfectly feasible to observe a number of different mechanisms occurring in the same sample/test.

The erosion mechanisms of PS TBCs is slightly more difficult to qualify as, unlike EB PVD TBCs as illustrated in Fig. 2.8, damage mechanisms are not often observed in PS TBCs. This is due to the fact that once damage is introduced into the PS TBC system, the material is usually removed from the surface and there is little evidence to be observed post impact. However, it is generally accepted that PS TBCs erode via the initiation and propagation of cracks along the splat boundaries resulting in the loss of the splat from the surface of the coating [58–60]. Under high energy impacts, Eaton and Novak [58] proposed that pre-existing pores within the coating would connect through a crack network resulting in the loss of several splats on impact. This was supported by their observations that the erosion rate of PS TBCS increased with increasing porosity.

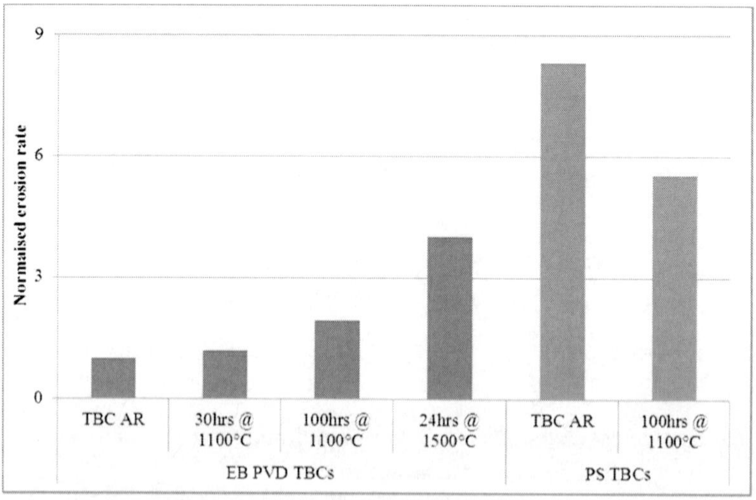

Figure 2.9 Relative erosion rated of EB PVD and PS TBCs after aging under different conditions. Graph replotted from Refs. [13, 61].

2.6.1 *Effect of Aging and Sintering on the Erosion of TBCs*

It is these different erosion mechanisms that are responsible for the different responses in erosion rates between EB PVD and PS TBCs after aging and the associated sintering that occurs during the aging of TBCs. As can be seen in the graph in Fig. 2.9, while the erosion rate of EB PVD TBCs increases with degree of aging and sintering, the erosion rate of PS TBCs decreases. This is a direct result of the different erosion mechanisms that are operating. In EB PVD TBCs, the sintering results not only in changes to pore geometry, which affect thermal conductivity, but also in necking between previously discrete columns. The result is that now cracks, generated during impact of the individual columns, will propagate into neighbouring columns that have sintered together, resulting in a greater mass loss per impact event than previously would have occurred. As the aging condition increases time and/or temperature, the degree of sintering increases and hence the erosion rate, as illustrated in the graph.

On the other hand in PS TBCs, aging results in sintering of the splats at the boundaries thus improving the inter-splat bonding and cohesion of the TBC. Since erosion of PS TBCs occurs through the propagation of cracks along splat boundaries, any increase in inter-splat cohesion will result in a decrease in the erosion rate, as illustrated in Fig. 2.9.

The erosion rates in the graph in Fig. 2.9 have all been normalised to an EB PVD TBC eroded at room temperature at an impact angle of 90°. All erosion rates in this chapter will be normalised in a similar manner.

2.6.2 *Effects of Dopants on the Erosion Rates of TBCs*

As mentioned previously, the addition of rare earth oxides can result in a significant reduction in the thermal conductivity of the TBC. Unfortunately, these additions also can have a negative effect on the erosion resistance of the TBCs. As demonstrated by Steenbakker et al. [62], the erosion rates of EB PVD TBCs increases quite dramatically with the addition of relatively small amounts (2–4 wt.%) of dopants as illustrated by the graph in Fig. 2.10; furthermore, the graph also illustrates how this effect is exacerbated

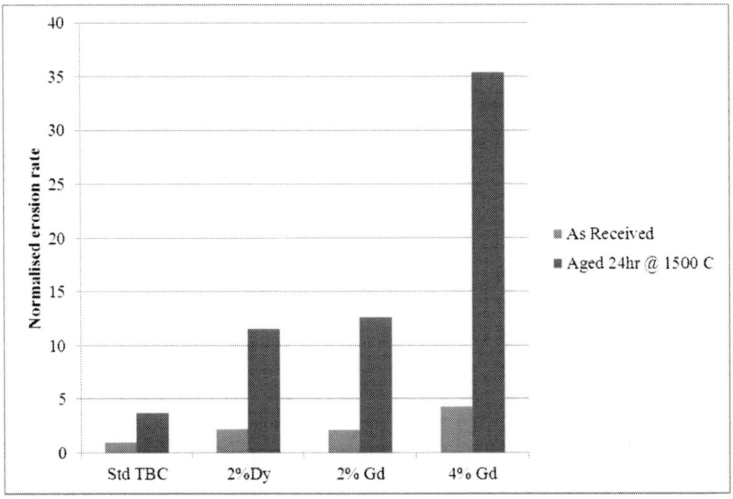

Figure 2.10 Effect of dopant additions on the erosion rates of EB PVD TBCs. Graph replotted from Ref. [62].

by the effect of aging, admittedly, in this case, under the relative harsh aging conditions of 24 hr at 1500°C.

This loss of erosion resistance has to be weighed against the reduction in thermal conductivity, which can see the conductivity of EB PVD TBC systems approach that of their std 7 YSZ PS TBCs counterparts. Although similar additions of the dopants to a PS TBC would result in a reduction of their thermal conductivity as well.

2.6.3 *Effect of Microstructure on the Erosion Rates of TBCs*

The effect of microstructure has already been discussed with respect to the difference between EB PVD and PS TBCs under aging and sintering effects; however, what about other microstructural changes, for example, porosity or column diameter? Recent developments in TBC technology has resulted in a number of changes to the microstructure of both families of TBCs. Strain-tolerant vertically cracked (VC) PS TBCs have been developed and more recently PS PVD TBCs, which have a low density EB PVD TBC type of microstructure but are deposited via plasma spraying, have entered the market. So how do all these changes in microstructure affect the erosion resistance of the various types of TBCs? The graph in Fig. 2.11 of the normalised erosion rates of various types of TBCs illustrates a number of issues.

In both EB PVD and PS TBCs an increase in porosity or a decrease in density will result in an initial lower thermal conductivity, it will also result in a higher erosion rate. On the other hand, the VC PS TBCs have been shown to have a lower erosion rate than std PS TBC and approaching that of EB PVD TBCs [29], although other authors have reported that highly segmented (VC) PS TBCs have an erosion rate slightly lower than standard EB PVD TBCs [60].

2.7 Corrosion of TBCs (CMAS Attack)

This chapter would not be complete without considering the concept of erosion–corrosion of TBCs, or more specifically the erosion of CMAS infiltrated TBCs. CMAS attack is a thermo-chemical reaction that occurs between the YSZ and atmospheric dusts and/or

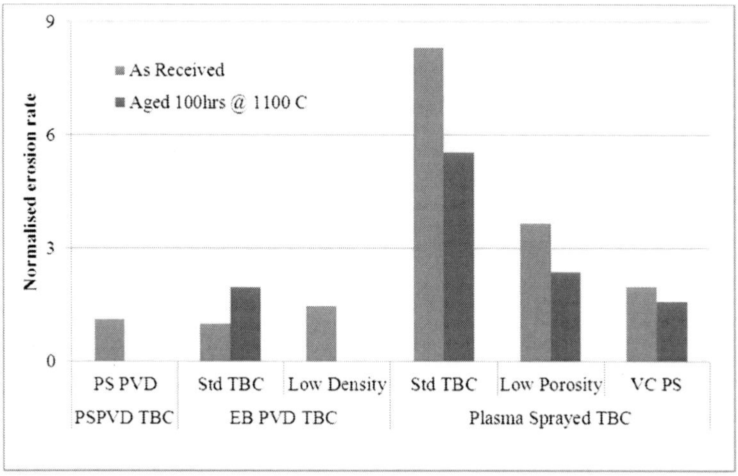

Figure 2.11 Effect of microstructure on the erosion rates of TBCs. Graph replotted from [54, 60, 63].

volcanic ash that are ingested, pass through the engine and deposit on the surface of coated components. When the surface temperature approaches or exceeds the melting point of the deposit, chemical interactions between the deposit and the YSZ can occur resulting in degradation of the TBCs.

There are essentially two main degradation routes, one mechanical and one chemical, that are not mutually exclusive. Mechanical degradation involves CMAS infiltration of the columnar microstructure of EB PVD TBCs when debris deposits on the coating and the surface temperatures are above the melting point of the deposited CMAS. Upon cooling, the CMAS solidifies and the resulting high in plane stiffness due to the loss of the strain compliance can initiate Mode I delaminations parallel to the surface [14]. Both EB PVD and APS TBCs are susceptible to loss of strain tolerance, due to the porous structures, which in the case of EB PVD TBCs are designed specifically to be effective in reducing thermal conductivity as well as accommodating the thermal expansion mismatch between the TBC and substrate.

Chemical degradation involves a combination of the melting and re-solidification of the ceramic topcoat and the chemical reaction of the CMAS and the ceramic topcoat. Research by Stott et al. was one of the first studies to show that TBC coatings were degraded readily by the molten sands and molten glass debris [64]. The zirconia and yttria were absorbed from the ceramic during attack preferentially at grain boundaries. It was noted that when the solubility limit of the Zirconia was exceeded that the phase was re-precipitated as a monoclinic fine grain structure depleted in yttria. This has also been reported in other studies; Kramer in particular has studied the effects of CMAS attack in great depth [65] and found that CMAS rapidly penetrates the open structure of an EB PVD TBC upon melting, which leads to changes in chemistry and microstructure. The extent of penetration is controlled by the interaction between the thermal gradient through the TBC with the fluidity of the melt. Recent work by Ndamka [66] has concentrated on finding the correlation between CMAS composition, viscosity and wettability on the extent of CMAS attack and has identified that the basicity index, a ratio of CaO to SiO_2, will have a significant effect on the extent of the attack and the melting point of the CMAS.

One of the drawbacks of most of the laboratory testing to date is the fact that it is generally conducted under isothermal conditions. However, a burner rig that operates in a thermal gradient has been developed at both Southside Thermal Sciences and at the Forschungszentrum Julich, which has been used to study the effects of CMAS attack under a thermal gradient and a thermal cyclic environment [67, 68]. More recently, information is becoming available in the open literature on the effects of CMAS attack on engine run hardware. Vidal-Setif et al. [69] have related the depth of CMAS penetration to the temperature of the component noting that the penetration drops off rapidly away from the hot zone of the aerofoils. Some of the features that they observed in engine run hardware, for example, the dissolution of the YSZ and the re-precipitation of an yttria-depleted monoclinic phase are in good agreement with the isothermal laboratory testing published in the literature. Meanwhile, Drexler et al. [26, 70, 71] have been studying the depth of penetration of CMAS into potential TBC materials including gadolinia zirconate, Al and Ti doped YSZ and various zirconia-based systems containing different levels of (Y^{3+}, Gd^{3+} or Yb^{3+}) solutes.

Currently there are three possible routes for dealing with the issue of CMAS attack of TBCs. The first is to ensure that the surface temperature of the coating never exceeds that of the melting point of the CMAS; however, this has the obvious drawback of limiting the operational temperature of the engine and hence is not a real option. The second is to apply a CMAS resistant top layer to protect the underlying YSZ, one of the current options being investigated is gadolinia zirconate. The final option is to develop a whole new ceramic topcoat that is CMAS resistant but that also has a good sintering resistance and phase stability.

2.8 Summary

This chapter was aimed at illustrating the multifunctional nature of TBCs highlighting the various aspects of their engineered functionality. As in nearly all multifunctional surface engineering solutions, and TBCs are no exception, there will always be a conflict between the various functional requirements and a compromise will need to be reached. For example, in current TBC systems, all the options that are available to lower the thermal conductivity result in a reduction in the erosion resistance. Although, various lifing models have been developed to predict service life of current TBC systems, these have been built up over numerous years of both engine run data and laboratory testing and are empirically based. Thus any new TBC system developed has to undergo rigorous testing under various conditions before it can be incorporated into engines.

2.9 Future TBC Systems

So what will future TBC systems be capable of from a surface engineering perspective and what degree of multifunctionality will be incorporated into the ceramic topcoat? One thing is certain, and that is that, as the drive for higher efficiencies and hence higher operating temperatures continues, TBC systems will have to evolve and develop to be able to accommodate the harsher environment in which they perform. As Vassen et al. [22] observed, the current standard 7 YSZ is fast approaching its operational limitations due to a combination of sintering, phase transformations, oxidation of the BC and the more recent issue of CMAS and VA attack. There are a number of alternatives that are currently being investigated as possible YSZ replacements (advanced TBCs), which include various doped zirconias, pyrochlores, hexaaluminates and zirconate perovskites [22]. While modelling is always a useful tool, TBCs are applied to prime reliant components and thus, for aero engine applications, rigorous testing will always be required. So while modelling TBCs will certainly inform the future direction of TBC development, it will never replace laboratory and engine testing.

As TBC systems evolve, it is certain that they will maintain their multilayer structure while becoming more complex in both the chemistry and structure of the insulating ceramic topcoat. Future systems will have to maintain their multifunctionality in terms of oxidation, sintering, CMAS and erosion resistance, while maintaining their microstructural stability at high operating temperatures. The possibility of including sensor systems in future generations of TBCs allowing both condition monitoring and accurate determination of the surface temperature will enable the systems to be safely operated at their capability limits without compromising safety or component integrity, thus resulting in higher efficiency engines.

References

1. R.G. Wellman, M.J. Deakin, J.R. Nicholls, The effect of TBC morphology and aging on the erosion rate of EB-PVD TBCs. *Tribol. Int.*, 38 (2005), 798–804.
2. T.A. Cruse, S.E. Stewart, M. Ortiz, Thermal barrier coating life prediction model development. *Transactions of the ASME*, 110 (1988), 610–616.
3. B.C. Wu, E. Chang, S.F. Chang, C.H. Chao, Thermal cyclic response of yttria-stabilised zirconia/CoNiCrAlY thermal barrier coatings. *Thin Solid Films*, 172 (1989), 185–196.
4. J.H. Sun, E. Chang, B.C. Wu, C.H. Tsai, The properties and performance of (ZR028wt%Y2O3) / (CVD Al2O3) / (Ni-22wt%Cr-10wt%Al-1wt%Y) thermal barrier coatings. *Surf. Coat. Technol.*, 58 (1993), 93–99.

5. F.C. Toriz, A.B. Thakker, S.K. Gupta, Flight service evaluation of thermal barrier coatings by physical vapor deposition at 5200H. *Surf. Coat. Technol.*, 39/40 (1989), 161–172.
6. W. Tabakoff, Investigation of coatings at high temperature for use in turbomachinery. *Surf. Coat. Technol.*, 30/40 (1989), 97–115.
7. J.R. Nicholls, Y. Jaslier, D.S. Rickerby, Erosion of EB-PVD thermal barrier coatings. *Mater. High Temp.*, 15 (1998), 15–22.
8. A.G. Davis, D.H. Boone, A.V. Levy, Erosion of ceramic thermal barrier coatings. *Wear*, 110 (1986), 101–116.
9. T. Rhys-Jones, F.C. Toriz, Thermal barrier coatings for turbine use in aero engines. *High Temperature Technology*, 7 (1989), 73–81.
10. R.G. Wellman, J.R. Nicholls, Some observations on the erosion mechanism. *Wear*, 242 (2000), 89–96.
11. R.G. Wellman, M.J. Deakin, J.R. Nicholls, The effect of TBC morphology on the erosion rate of EB PVD TBCs. *Wear*, 258 (2005), 349–356.
12. X. Chen, M.Y. He, I. Spitsberg, N.A. Fleck, J.W. Hutchinson, A.G. Evans, Mechanisms governing the high temperature erosion of thermal barrier coatings. *Wear*, 256 (2004), 735–746.
13. R.G. Wellman, J.R. Nicholls, On the effect of aging on the erosion of EB PVD TBCs. *Surf. Coat. Technol.*, 177–178 (2004), 80–88.
14. C. Mercer, S. Faulhaber, A.G. Evans, R. Darolia, A delamination mechanism for thermal barrier coatings subject to calcium-magnesium-alumino-silicate (CMAS) infiltration. *Acta Mater.*, 53 (2005), 1029–1039.
15. J.R. Nicholls, Designing oxidation-resistant coatings. *JOM*, 52 (2000), 28–35.
16. J. Stringer, R. Viswananthan, Gas Turbine Hot Section Materials and Coatings in Electric Utility Applications. *Proc. ASM 'Materials Week'* (1993), pp. 1–21.
17. G.W. Goward, Progress in coatings for gas turbine airfoils. *Surf. Coat. Technol.*, 108–109 (1998), 73–79.
18. J.R. Nicholls, Advances in coating design for high-performance gas turbines. *MRS Bull.*, 28 (2003), 659–670.
19. D.M. Nissley, Thermal Barrier Coating Life Modelling in Aircraft Gas Turbine Engines. *Thermal Barrier Coatings Workshop NASA Conference Publication*, 3312 (1995), 265–281.
20. R. Taylor, J.R. Brandon, P. Morrell, Microstructure, composition and property relationships of plasma-sprayed thermal barrier coatings. *Surf. Coat. Technol.*, 50 (1992), 141–149.

21. S. Bose, J. Demasi-Marcin, Thermal Barrier Coating Experience in the Gas Turbine Engine at Pratt & Whitney. *Thermal Barrier Coatings Workshop NASA Conference Publication*, 3312 (1995), 63–78.
22. R. Vaßen, M.O. Jarligo, T. Steinke, D.E. Mack, D. Stöver, Overview on advanced thermal barrier coatings. *Surf. Coat. Technol.*, 205 (2010), 938–942.
23. J.R. Nicholls, K.J. Lawson, D.S. Rickerby, P. Morrell, Advanced Processing of TBCs for Reduced Thermal Conductivity. *Thermal Barrier Coatings Agard Report*, 823 (1998), Paper 6.
24. D. Zhu, R. A. Miller, Thermal conductivity and sintering behavior of advanced thermal barrier coatings. *Ceram. Eng. Sci. Proc.*, 23 (2002), 457–468.
25. J.R. Nicholls, K.J. Lawson, A. Johnson, D.S. Rickerby, Methods to reduce the thermal conductivity of EB-PVD TBCs. *Surf. Coat. Technol.*, 151–152 (2002), 383–391.
26. J.M. Drexler, C.-. Chen, A.D. Gledhill, K. Shinoda, S. Sampath, N.P. Padture, Plasma sprayed gadolinium zirconate thermal barrier coatings that are resistant to damage by molten Ca-Mg-Al-silicate glass. *Surf. Coat. Technol.*, 206 (2012), 3911–3916.
27. D. Zhu, J. A. Nesbitt, T. R. McCue, C. A. Barrett, R. A. Miller, Furnace cyclic behavior of plasma-sprayed zirconia-yttria and multi-component rare earth oxide doped thermal barrier coatings. *Ceram. Eng. Sci. Proc.*, 23 (2002), 533–545.
28. M. Watanabe, C. Mercer, C. G. Levi, A. G. Evans, A probe foe the high temperature deformation of thermal barrier oxides. *Acta Mater.*, 52 (2004), 1479–1487.
29. R.G. Wellman, J.R. Nicholls, A review of the erosion of thermal barrier coatings. *J. Phys D: Appl. Phys.*, 40 (2007), 293–305.
30. A.F. Renteria, B. Saruhan, U. Schulz, H.-. Raetzer-Scheibe, J. Haug, A. Wiedenmann, Effect of morphology on thermal conductivity of EB-PVD PYSZ TBCs. *Surf. Coat. Technol.*, 201 (2006), 2611–2620. doi: 10.1016/j.surfcoat.2006.05.003.
31. L. Wang, Y. Wang, X.G. Sun, J.Q. He, Z.Y. Pan, C.H. Wang, A novel structure design towards extremely low thermal conductivity for thermal barrier coatings – Experimental and mathematical study. Mater Des 35 (2012), 505–517. doi: 10.1016/j.matdes.2011.09.031.
32. T.W. Clyne, S.C. Gill, Residual stresses in thermally sprayed coatings and their effect on interfacial adhesion: A review of recent work. *J. Therm. Spray Technol.*, 5 (1996), 1–18.

33. J.A. Thompson, T.W. Clyne, The effect of heat treatment on the stiffness of zirconia top coats in plasma-sprayed TBCs. *Acta Mater.,* 49 (2001), 1565–1575.
34. R.L. Jones, Thermal barrier coatings. In *Metallurgical and Ceramic Protective Coatings*, K.H. Stern, ed. (1996), 194–235.
35. L. Xie, M.R. Dorfman, A. Cipitria, S. Paul, I.O. Golosnoy, T.W. Clyne, Properties and performance of high-purity thermal barrier coatings. *Proceedings of the International Thermal Spray Conference*, (2007), 804–808.
36. X. Zhao, X. Wang, P. Xiao, Sintering and failure behaviour of EB-PVD thermal barrier coating after isothermal treatment. *Surf. Coat. Technol.*, 200 (2006), 5946–5955. doi: 10.1016/j.surfcoat.2005.09.006.
37. A.F. Renteria, B. Saruhan, Effect of ageing on microstructure changes in EB-PVD manufactured standard PYSZ top coat of thermal barrier coatings. *J. Eur. Ceram. Soc.*, 26 (2006), 2249–2255. doi: 10.1016/j.jeurceramsoc.2005.04.027.
38. J. Ruud, Y.C. Lau, V. Kwasniewski, ATP Status Report 95–07-0018, NIST (2003).
39. K.L. Choy, J. Mei, J.P. Feist, A.L. Heyes, Microstructure and thermoluminescent properties of ESAVD produced Eu doped Y 2O 3-ZrO 2 coatings. *Surf. Eng.*, 16 (2000), 469–472.
40. J.P. Feist, A.L. Heyes, Europium-doped yttria-stabilized zirconia for high-temperature phosphor thermometry. *Proc. Inst. Mech. Eng. L J. mater. Des. Appl.*, 214 (2000), 7–12.
41. J.P. Feist, A.L. Heyes, Coatings and an optical method for detecting corrosion process in coatings (2003), GB Patent 0318929.7.
42. A.M. Srivatave, A.A. Setler, H.A. Comanzo, J.W. Devitt, J. Ruud, L.N. Brewer, Apparatus for determineing past-service conditions and remaining life of thermal barrier coatings and components having such coatings (2003), US Patent 6730918B2.
43. J.I. Eldridge, T.J. Bencic, C.M. Spuckler, J. Singh, D.E. Wolfe, Delamination-indicating thermal barrier coatings using YSZ:Eu sublayers. *J. Am. Ceram. Soc.*, 89 (2006), 3246–3251.
44. C.C. Pilgrim, S. Berthier, J.P. Feist, R.G. Wellman, A.L. Heyes, Photo-luminescence for quantitative non-destructive evaluation of thermal barrier coating erosion. *Surf. Coat. Technol.*, 209 (2012), 44–51. doi: 10.1016/j.surfcoat.2012.08.025.
45. J.I. Eldridge, J. Singh, D.E. Wolfe, Erosion-indicating thermal barrier coatings using luminescent sublayers. *J. Am. Ceram. Soc.*, 89 (2006), 3252–3254.

46. R.J.L. Steenbakker, J.P. Feist, R.G. Wellman, J.R. Nicholls, Sensor thermal barrier coatings: Remote in situ condition monitoring of EB-PVD coatings at elevated temperatures. *J. Eng. Gas Turbines Power*, 131 (2009). doi: 10.1115/1.3077662.
47. M.P. Boyce, *Gas Turbine Engineering Handbook*. Gulf Professional Publishing, USA, (2002).
48. B.W. Noel, H.M. Borella, W. Lewis, W.D. Turley, D.L. Beshears, G.J. Capps, M.R. Cates, J.D. Muhs, K.W. Tobin, Evaluating thermographic phosphors in an operating turbine engine. *J. Eng. Gas Turbines Power*, 113 (1991), 242–245.
49. K.L. Choy, J.P. Feist, A.L. Heyes, B. Su, Eu-doped Y2O3 phosphor films produced by electrostatic-assisted chemical vapor deposition. *J. Mater. Res.*, 14 (1999), 3111–3114.
50. B.R. Lawn, A.G. Evans, D.B. Marshall, Elastic/Plastic Indentation Damage in Ceramics: The Median/Radial Crack System. *J. Am. Ceram. Soc.*, 63 (1980), 574–581.
51. B.R. Lawn, M.V. Swain, Microfracture beneath point indentations in brittle solids. *J. Mater. Sci.*, 10 (1975), 113–122.
52. D.B. Marshall, B.R. Lawn, A.G. Evans, Elastic/Plastic Indentation Damage in Ceramics: The Lateral Crack System. *J. Am. Ceram. Soc.*, 65 (1982), 561–566.
53. B.R. Lawn, E.R. Fuller, Equilibrium Penny-Like Cracks in Indentation Fracture. *J. Mater. Sci.*, 10 (1975), 2016–2024.
54. R.G. Wellman, J.R. Nicholls, Erosion, corrosion and erosion-corrosion of EB PVD thermal barrier coatings. *Tribol. Int.*, 41 (2008), 657–662.
55. R.G. Wellman, A. Dyer, J.R. Nicholls, Nano and Micro Indentation Studies of Bulk Zirconia and EB PVD TBCs. *Surf. Coat. Technol.*, 176 (2004), 253–260.
56. R.G. Wellman, J.R. Nicholls, The design of thermal barrier coating to resist erosion and FOD. Thermal Barrier Coating Workshop, University of California, Santa Barbara (2003).
57. R.G. Wellman, J.R. Nicholls, Observed damage mechanisms in the erosion and FOD of EB PVD TBCs. Thermal Barrier Coatings Conference, Irsee, Germany (2003).
58. H.E. Eaton, R.C. Novak, Particulate erosion of plasma-sprayed porous ceramic. *Surf. Coat. Technol.*, 30 (1987), 41–50. doi: 10.1016/0257-8972(87)90006-5.

59. C. Li, G. Yang, A. Ohmori, Relationship between particle erosion and lamellar microstructure for plasma-sprayed alumina coatings. *Wear*, 260 (2006), 1166–1172. doi: 10.1016/j.wear.2005.07.006.

60. F. Cernuschi, L. Lorenzoni, S. Capelli, C. Guardamagna, M. Karger, R. Vaßen, K. von Niessen, N. Markocsan, J. Menuey, C. Giolli, Solid particle erosion of thermal spray and physical vapour deposition thermal barrier coatings. *Wear*, 271 (2011), 2909–2918. doi: 10.1016/j.wear.2011.06.013.

61. R.G. Wellman, E. Besnard, T.W. Clyne, J.R. Nicholls, Erosion of Aged and Vertically Cracked Plasma Sprayed TBCs, Euromat Conference Proceedings, Rimini, Italy (2005).

62. R.J.L. Steenbakker, R.G. Wellman, J.R. Nicholls, Erosion of gadolinia doped EB PVD TBCs. *Surf. Coat. Technol.*, 201 (2006), 2140–2146.

63. R.G. Wellman, J.R. Nicholls, K. Murphy, Effect of microstructure and temperature on the erosion rates and mechanisms of modified EB PVD TBCs. *Wear*, 267 (2009), 1927–1934. doi: 10.1016/j.wear.2009.04.002.

64. F.H. Stott, D.J. de Wet, R. Taylor, Effects of molten silicate deposits on the stability of thermal barrier coatings for turbine applications at very high temperatures. International SAMPE Metals and Metals Processing Conference, 3 (1992), 92–101.

65. S. Kramer, J. Yang, C.G. Levi, Thermochemical interaction of thermal barrier coatings with molten CaO-MgO-Al2O3-SiO2 (CMAS) deposits. *J. Am. Ceram. Soc.*, 89 (2006), 3167–3175.

66. N.L. Ndamka, CMAS Attack of TBCs: The Effect of Basicity. Thermal Barrier Coatings III Conference, Irsee, Germany, (August 2011).

67. T. Steinke, D. Sebold, D.E. Mack, R. Vaßen, D. Stöver, A novel test approach for plasma-sprayed coatings tested simultaneously under CMAS and thermal gradient cycling conditions. *Surf. Coat. Technol.*, 205 (2010), 2287–2295.

68. R. Vassen, F. Cernuschi, G. Rizzi, A. Scrivani, N. Markocsan, L. Ostergren, A. Kloosterman, R. Mevrel, J. Feist, J. Nicholls, Recent activities in the field of thermal barrier coatings including burner rig testing in the european union. *Adv. Eng. Mater.*, 10 (2008), 907–921.

69. M.H. Vidal-Setif, N. Chellah, C. Rio, C. Sanchez, O. Lavigne, Calcium-magnesium-alumino-silicate (CMAS) degradation of EB-PVD thermal barrier coatings: Characterization of CMAS damage on ex-service high pressure blade TBCs. *Surf. Coat. Technol.*, 208 (2012), 39–45.

70. J.M. Drexler, A.L. Ortiz, N.P. Padture, Composition effects of thermal barrier coating ceramics on their interaction with molten Ca-Mg-Al-silicate (CMAS) glass. *Acta Mater.*, 60 (2012), 5437–5447.
71. J.M. Drexler, A. Aygun, D. Li, R. Vaßen, T. Steinke, N.P. Padture, Thermal-gradient testing of thermal barrier coatings under simultaneous attack by molten glassy deposits and its mitigation. *Surf. Coat. Technol.*, 204 (2010), 2683–2688.

Chapter 3

Scratch Damage in Coated Glass

Stephen J. Bull
School of Chemical Engineering and Advanced Materials, University of Newcastle, Newcastle upon Tyne, NE1 7RU, UK
steve.bull@ncl.ac.uk

The major in-service failure mechanism of modern solar control coatings for architectural glass can be mechanical (e.g., scratch damage). Many of these coatings are multilayer structures of less than 100 nm thickness and different coating architectures are possible (i.e., different layer materials, thickness and stacking order). For high-performance solar control coatings deposited by physical vapour deposition processes, the active layer is a thin silver coating (~8 nm thick) surrounded by anti-reflection coatings (e.g., ZnO, SnO_2) and barrier layers (e.g., TiO_xN_y). Scratches are often found during delivery of the coated glass (called transit scratches) and it has been determined that the cause of the scratches was the polymer balls sprayed onto the glass to separate sheets while in transportation. This chapter highlights the importance of using an appropriate simulation test for the transit scratches and has determined that the polymer/coating friction and how it controls through-thickness fracture and adhesion of layers within the multilayer stack is critical in determining performance. To test the adhesion of the coatings, coated samples have been subjected

Multifunctional Materials for Tribological Applications
Edited by Robert J. K. Wood

ISBN 978-981-4463-90-4 (Hardcover), 978-981-4463-91-1 (eBook)
www.panstanford.com

to scratch tests using a range of indenters and the most visible damage has been characterised. Through-thickness cracks were observed and it was seen that the coating was stripped by the balls at the weakest point in the coating stack. Microanalysis reveals this weakest point to be the silver/zinc oxide interface in the materials analysed in this study.

Nomenclature

A, a_c	Contact radius
c_{int}	Interfacial crack length
δ	Indenter displacement
E	Young's Modulus (the subscripts c and s refer to coating and substrate, respectively)
E_{int}	Interfacial Modulus
G_{int}	Interfacial strain energy release rate
h	Bulge height
H	Hardness (the subscripts c and s refer to coating and substrate, respectively)
K_{int}	Interfacial toughness
ν	Poisson's ratio
P	Indenter load
r_d	Delamination radius
σ_c	Critical buckling stress
σ_r	Residual stress
t	Coating thickness
τ_c	Interfacial shear strength
θ	Half-included angle of the indenter
u_0	Maximum film deflection
W	Work of indentation

3.1 Introduction

Due to the increasing usage of glass as a modern construction and design material, controlling the sunlight that comes in through architectural glass and the building heat lost through it has become increasingly important. More recently, the drive to reduce the energy consumption of building air conditioning or heating systems has given additional importance to high-performance thin film solar control coatings for architectural glass.

Nearly all coated glass applications require multifunctional behaviour. For instance, in a typical physical vapour deposited (PVD) solar control coating the active layer of silver that acts as a barrier to heat is surrounded by anti-reflection coatings and barrier layers designed to improve the optical performance, environmental stability and scratch resistance of the multilayer coating stack. This may also be combined with thin titania layers to produce a self-cleaning product. Anti-reflection coatings and barrier layers are generally metal oxides with thicknesses in the range of tens of nanometres. Generally such multilayer coatings are designed on the basis of their optical performance (i.e., on the thickness of each layer and the layer materials) and then materials are changed or other coating layers are added to improve processability and damage resistance. For instance different materials may be selected for anti-reflection coatings depending on the underlying layer material on which they are deposited in order to improve coating adhesion. It is the properties and durability of the complete coating stack that determines whether it will be successful in the intended application.

One of the problems faced by the glass industry is during the delivery of the coated glass. The glass sheets that are cut to size (usually $3\,m \times 6\,m$) prior to coating are sprayed with polymer balls on the coated surface to protect them during transit or transportation. The polymer balls separate the glass sheets when they are stacked almost vertically on a steel A-frame support called a stillage for transportation. In combination with cardboard spacers, this is designed to reduce the contact stresses on the coated glass surface and prevent damage.

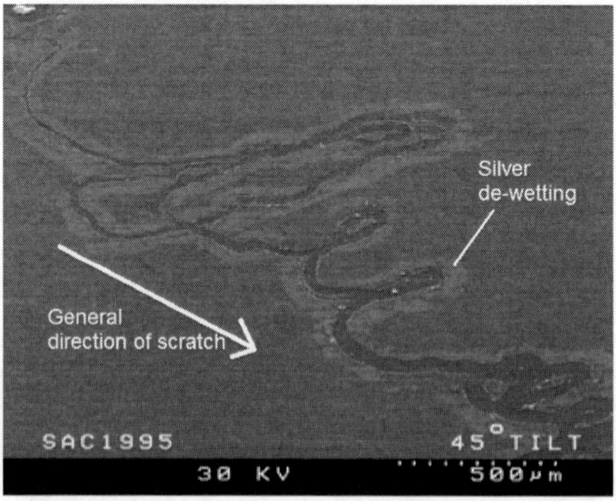

Figure 3.1 Typical transit scratch in Pilkington low-emissivity glass.

However, despite this scratches are still observed on the coated surface during the transportation of the glass from the manufacturing plant to the final customer who usually assembles the coated glass into double glazing units with the easily damaged coating on the inside. Scratched glass is rejected and can lead to costly returns (Fig. 3.1). Small grooves do not generally result in visible scratches and thus plastic deformation of the coating stack is a relatively minor failure mode. However, coating detachment is generally very visible and is to be avoided. Thus, the fracture toughness and adhesion strength are important aspects in evaluating the quality of coating layers. Most solar control coatings are multilayer structures. The reliability and the effectiveness of the coated glass thus depend on how well the thin films adhere to each other as well as to the substrate.

Many tests can be used to assess adhesion of coatings on different substrates [1] but perhaps the most widely used techniques are the indentation and scratch tests when the coating and substrate are hard [2–3]. Such tests usually provide a measure of 'practical adhesion,' which may be test specific and different from the basic adhesion between coating and substrate [4–5]. Scratch testing tends to induce failure in a coated systems at a lower load than in static indentation testing and is therefore often the preferred test, particularly when trying to assess the failure of surfaces in relative motion, such as occurs in a transit scratch.

Scratch adhesion testing is a process of displacing the coating at a constant velocity under an indenter whilst applying a constant or an increasing load to the coating surface. As the load increases, the contact generates increasing stresses and at a given load damage of the coating surface can occur. The minimum load at which the coating fails is termed the critical load [6]. This damage may be plastic deformation, through-thickness fracture or detachment of the coating. Scratch tests are usually carried out with styli that are harder and stiffer than the surface under test and a considerable amount is known about the assessment of the tribological performance of a material tested with a hard counter face [7–9]. However, when looking to mimic transit scratches, a softer stylus is required and the damage of a hard coating by a softer slider is much less well understood.

Multilayer thin films are now widely used in modern technology due to their unique combinations of physical properties, which differ significantly from those of single materials [8, 9]. One possible reason for this difference can be the different microstructures created by deposition onto different underlying layers that may be different for layers of the same material at different levels in the coating stack [10]. Hence it is very important that we understand and take into account not only the coating–substrate adhesion but also the adhesion at different layers in the coating stack.

In this chapter the cause of transit scratches in a commercial solar control coating is investigated. The active layer in a solar control coating is a thin layer of pure silver ($\sim$10 nm thick) surrounded by various anti-reflection and barrier layers that are usually oxides. Transit scratches are produced by the sliding of the polymer balls during transportation of the glass; this leads to through-thickness fracture which is quickly followed by adhesion failure—typically this occurs within the coating stack at the weakest interface. For the materials tested here this is the Ag/ZnO interface, which is the interface of the weak silver layer closest to the glass substrate. A test to simulate these scratches is also described. It has been shown that under similar experimental conditions, the occurrence of transit scratches was higher on a clean sample than on a sample with a thin layer of atmospheric contamination; this is due to the reduction in friction on the contaminated sample because of the presence of adsorbed hydrocarbons.

3.2 A Review of Test Methods for Adhesion Assessment

3.2.1 *Measurement of Adhesion*

Adhesion tests aim to introduce and propagate a mechanically stable crack at the interface; in practice, this depends on the interfacial flaw distribution as well as the interfacial bonding. The resistance to crack initiation or propagation is a measure of adhesion, which is best characterised in terms of the critical strain energy release rate for interfacial crack propagation, G_{int}. The extent of delamination can be large and there are lots of failure models developed based on linear elastic fracture mechanics (LEFM). However, the delamination may be of limited extent, confined within in a plastically deforming region and these models are no longer valid. Some different approaches to address this problem are outlined in this chapter.

There are numerous methods [e.g., 11–12] to assess the interfacial toughness that are summarised in the following section.

3.2.1.1 Techniques to measure adhesion

Pull-off methods: This group includes tests where the coating is removed by another material bonded to it using either shear or tensile forces. The techniques are most commonly used when the adhesion is poor since they are limited by the strength of available adhesives. Shear tests include the tape peel test or the lap shear technique [e.g., 13, 14]. The tape test uses a pressure-sensitive tape applied to the coating surface, which is pulled off to determine the coating detachment load per unit tape width. It is quite easy to apply but its main disadvantage is that it can be only applied to thick and rather weakly bonded coatings. Alternatively, one may bond studs to coating and substrate by an adhesive and then perform a tensile test to pull off the coating. However, perfect alignment to ensure uniform loading across the interface is difficult to achieve, in particular for small samples. The mixture of tensile and shear stress during the test leads to difficulty in interpretation of the results. Also, when the coating is very thin or porous, penetration of the adhesive through the coating to the interface may affect results.

Superlayer test: This test is based on the observation that a compressively stressed film can spontaneously delaminate from the substrate. In many cases, the stresses in the film are not high enough to achieve this but the addition of another stressed layer on top of the coating of interest can generate delamination. A well-adherent, thick superlayer is added on the top of the film so that it increases the apparent film thickness and residual stress without altering the tested interface [15, 16]. This method is nearly always applied to metal–ceramic systems. Although accurate adhesion measurements can be made by this technique, it requires tedious tests changing the thickness and process conditions of the superlayer until delamination occurs and it requires careful choice of a suitable material for the superlayer, which is compatible with the coating of interest.

Bending test: This is usually designed for metal–ceramic systems where through-thickness fracture in a brittle coating is diverted along the interface by a more ductile metal substrate to cause interfacial failure [e.g., 17, 18]. This method requires a relatively big sample and cannot directly account for residual stress in the coating, which may be a major driver for interfacial failure [19]. Also, this method may not work for ceramic–ceramic systems since it is more likely to cause catastrophic brittle fracture independent of the interface between the brittle ceramic layers.

Bulge test: This is often used to assess the adhesion for thin, ductile metal films on harder substrates. To achieve the test, it is necessary to etch the substrate from behind to obtain a free-standing thin film window. Pressure is applied to the rear of the film to cause it to deflect – tensile stresses at the edge of the bulge can drive delamination and are relatively easy to analyse. However, etching may lead to contamination of the interface. Also, the test may not work for thin films (<2 μm) due to possible wrinkling effects [20]. In some cases, the removal of the substrate causes the film to fragment and porous films may not be sufficiently gas or liquid tight to deflect.

Shockwave tests: These produce delamination by a rapidly applied external stimulus, which creates a shockwave [21, 22]. The stimulus could be the impact of erosive particles or a laser pulse. There can be problems in coupling the stimulus to the coating and interpretation of the results is difficult. The test may also influence the properties of the surface, particularly for thin films.

Double cantilever beam test: In this test, controlled interfacial cracking is introduced for a coating between two rigid plates. The energy release rate can be determined by measuring the fracture load, precrack length and the height of the specimen [23]. The advantage is that this method leads to almost pure mode I fracture, which makes it easy to analyse. It works well for thick coatings; however, it is not suitable for thin films due to the need to glue the coating to one of the plates and the difficulty of introducing a stable interfacial crack.

Scratch test: This is a simple way to estimate adhesion; however, it is usually regarded as semi-quantitative because there are a lot of intrinsic (such as, instrument specific) and extrinsic parameters (e.g., coating thickness, substrate hardness), which control test results [16]. It is most useful for hard coatings on softer substrates.

Indentation: For thin films, nanoindentation is a very effective technique to measure the mechanical properties of the coatings [e.g., 24]. If delamination occurs, the interfacial toughness can be measured. Given the advantages of simplicity and convenience, the use of the indentation test to assess thin film adhesion is the emphasis of this chapter. Many models have been developed to extract quantitative toughness values from indentation-induced delamination. In the following section, the existing analytical models are discussed.

3.2.2 *Models for Interfacial Fracture during Indentation*

Models based on LEFM: For a soft coating on a hard substrate, approaches based on initiation of delamination under the indenter have been proposed [25, 26]. The expression linking interfacial shear strength to coating hardness is [31],

$$\tau_c = \frac{0.56 H_c}{K_{1'}(\varphi a_c/t)/\varphi K_1(\varphi a_c/t) + \nu t/a_c\varphi^2} \tag{3.1}$$

where H_c is the hardness of the coating, K_1 is a modified Bessel function of the second kind, $K_{1'}$ is the derivative of the function, K_1; a_c is the contact radius, and

$$\phi = [6(1-\nu)/(4+\nu)]^{1/2} \tag{3.2}$$

The interfacial toughness, K_{int} is then given by [27],

$$K_{int} = \tau_c\sqrt{\pi c_{int}} \tag{3.3}$$

where interfacial failure is treated as mode I failure and c_{int} is the appropriate flaw size at the interface. In this way, Malzbender et al. [28] obtained an average interfacial toughness of 0.18 $MPam^{0.5}$ for their hybrid coatings, which falls into the range determined previously based on analysis by He et al. [29]. This indicates that this model works well, but c_{int} is not easy to determine accurately. Therefore, it may be more convenient to relate the interfacial toughness to a crack dimension that can be readily measured.

Based on the delamination of films with axisymmetric geometries, Rosenfeld et al. [30] proposed the following expression that relates interfacial fracture toughness to coating hardness H, applied load P, and delamination size c,

$$K_{int} = \frac{0.792H\sqrt{(1-\nu^2)t}}{[1+\nu+2(1-\nu)Hc^2/P]} \tag{3.4}$$

This assumes that the pressure at the interface is approximately equal to the hardness of the coating, which is reasonable for a soft coating or a sufficiently thick coating on the substrate under the condition that the effect of cracking on hardness can be ignored. This simple form makes it easy to apply, but it will considerably overestimate or underestimate the adhesion depending on the actual coated system. For instance, overestimation was found when applying Eq. (3.5) to sol–gel coatings (thickness of several micrometres) on float glass [33] and CN_x coatings on ceramic substrates [31]. In addition, friction between fracture faces may also contribute to the overestimation of the interfacial fracture toughness for mode II fracture [32].

An alternative approach was proposed by Thouless [3] and further developed by Toonder et al. [33]. It was assumed that the delaminated coating is a part of a circular disk extending in radius with increasing load and it will buckle when the in-plane

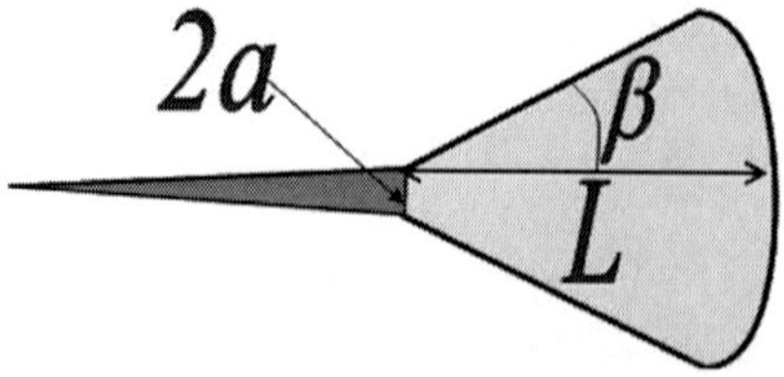

Figure 3.2 Schematic of the chipped area in the analysis by Thouless (1998).

stress reaches a critical value. Supposing that the driving force of delamination equals the critical load for buckling, Thouless derived the following equation for interfacial fracture energy [38],

$$G_{\text{int}} = 0.35\frac{Et^5}{L^4}\left(\frac{\tan\beta + 2a/L}{\tan\beta + a/L}\right)^2 \tag{3.5}$$

where L, a, β are the geometric parameters for a chip as depicted in Fig. 3.2. By considering the curved geometry and the residual stress, Toonder et al. modified the model to generate a more complex expression [39],

$$G_{\text{int}} = 1.42\frac{Et^5}{L^4}\left(\frac{\frac{a}{L}+\frac{\beta\pi}{2}}{\frac{a}{L}+\beta\pi}\right)^2 + \frac{t(1-\nu)\sigma_r^2}{E} + \frac{3.36(1-\nu)t^3\sigma_r}{L^2}\left(\frac{\frac{a}{L}+\frac{\beta\pi}{2}}{\frac{a}{L}+\beta\pi}\right) \tag{3.6}$$

Eq. (3.6) provides similar results for interfacial toughness of sol–gel coatings on glass where failure is caused by indentation or scratch [39]. The method gives reasonable results (e.g., Ref. [36]).

All the models above are based on LEFM, however, when the fracture is within the plastic deformation zone, LEFM becomes invalid. Therefore, a model which accounts for the plastic deformation is needed.

Models considering plastic deformation: A so-called plastic hinge model was proposed and applied to assess indentation-induced delamination for a ductile film on an elastic substrate to deal with the problem that buckling is inside the plastic deformation zone. Instead of elastic plate theory, a model of a travelling plastic hinge at the apex of the buckled coating [34] was adopted for stress analysis for the delaminated film. The detailed stress analysis is given by Li and Siegmund [35].

Compared to the models based on LEFM (e.g., Marshall et al. [36] and Hutchinson and Suo [37]) at intermediate interfacial toughness where the plastic deformation zone does not extend beyond the buckling height, the travelling plastic hinge model gives the best prediction of energy release rate by delamination, G_{int}. However, in the cases of very low G_{int} and high G_{int}, it leads to deviation when compared to numerical simulations (see details in Ref. [41]).

Alternative energy based methods: There are several energy based models to assess the fracture toughness of the coating (e.g., described in [38] and Wt-dp method in [15, 39]). If a feature in the indentation load-displacement curve is related to delamination, these energy based models are still valid and can be used to assess interfacial toughness. For example, Wt-dp method gives the energy release rate of $30 \pm 14\,\text{J}\,\text{m}^{-2}$ and $19 \pm 9\,\text{J}\,\text{m}^{-2}$ for a 400 nm TiO_xN_y/glass interface and 240 nm indium tin oxide (ITO) multilayer on glass, respectively, which is a reasonable result for such coated systems.

Another alternative energy based model was presented by Malzbender and de With [40] based on plotting irreversible work W_{irr} versus applied load. For the thicker sol–gel coatings in Ref. [46], a slope change of the curve of W_{irr} versus applied load (W_{irr}–P curve) was observed when delamination occurred. Thus the energy $\Delta W_{\text{irr}}^{\text{d}}$ (see Fig. 3.3) was assumed to be the energy dissipated by delamination, which is the difference between the energy at a given delamination dimension r_{d} and the extrapolated energy for the plastic deformation of coatings before delamination. The energy release rate for delamination, G_{int}, is then given by,

$$G_{\text{int}} = \frac{\Delta W_{\text{irr}}^{\text{d}}}{\pi r_{\text{d}}^2} \tag{3.7}$$

Thus the interfacial toughness K_{int} is given by,

$$K_{\text{int}} = \sqrt{G_{\text{int}} E_{\text{int}}} \tag{3.8}$$

where E_{int} is a so-called interfacial modulus which is given by Ref [43],

$$\frac{1}{E_{\text{int}}} = \frac{1}{2}\left(\frac{1}{E_{\text{c}}} + \frac{1}{E_{\text{s}}}\right) \tag{3.9}$$

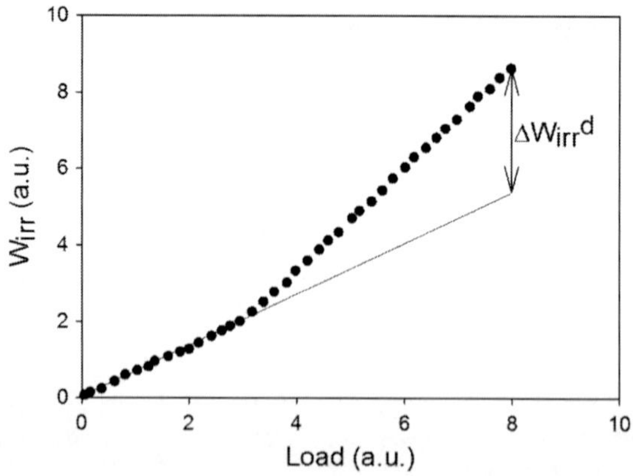

Figure 3.3 Schematic of irreversible energy dissipated as a function of load during indentation [33].

where E_c and E_s are the Young's modulus of coating and substrate, respectively.

This method is more accurate than an alternative method based on extrapolating the loading curve (ld–dp method [44]). When extrapolating the irreversible work (i.e., plastic deformation in the initial paper [46]), it was assumed the relationship is independent of applied load if no delamination occurs. However, the radial cracking occurring prior to delamination may change the W_{irr}–P relation. Furthermore, as the load increases the role of the substrate will also change significantly introducing inaccuracy. The energy release rate for delamination of sol gel coatings coded A and B [46] was 11 to 290 J m^{-2} by this method, which is unrealistically high for such coatings with poor adhesion and it was much higher than previous results (<4 J m^{-2}) reported elsewhere [41].

In order to eliminate the substrate influence and the change of apparent Young's modulus and hardness caused by cracking and delamination, the fracture dissipated energy can be assessed by extrapolating the ΔW_{irr}^{d} versus the reciprocal of coating thickness to infinite coating thickness. In this way, a reasonable approximation (same order of magnitude) of interfacial toughness can be obtained. However, different fracture behaviour in the same coating composition will occur at different thickness and this may vary significantly for different coatings and deposition technologies (e.g., sol–gel coating [33, 46], and PVD ceramic coatings [15, 45] behave very differently). Thus the extrapolation method is not expected to be valid in all cases, nor agree in different coating systems.

It is necessary to point out that the W_{irr} method is also only valid when there is a delamination feature in the load–displacement curve [42]. Fracture does not always lead to an obvious excursion in load–displacement curve (P–δ curve). In this case, the fracture toughness can still be measured by an energy-based model proposed by Chen and Bull [43]. Combining the numerical simulations and calibration on realistic materials, the plastic work can be estimated and the irreversible work can be easily calculated from the load–displacement curve. Then the difference between plastic work and irreversible work is regarded as the fracture dissipated energy. The interfacial toughness for 400 nm ITO/glass is 0.64 ± 0.07 MPam$^{1/2}$ given by this model, which is very close to 0.44 MPam$^{1/2}$ determined from the geometrical lateral cracking along the interface in higher load indentation tests seen later.

In fact, under certain conditions, that is, rebound after delamination leading to extra linear recovery in the unloading part of P–δ curve, the adhesion can be assessed from the difference in the elastic work recovered on unloading. The linear part (as marked in Fig. 3.4) occurring near the end of unloading procedure is believed to be accounted for by debonding during unloading. The area of the interfacial crack is πC_{R}^2. If extrapolating the initial part of the unloading curve to the complete unload point, the difference between the extrapolated unloading curve and the actual unloading curve can be regarded as an approximation of the fracture dissipated energy in delamination. The mechanism can be explained as follows (see Fig. 3.5). The coating underneath the indenter deflects into the substrate, due to the tensile stress at the edges caused by the pre-existing buckling and the compressive stress at the tip. On unloading the coating within the region, A-A (see Fig. 3.5) will rebound thus resulting in an extra recovery in the load versus displacement (P–δ) curve.

Another adhesion model based on equilibrium theory and the solution for the deflection of a centrally loaded disc was proposed to deal with the same phenomena by Hainsworth et al. [44]. However,

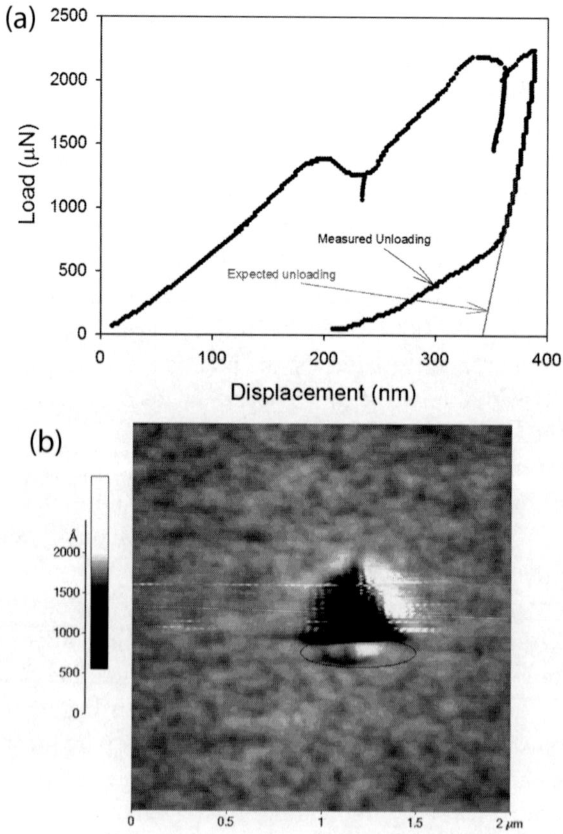

Figure 3.4 Extra linear elastic recovery in load–displacement curve (a) is observed during unloading for a 400 nm TiO_xN_y on commercial soda-lime glass and its corresponding AFM image (b).

it dramatically overestimated the adhesion. In contrast, the method presented here provides reasonable approximation giving the fracture dissipated energy as 79.7 ± 17.6 J m^{-2} for the TiO_xN_y/glass interface. Thus the interfacial toughness is approximately 2.8 ± 0.3 MPam$^{1/2}$. For the ITO/Ag interface, the values of fracture dissipated energy and toughness are 35.8 ± 14.1 J m^{-2}, and 1.9 ± 0.3 MPam$^{1/2}$, respectively. These results agree reasonably with those determined by a different method as discussed in the previous section. For a 240 nm ITO coating, it is observed that the high pre-existing residual

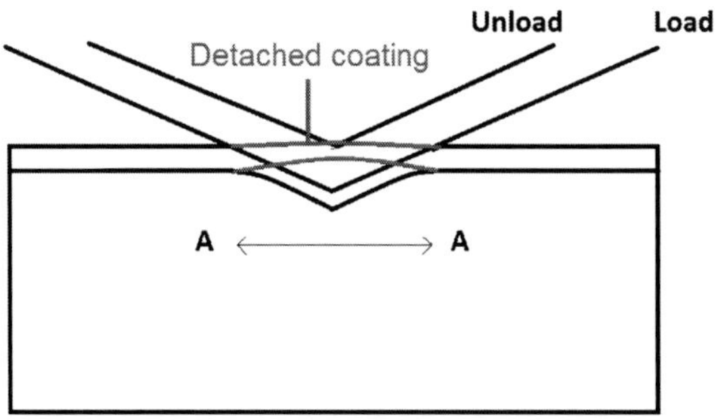

Figure 3.5 Schematic of double blister formation during loading. The coating within region A-A can become detached and rebound on unloading giving the results shown in Fig. 3.4.

stress in the coating makes it more likely to fail at the ITO/Ag interface rather than at a lower interface in a multilayer coating stack. Accurate metrology of the spallation at high loads by AFM confirms the delamination at this interface [49].

Some special indentation techniques: In addition to the development and improvement of assessment models, some special indentation techniques have also been proposed, for instance, the use of tests with special indenters, such as the micro-wedge indentation test to measure adhesion [45]. By extending the model given by Marshall and Evans [46] to the plane strain situation, the energy release rate can be determined. Such a method is less sensitive to the measurement accuracy of crack length and it is easier to induce interfacial failure compared to the normal indentation test by conical indenters or pyramids. The disadvantage is the difficulty to align the wedge perpendicular to the plane of a thin film [24]. This model ignores the bending effect of film and substrate during the indentation cycle and this leads to deviation of its results from data from other methods, particularly, for ductile substrates.

In order to eliminate the influence of plastic deformation, in particular, for a ductile coating on a much stiffer substrate, a cross-sectional indentation test was suggested [47, 48]. The interfacial toughness can be estimated by treating the interface as a homogeneous material with the Young's modulus and hardness as composite parameters of the coating and substrate so that linear fracture mechanics for brittle materials can be applied. Then [40],

$$K_{\text{int}} = 0.015(E/H)_{\text{int}}^{1/2} P/a^{3/2} \tag{3.10}$$

where P and a are the critical load and critical contact radius for the delamination at the interface, respectively.

Here,

$$(E/H)_{\text{int}}^{1/2} = \frac{\left(\frac{E}{H}\right)_{\text{c}}^{1/2}}{1 + \left(\frac{H_{\text{c}}}{H_{\text{s}}}\right)^{1/2}} + \frac{\left(\frac{E}{H}\right)_{\text{s}}^{1/2}}{1 + \left(\frac{H_{\text{s}}}{H_{\text{c}}}\right)^{1/2}} \tag{3.11}$$

This model provides reasonable results for very thick ceramics (hundreds of micrometres) such as CVD chromium carbide and thermal sprayed coatings on metal substrates. It requires one diagonal of a Vickers indentation parallel to the interface, which may not be easily achieved for thin films. The assumption of Eq. (3.11) that the equivalent radius of the deformation zone at the interface (the materials is assumed to be homogeneous) is the average of those in film and substrate is only reasonable if the mismatch between the coating and substrate is negligible; otherwise, significant deviation will occur.

Based on the plate model, the energy release rate for the interface can be determined by [54],

$$G_{\text{int}} = \frac{Et^3u_0^2}{12(a-b)^2}(1-\lambda)^4(2F+\lambda F') \tag{3.12}$$

$$F(\lambda) = \frac{2\ln\lambda + \frac{1+\lambda}{1-\lambda}\ln^2\lambda}{[(1+\lambda)\ln\lambda + 2(1-\lambda)]^2} \tag{3.13}$$

where a, b are the delamination radius and contact radius, respectively; u_0 is the maximum film deflection, t is film thickness, $\lambda = a/b$, $F' = dF/d\lambda$. The latter model [i.e., Eq. (3.12–3.13)] is preferred to assess the adhesion rather than Eq. (3.11).

The cross-section method is particularly useful for a thick brittle coating or a ductile coating on a brittle substrate. In this case, the crack initiates in the brittle substrate and it will propagate along the weaker interface if this is less tough than the coating itself. The nature of this method is that it induces a pre-crack at the interface. The disadvantage of this method is that there are some critical parameters that are difficult to control experimentally such as the orientation of the indenter tip and the indentation position with respect to the interface for controlled delamination. Also, if the interface is not brittle enough, the relatively low stress (compared to indenting direction) in the lateral direction may not lead to delamination.

As discussed previously, superlayer test can make it easier to cause blistering. Therefore, indentation in combination with superlayer techniques is a very useful technique [e.g., Ref. 49] especially for the assessment of thin metallic films on ceramic substrates.

Lateral cracking at the interface: The models above deal with blistering or delamination, while, interfacial failure can be also caused by lateral cracking at or adjacent to the interface. Marshall et al. [50] proposed a model to relate the toughness of brittle bulk materials to the depth where lateral cracks occur and the crack driving force P_{r} due to the residual stress during unloading. This model can be further developed to link the toughness to the threshold load and the other basic parameters such as hardness, elastic modulus, giving [56],

$$K_{\mathrm{c}}^{4}=\frac{P_{0}A_{0}^{2}}{\delta_{0}}(\cot\theta)^{\frac{2}{3}}\frac{H}{E}H^{3} \tag{3.14}$$

where P_0 is threshold load for lateral cracking; A_0 and δ_0 are dimensionless constants ($A_0 = 3/4$ and $\delta_0 = 1200$), and ϑ is the half-included angle for the indenter. With the aid of nano-indentation, this method can be applied to coated systems. Under certain circumstances, the method can be extended to assess interfacial failure of a coated system if the lateral crack occurs at a relatively weak interface compared to the coating and substrate.

It has been shown that such a lateral crack may occur at the ITO/glass interface during indentation by a blunt Berkovich tip [49]. Therefore, Eq. (3.15) can be applied. Since the lateral cracking tends to occur near the elastic-plastic boundary. It is reasonable to assume that the critical load for the lateral cracking can be obtained when the plastic deformation zone reaches the interface. Here, the hardness and elastic modulus are composite parameters resulting from the contribution of coating and substrate, which can be estimated based on an energy-based model developed by Bull

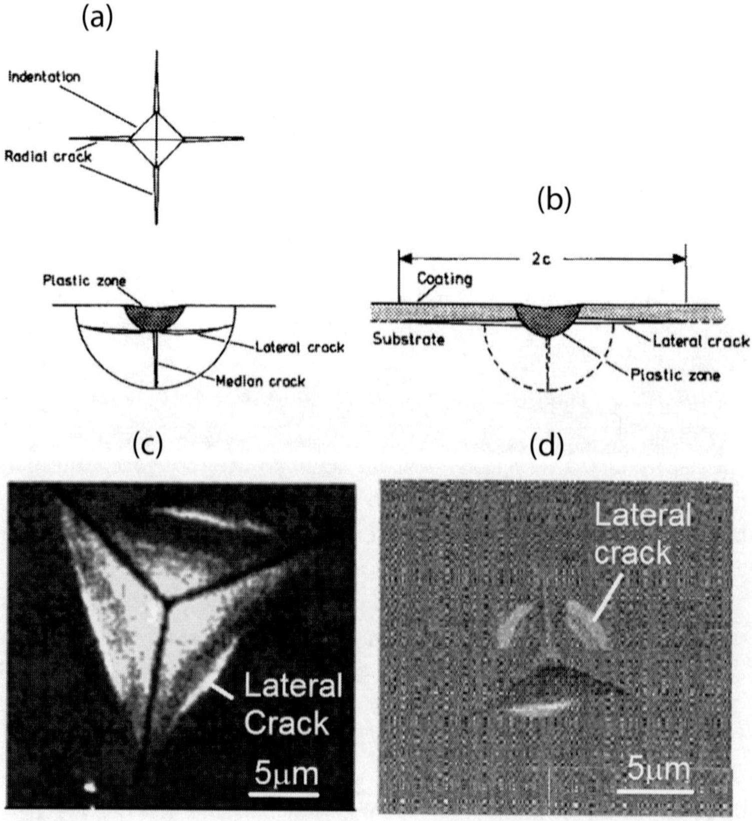

Figure 3.6 (a) Schematic of cracking around a Vickers indent in a bulk material and (b) lateral cracking at the interface in a coated system (c) scanning electron micrograph (SEM) and (d) reflected light micrograph showing lateral cracks for a 400 nm ITO layer on glass.

et al. [51–52]. Thus we can get a first approximation of interfacial toughness for ITO/glass, 0.44 $\text{MPam}^{1/2}$, which is consistent with results determined by different methods as discussed in the previous section.

Spontaneous blistering: If spontaneous blistering occurs in a coated system, the interfacial toughness can be directly estimated without additional tests. The approach considers the contribution from the critical buckling stress, σ_c, and the driving residual stress σ_r, of the film. The relationship between them was developed by

Hutchinson and Suo [42]:

$$G_{\text{int}} = \left[\frac{(1-\nu^2)t}{2E}\right](\sigma_r - \sigma_c)(\sigma_r + 3\sigma_c) \quad (3.15a)$$

where $$\sigma_c = \frac{kE}{12(1-\nu^2)}\left(\frac{t}{a}\right)^2 \quad (3.15b)$$

and $$\sigma_r = \sigma_c\left[c_1\left(\frac{h}{t}\right)^2 + 1\right] \quad (3.15c)$$

where h is the bulge height, $c_1 = 0.75$, and, for a spontaneous blister, $k = \pi^2$.

For CN_x coatings (1 μm in thickness) on M2 steel, Eq. (3.15) leads to the energy release rate of $3.3 \pm 0.1\,\text{J m}^{-2}$ [53]. For a 400 nm ITO coating on glass, the energy release rate for adhesion is around $2.7 \pm 0.1\,\text{J m}^{-2}$ given by Eq. (3.16), similar to what is derived by other methods. Therefore, an alternative method to assess adhesion is to cause spontaneous blistering by varying the coating thickness.

3.2.3 *Choosing an Appropriate Test*

The selection of an appropriate adhesion test depends on what the test is designed to achieve. If design data for mechanical modelling of the coating/substrate system is required, then many of the methods outlined in the previous section are appropriate because they give quantitative adhesion energies. However, the mechanics of adhesion failure may not be the same as that encountered in service. To predict in-service performance, it is better to choose a test that simulates the operating failure mode and use an adhesion-based failure criterion to rank different coating systems under investigation. This sort of test is useful for both process development and quality control.

It is often assumed that the damage to optical coatings arises from contact with hard debris such as sand or glass cullet and indentation or scratch tests using hard diamond indenters are suitable for coating assessment. However, this is not suitable for all types of failure, such as the transit scratches mentioned previously. In this case, a laboratory simulation test must be developed that

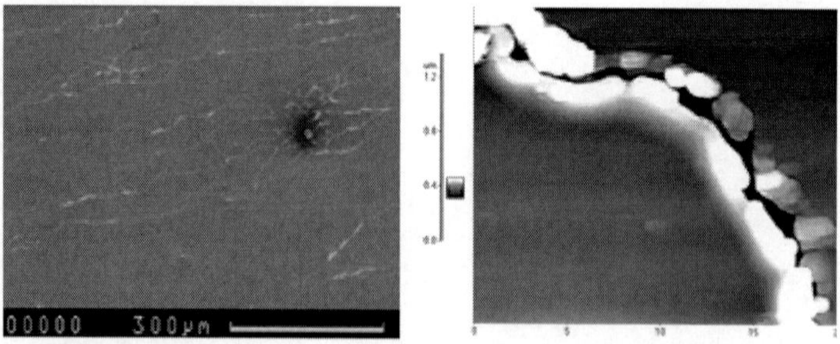

Figure 3.7 SEM (a) and AFM (b) image of spontaneous blister in 400 nm ITO on glass.

more close approximately real service conditions. This is discussed in more detail in the next section.

3.3 Simulation of Scratch Damage in Low-Emissivity Glass

3.3.1 *Coated Glass Samples Investigated*

Experiments were carried out on float glass, coated with a multilayer stack of silver and metal oxides in a solar control configuration; the total thickness of the coating stack is approximately 100 nm. The coating consists of a 10 nm silver layer (which is the active layer in the stack) surrounded by SnO_2 and ZnO anti-reflection coatings and TiO_xN_y barrier layers; the layer structure and nominal coating thicknesses are presented in Table 3.1. A thin conducting ITO layer is used to prevent the silver layer from oxidation during the subsequent deposition of tin oxide. The coatings were produced

Table 3.1 Structure of the multilayer coatings investigated

Position in the coating stack	Layer Material	Nominal thickness (nm)
Outermost barrier coating	TiO_xN_y	3
Outer AR coating	SnO_2	40
Protective layer	ITO	2
Wavelength selective layer	Silver	10
Inner AR layer	ZnO	10
Inner barrier layer	TiO_xN_y	20
Substrate	Float glass (air side coated)	42000000

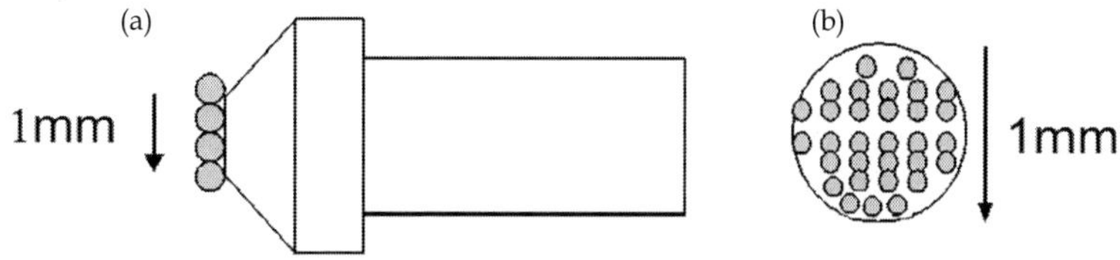

Figure 3.8 (a) Geometry of the scratch stylus and (b) the arrangement of balls on the scratch stylus tip.

by sputtering in a commercial coating plant at Pilkington Technical Centre (Lathom, UK).

3.3.2 *Simulation Test*

A steel scratch stylus was designed and constructed to simulate the transit scratches, which had a flat tip of 1 mm diameter at the end which was repeatedly reciprocated on an unimportant area of the glass surface prior to testing to ensure the flat was parallel to the test surface. Polymer balls were glued to the end of this tip after this process (Fig. 3.8)—the surface roughness of the flat was less than 1 μm R_a and the ball diameter was about 80 μm; so, provided that a good surface coverage is maintained and the tip alignment is good, the load is only carried on the balls and is uniformly distributed among them.

Scratch tests were performed on the specimen using a commercial scratch tester (Teer Coatings, Worcester, UK) at 8 N load and the table speed was set to 10 mm min^{-1}. Prior to commencing the scratch the static stylus was loaded to the test load at a rate of 100 N min^{-1}. Sieved poly methyl methacrylate (PMMA) balls with a diameter of 80 ± 20 μm were used; the balls were bonded carefully onto the indenter tip using a cyanoacrylate adhesive and their number was determined using a light microscope. The balls were counted before and after the experiments were performed so as to know the average load on each ball and the number of balls causing the damage. A reflected light microscope was used initially to identify surface damage. SEM and optical profilometry were used to analyse the scale of the damage and to determine which interfaces had failed.

Table 3.2 Mechanical properties used for contact mechanics calculations

Material	E (GPa)	υ
Soda-lime glass	70	0.18
PMMA	1	0.3

3.3.3 *Measurements and Calculations*

In order to calculate the contact stresses, a simple Hertz stress analysis was used. The contact modulus was calculated using the equation [54]

$$\text{Contact modulus} \quad \frac{1}{E^*} = \frac{1-v_1^2}{E_1} + \frac{1-v_2^2}{E_2} \tag{3.16}$$

where E_1, E_2 and υ_1, υ_2 are the Young's moduli and Poisson's ratios of the ball and coated surface, respectively. Since the load is very high and the mean load per ball, P, is a substantial fraction of a Newton, the substrate dominates the stress response and the values for glass have been used in the calculations here (Table 3.2).

The maximum contact pressure and semi-contact width were calculated using the Hertz equation for elastic contact for a sphere of radius, R [61]

$$\text{Maximum contact pressure} \quad P_0 = \frac{1}{\pi}\left(\frac{6PE^{*2}}{R^2}\right)^{1/3} \tag{3.17}$$

$$\text{Contact radius} \quad a = \left(\frac{3PR}{4E^*}\right)^{1/3} \tag{3.18}$$

3.4 Test Conditions Investigated

3.4.1 *Contaminated Coated Glass*

The coated specimen was exposed to the environment for few days as they do in practice in industry, a contamination layer was formed on the surface and the scratch tests were carried out on the contaminated surface. The contamination layer consists of adsorbed water and hydrocarbons.

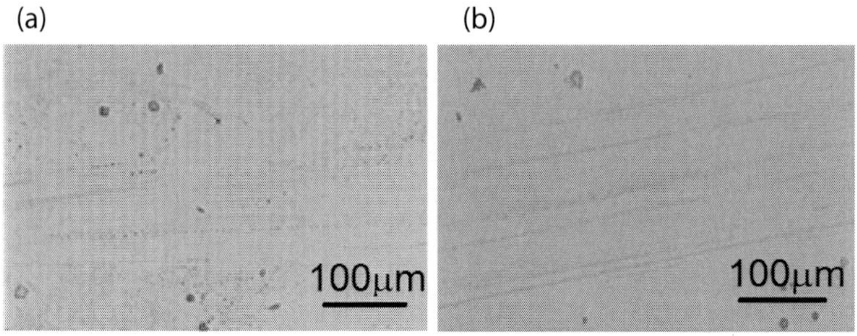

Figure 3.9 Reflected light micrographs of (a) Contaminated surface prior to testing and (b) scratch tracks at 7.5 N on a contaminated surface.

As seen in Fig. 3.9, no damage has occurred on the surface; this may be due to the low tensile stress in the sliding contact caused by the lubricating effect of the contamination layer.

3.4.2 *Cleaned Coated Glass*

The specimen was ultrasonically cleaned using industrial methylated spirit and the experiments were performed immediately. This removes the contamination layer and does not allow sufficient time for it to re-form. Damage was seen and what appeared to be a transit scratch occurred on this sample (Fig. 3.10).

From Figs. 3.9 and 3.10, it is clearly visible that it is possible for the polymer balls to damage the surface of the coated glass when the sliding friction is high (typically the coefficient of friction is greater than 0.5 in such cases).

3.4.2.1 Scratch test on cleaned coated glass without the polymer balls on the tip

Experiments were also carried on the clean coated glass at the same load of 8 N to confirm that the damage caused to the surface was by the polymer balls and not by the tip. Very wide scratch tracks were produced with a large amount of debris collected at the end. In fact the coating was completely stripped, which does not match with what is observed in transit scratches by industry.

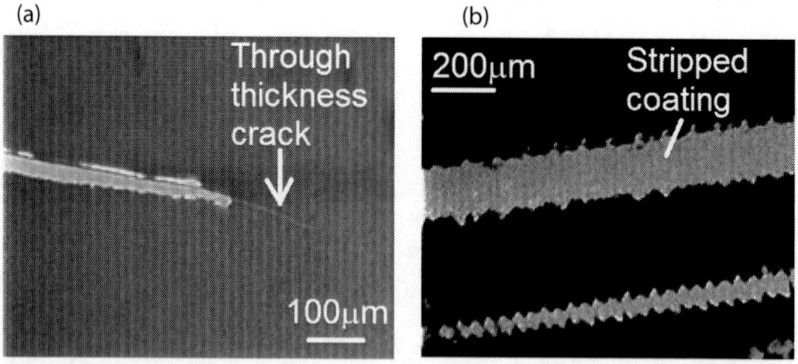

Figure 3.10 (a) Through thickness crack and the initiation of a transit scratch. The uncovered surface is shiny and metallic which suggests it is silver (b) well-established transit scratch.

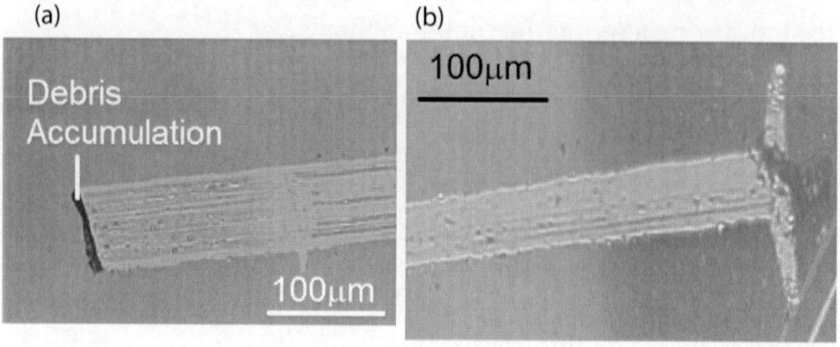

Figure 3.11 (a) Complete stripping of the coating and the pile up of debris at the track end. (b) The starting point of the scratch track, which is clearly different from that shown in Fig. 3.10.

From Figs. 3.11(a and b) it can be clearly seen that the scratch formation and also the starting and the end points of the scratch are totally different from the damage when polymer balls were used on the tip. This type of damage is nothing like a transit scratch.

3.5 Analysis of Failures

Since the cleaned surface scratched by the polymer ball generated what appeared to be transit scratches in the scratch test, this

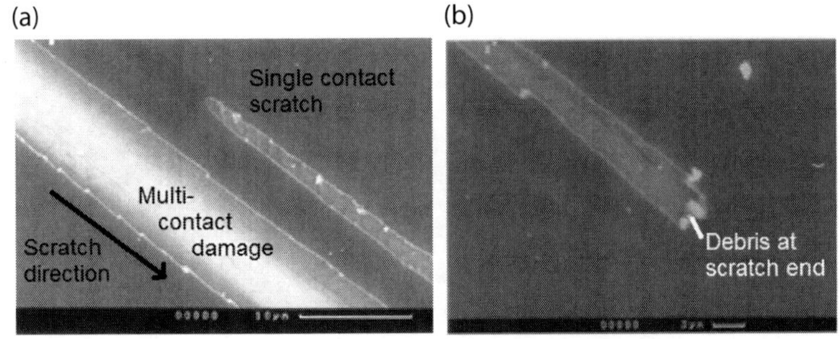

Figure 3.12 (a) SEM micrographs of the scratch track showing the presence of metal debris to the sides and in the middle. (b) The end of the scratch where a pile up of metal debris occurs.

sample was analysed in detail. When observed by SEM (Fig. 3.12), the micrographs clearly show the presence of small pieces of metal debris within the scratch track which were suspected to be silver.

To identify the interface which had failed, energy dispersive X-ray (EDX) microanalysis was done on the unscratched surface and in the scratch track and by subtracting the spectra, the difference spectrum obtained is representative of the material removed (Fig. 3.13). This procedure is necessary because the electron penetration depth and X-ray escape depth are much greater than the total coating thickness and the EDX signal is dominated by the substrate. The difference spectrum shows peaks for Ti, Sn and Ag but no peaks for Zn. This suggests that the top layers of the coating have been removed and failure has occurred at the Ag/ZnO interface. In this coating design, the outermost layer is a protective TiO_xN_y film and beneath this is a tin oxide anti-reflection coating on top of the active silver layer. The anti-reflection layer below the silver is zinc oxide.

Optical profilometry was performed to determine the depths at which the damage had occurred and to confirm which interface had failed.

From Fig. 3.14, the depth of the scratch was seen to be 53.2 nm and from the nominal coating thickness data this is close to the depth of the silver/ZnO interface below the coating surface.

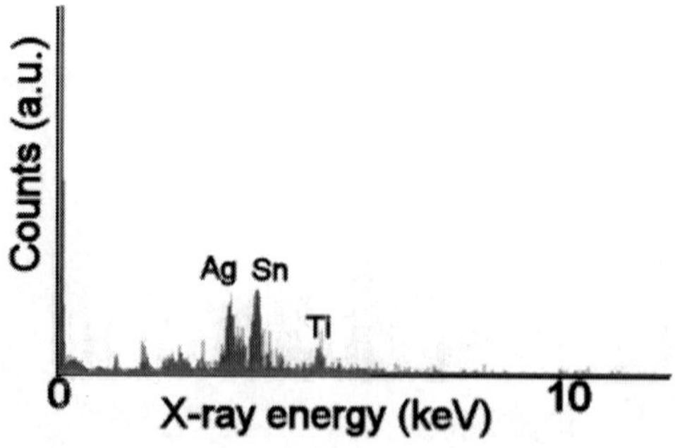

Figure 3.13 EDX difference spectrum of the scratch track.

3.6 Discussion

In this study, we have simulated a transit scratch in coated glass when the surface is not contaminated. The main reason for the damage is due to the stress field created by a circular sliding contact with the ball. Up to the point of failure, there is no evidence for any scratch track being produced, so plastic deformation of the coating/substrate system has not occurred as expected from the low hardness of the ball. This is exactly the same as is observed in real transit scratches and it shows that these scratches are not caused by hard debris particles (e.g., glass cullet) accidentally trapped between the coated glass panels. Rather it is the tensile stress generated at the trailing edge of the contact with a sliding sphere which is responsible for the observed damage (Fig. 3.15).

Figure 3.16 explains the failure mode. Initially the polymer balls are loaded onto the surface where they deform and flatten, establishing a circular contact patch. Movement of the glass (or stylus) causes the ball to slide across the surface. Since the friction coefficient is not zero, tensile stresses are generated at the trailing edge of the contact [55] and these are sufficient to generate through-thickness cracks at this trailing edge. Once this occurs, the coating can be dragged forward by the indenter and failure will occur at the weakest interface in the coating stack. This failure may propagate ahead of the moving stylus and coating debris is folded and concentrated ahead of the moving stylus. The effect is to strip

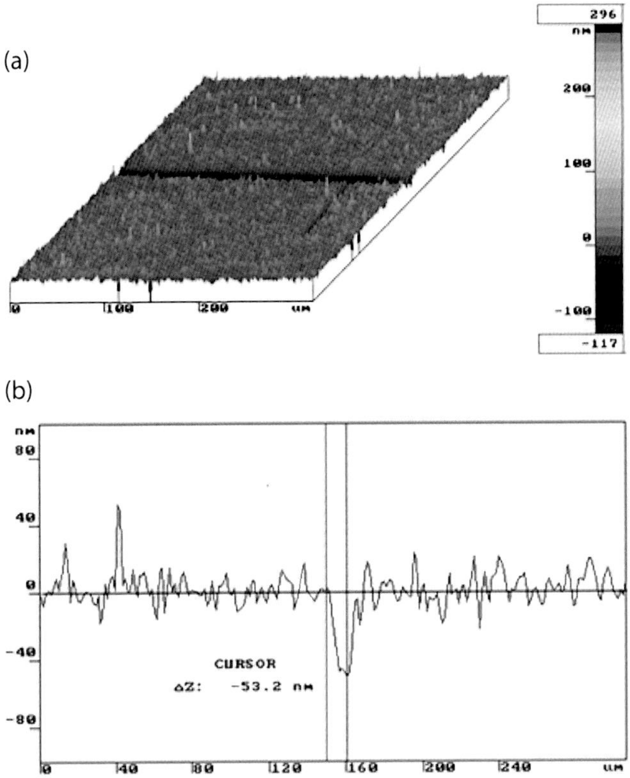

Figure 3.14 Optical profilometry results on the simulated transit scratch (a) aerial view and (b) profile.

the coating at the weakest interface and produce a scratch track with a ragged edge as observed.

As stated by Hamilton and Goodman [61] with brittle materials, the appearance of tensile stresses is more important than the yield parameters since these stresses act in the radial direction and drive the initiation of partial ring cracks as noticed in our experimental analysis. The fracture is critically dependent on the friction coefficient between the ball and the coated glass surface. In order to determine this, a single ball was glued to the end of a nanoscratch stylus and measurements of friction were made as a function of load. The friction coefficient is high at low loads and

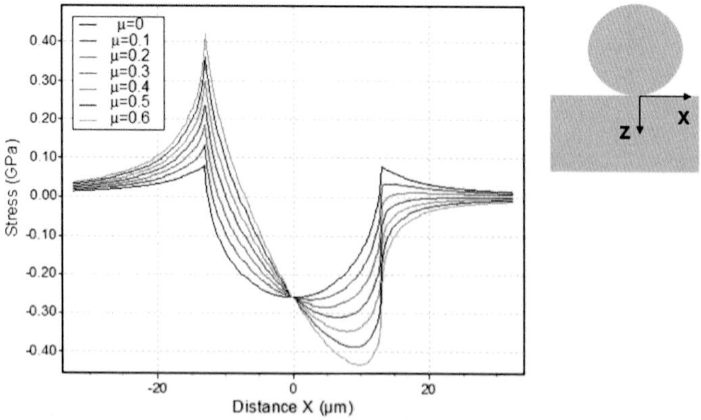

Figure 3.15 Schematic of the stress along the centreline of a sliding Hertzian contact. As the coefficient of friction increases, the stress in the x-direction becomes increasingly compressive at the leading edge of the contact and tensile at the trailing edge.

reduces as the load increases to a steady-state value which depends on the surface condition of the coated glass. The outer coating layer, TiO_xN_y, has a steady-state friction coefficient of ~0.4 for the contaminated surface rising to 0.6 when the surface is cleaned. This confirms that the frictional tractions are important in producing the transit scratch failure.

The frtictional stresses for a titanium oxide top layer on a multilayer solar control coating are typically about 0.45 GPa tensile. The stress which causes failure is a combination of the frictional stresses and the residual stress in the coating; wafer curvature methods give an average residual stress of 1.5 GPa so the failure stress is just less than 2 GPa. The typical interface toughness is 1.9 $MPam^{0.5}$ (from extra recovery in nanoindentation unloading curve as described previously).This implies that the critical interfacial defect size is estimated to be around 300 nm which is similar to the size of the initiating defects in Fig. 3.12a.

In the actual environment, a thin film of polymer balls is sprayed onto the coated glass which separates it from the other glass before they are wrapped tightly into a stack. The balls are compressed during the wrapping process. During transportation, the glass sheets can move with respect to each other due to the local accelerations

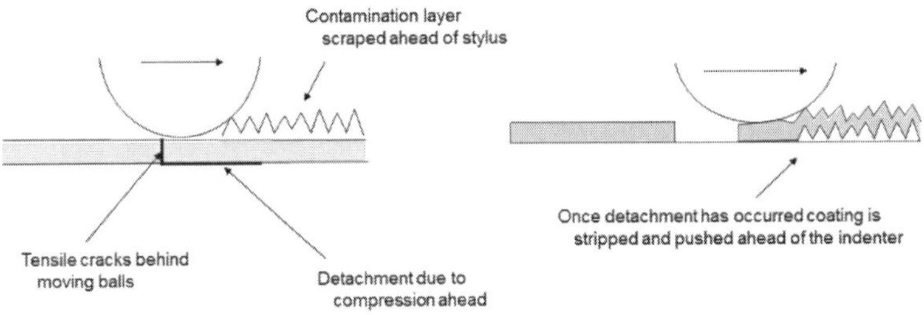

Figure 3.16 Types of failures occurring in the coating.

and decelerations associated with road conditions, unloading and loading. The polymer balls are intended to stop the glass sheets from touching each other, which they do successfully. However, the glass sheets still move with respect to each other and in the process the polymer balls slide against the glass; the frictional force which results generates tensile stresses. These stresses are not sufficient to cause failure in most cases but if there is a weak point in the coating stack the frictional force generated is just enough to initiate the damage of the coating and during the rest of the course of the transportation the scratch elongates. In a real transit scratch, the direction of sliding will vary depending on the movement of the glass sheets which is critically dependent on the route taken by the delivery lorry.

A typical transit scratch was shown in Fig. 3.1. In our experimental analysis, the track of the transit scratch was different from that shown here and this is because we made our scratch in a more controlled fashion. Given the density of balls on the glass sheets and typical acceleration forces which might be encountered in transportation, the load on an individual ball is expected to be of the order of 1 mN.

We have performed the experiments at 8 N using 85 PMMA balls stuck at the tip with an average diameter of 78.9 μm. The compression forces were divided equally among the balls as all the balls were loaded at this load. Thereby the load on each ball can be calculated as 9.4 mN. The majority of the damage was done by only eight PMMA balls: the number of balls bonded to the stylus was counted before and after the experiments (from the pictures shown

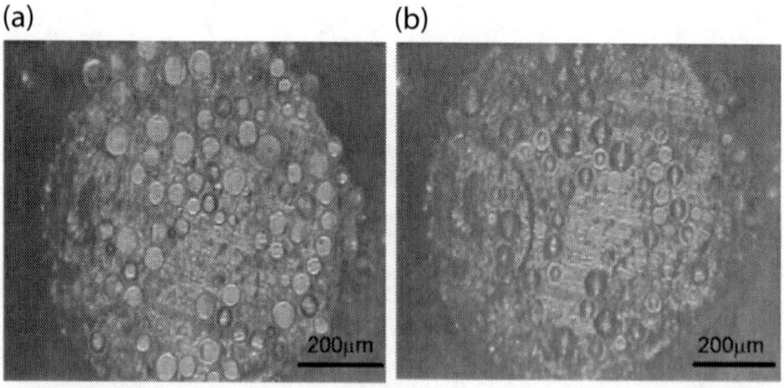

Figure 3.17 Reflected light micrographs of the indenter tip (a) before the scratch test and (b) after the scratch test.

in Fig. 3.17), and the missing balls were found on the test surface associated with the scratch tracks produced. However, since the load was applied on all the balls initially, the average load on a ball was used for calculation of contact stresses.

Using the average load on the balls calculated above, the data in Table 3.2 and the mean radius if the polymer balls the semicontact width calculated from Eq. (3.16) is 6 μm giving a possible scratch track width of 12 μ, if the coating is removed across the complete width of the contact patch. This is comparable to the track width in Fig. 3.10a confirming that the through-thickness crack is then likely to propagate along the Ag/ZnO interface (which adhesion tests show is the weakest in the coating stack with $G_{\text{int}} = 2.8\,\text{J m}^{-2}$) as the coating is dragged forward by the frictional traction. The maximum contact pressure from Eq. (3.17) is only about 130 MPa.

3.7 Conclusions

Scratches found during the delivery of coated glass, called transit scratches, are caused by the polymer balls sprayed to separate the glass sheets during transportation. Through-thickness cracks form initially and divert along the interface at the weakest point in the coating stack. Microanalysis revealed that this was the silver/zinc oxide interface.

The failure depends on the friction between the polymer ball and the coated glass surface. After cleaning the friction coefficient is high and the failure is more likely. Once the surface becomes contaminated by atmospheric exposure the friction coefficient reduces and the probability of transit scratch formation is decreased.

Even when the friction coefficient is high, in order for fracture to take place at the typical loads on a polymer ball, a defect is required in the coated glass surface. For this reason, the formation of transit scratches is not widespread; but given the total area of coated glass delivered, it is still an important failure mode.

Acknowledgements

This work was partly supported by Pilkington Technical Centre.

References

1. J. Chen, S. J. Bull, *J. Phys. D*, 44 (2011), 034001.
2. D. S. Rickerby, *Surf. Coat. Technol.*, 36 (1988), 541–557.
3. M. D. Thouless, *Eng. Fracture Mech.*, 61 (1998), 75–81.
4. S. Venkatraman, D. L. Kohlstedt, W.W. Gerberich, *J. Mater. Res.*, 11 (1996), 3133–3145.
5. T. W. Wu, *J. Mater. Res.*, 6 (1991), 407–426.
6. K. L. Mittal, *Polym. Eng. Sci.*, 17 (1977), 467–473.
7. K. L. Mittal, Adhesion Measurement of Thin Films, Thick Films and Bulk Coatings, (ed. K.L. Mittal), pp. 1–13, VSP, Utrecht (1995).
8. K. L. Mittal (ed.), STP No. 640, pp. 5–17, ASTM, Philadelphia, PA (1978).
9. S. J. Bull, D. S. Rickerby, *Surf. Coat. Technol.*, 42 (1990), 149–164.
10. T. Z. Kattamis, *J. Adhesion Sci. Technol.*, 7 (1993), 783–799.
11. M. J. Cordill, D. F. Bahr, N. R. Moody, W. W. Gerberich, *IEEE Trans. Mater. Reliab.*, 4 (2004), 163–168.
12. N. Sato, K. Habu, *J. Appl. Phys.*, 61 (1987), 4287–4289.
13. G. Binasch, P. Grünberg, F. Saurenbach, W. Zinn, *Phys. Rev. B*, 39 (1989), 4828–4830.
14. J. Zweck, M. Tewesy, H. Hoffmann, *J. Phys.: Condens. Matter*, 8 (1996), 649–663.
15. J. Chen, S. J. Bull, *This Solid Films*, 494 (2006), 1.

16. S. J. Bull, Interfaces and adhesion, (eds. W. Gissler, H. A. Jehn)Advanced techniques for Surface Engineering, 1992, ECSC, EEC, EAEC, Brussels and Luxembourg. 31–68.
17. S. J. Bull, E. G-Berasetegui, *Tribol. Int.*, 39 (2006), 99–114.
18. R. Jacobsson, *Thin Solid Films*, 34 (1976), 191–199.
19. G. Katz, *Thin Solid Films*, 33 (1976), 99–105.
20. A. Bagchi, G. E. Lucas, Z. Suo, A. G. Evans, *J. Mater. Res.*, 9 (1994), 1734–1741.
21. A. Bagchi, A. G. Evans, *Mat. Res. Soc. Symp. Proc.* 383 (1995), 183–196.
22. P. G. Charalambides, J. Lund, A. G. Evans, R. M. McMeeking, *J. Appl. Mech.*, 111 (1989), 77–82.
23. I. Hofinger, M. Oechsner, H. A. Bahr, M. V. Swain, *Int. J. Fract.*, 92 (1998), 213–220.
24. A. A. Volinsky, N. R. Moody, W. W. Gerberich, *Acta. Mater.*, 50 (2002), 441–466.
25. M. K. Small, W. D. Nix, *J. Mater. Res.*, 7 (1992), 1553–1563.
26. R. L. Loh, C. Rossington, A. G. Evans, *J. Am. Ceram. Soc.*, 69 (1986), 139–142.
27. V. Gupta, A. S. Argon, J. A. Cornie, D. M. Parks, *Mat. Sci. Eng.*, A 126 (1990), 105–117.
28. C. C. Berndt, R. McPherson, *Mater. Sci. Res.*, 14 (1981), 619–628.
29. W. C. Oliver, G. M. Pharr, *J. Mater. Res.*, 7 (1992), 1564–1583.
30. F. Ergogan, G. C. Sih, *J. Basic Eng.*, 85 (1963), 519–527.
31. D. K. Shetty, A. R. Rosenfeld, W. H. Duckworth, *J. Am. Ceram. Soc.*, 26 (1987), 825–840.
32. D. Singh, D. K. Shetty, *J. Am. Ceram. Soc.*, 72 (1989), 78–84.
33. J. Malzbender, G.de With, J. M. J. den Toonder, *Thin Solid Films*, 366 (2000), 139–149.
34. M. Y. He, A. G. Evans, J. W. Hutchinson, *Int. J. Solids Struct.*, 313 (1994), 3443–3455.
35. L. G. Rosenfeld, J. E. Ritter, T. J. Lardnerr, M. R. Lin, *J. Appl. Phys.*, 67 (1990), 3291–3296.
36. J. Fernandez-Palacio, I. Arce-Garcia, S. J. Bull, *Tribol. Int.*, 37 (2004), 929–940.
37. S. Suresh, C. F. Shih, A. Morrone, N. P. O' Dowd, *J. Am. Ceram. Soc.*, 73 (1990), 1257–1267.

38. J. den. Toonder, J. Malzbender, G. de With, R. Balkenende, *J. Mater. Res.*, 17 (2002), 224–233.
39. V. S. Deshpande, N. A. Fleck, *J. Mech. Phys. Solids.*, 51 (2003), 187–208.
40. W. Li, T. Siegmund, *Acta Mater.*, 52 (2004), 2989–2999.
41. D. B. Marshall, A. G. Evans, *J. Appl. Phys.*, 56 (1984), 2632–2638.
42. J. W. Hutchinson, Z. Suo, *Adv. Appl. Mech.*, 29 (1992), 63–191.
43. X. D. Li, D. F. Diao, B. Bhushan, *Acta. Mat.*, 45 (1997), 4453–4461.
44. J. Chen, S. J. Bull, *MRS Symp. Proc.*, 890 (2006), 57–62.
45. J. Malzbender, G. de. With, *Surf. Coat. Technol.*, 135 (2000), 60–68.
46. J. Malzbender, G. de With, *J. Non-Cryst. Solids* 265 (2000), 51–60.
47. S. Etienne-Calas, A. Duri, P. Etienne, *J. Non-Crystalline Solids*, 344 (2004), 60–65.
48. J. Chen, S. J. Bull, *J. Phys. D*, 40 (2007), 5401–5417.
49. S. V. Hainsworth, M. R. McGurk, T. F. Page, *Surf. Coat. Technol.*, 102 (1998), 97–107.
50. M. P. de Boer, W. W. Gerberich, *Acta Mater.*, 44 (1996), 3169–3175.
51. D. B. Marshall, A. G. Evans, *J. Appl. Phys.*, 56 (1984), 2632–2638.
52. D. Chicot, P. Démarécaux, J. Lesage, *Thin Solid Films*, 283 (1996), 151-157.
53. J. M. Sanchez, S. El-Mansy, B. Sun, T. Scherban, N. Fang, D. Pantuso, W. Ford, M. R. Elizalde, J. M. Martinez-Escanola, A. Martin-Meizoso, J. Gil-Sevillano, M. Fuentes, J. Maiz, *Acta Mater.*, 4 (1999), 4405–4413.
54. M. D. Kriese, W. W. Gerberich, N. R. Moody, *J. Mater. Res.*, 14 (1999), 3007–3018.
55. D. B. Marshall, B. R. Lawn, A. G. Evans, *J. Am. Ceram. Soc.*, 65 (1982), 561–566.
56. S. J. Bull, *J. Vac. Sci. Technol. A* 19, (2001), 1404–1414.
57. S. J. Bull, E. G-Berasategui, T. F. Page, *Tribology Int.*, 39 (2006), 99–114.
58. S. J. Bull, E. G-Berasetegui, T. F. Page, *Wear*, 256 (2004), 857–866.
59. J. F. Palacio, Ph. D thesis, University of Newcastle Upon Tyne, 2006.
60. H. Hertz, Gesammelte Werke, (ed. P. Lenard), pp. 174–196, J. A. Barth, Leipzig (1895).
61. G. M. Hamilton, L. E. Goodman, *J. Appl. Mech.*, 37 (1966), 371–376.

Chapter 4

Functionally Graded and Multilayered Carbon-Based Coatings: The Microstructure and Tribological Performances

Liping Wang, Shengguo Zhou, and Jibin Pu

State Key Laboratory of Solid Lubrication, Lanzhou Institute of Chemical Physics, Chinese Academy of Sciences, Lanzhou 730000, PR China

lpwang@licp.cas.cn

In many engineering applications, the carbon-based coatings demonstrate improved friction and wear responses for components and devices. However, due to the high level of residual stress and brittleness, their use has often been restricted, especially, when exposing to high loading. Design of functional gradient and multilayer architecture has been developed to overcome those limitations of carbon-based coatings. Especially, it is extremely beneficial to achieve high levels of performance coatings in a variety of tribological applications under severe serving environments. This critical review presents recent progress on functionally graded and multilayered carbon-based coatings, mainly focusing on their microstructure and tribological performances as well as some applications, which helps to progressively make their way into the

Multifunctional Materials for Tribological Applications
Edited by Robert J. K. Wood

ISBN 978-981-4463-90-4 (Hardcover), 978-981-4463-91-1 (eBook)
www.panstanford.com

commercial marketplace and meet the ever-increasing performance demands of more severe applications.

4.1 Introduction

Energy efficiency and mechanical service lifetime have a direct linkage with the friction and wear at surfaces in all moving systems. Statistics analysis shows that a significant part (about 30%) of the 20 world's energy resources is spent on friction in different forms; 80–90% of movable machinery joints become disabled because of wear. The annual cost of friction- and wear-related energy and material losses is estimated to be 5% to 7% of each country's gross national products, and the prospect of further reducing friction of all moving systems is extremely beneficial for conserving not only our ever-diminishing energy resources but also the environment [1–4]. Therefore, ever-increasing demands in reduced friction and wear situations have stimulated the development of a variety of innovative coatings such as TiN, CrN, ZrN, WC-Co, WC-Ni, ZrCN and TiCN coatings [5–10]. In particular, the solid lubrication coatings that exhibit lower friction and wear are of extremely importance to enhance surface lubrication characteristics, and thereby they are widely applied as an alternative lubricant in severe operating environments where lubricating oil cannot be used and have been found greater practical applications in the fields of vacuum equipment, aviation and space [11–15]. For example, CrB-MoS_x, TiB_2-MoS_2, TiN-MoS_2, TiCN-MoS_2 and Al_2O_3-MoS_2 solid lubricants have been observed to offer substantial friction coefficient reduction [16–22], and WS_2 and MoS_2 lamellar coatings have also been the most extensively used material due to both having friction and wear reduction effect [23, 24]. In recent years, diamond-like carbon (DLC) based coatings are recognised as excellent candidates of solid lubrication materials for tribological applications in a variety of environments. These are attributed to their remarkable mechanical and tribological properties (high hardness, low friction coefficient, high wear resistance and so on), together with its high surface smoothness, high chemical inertness and biocompatibility [25–32].

Recent work shows that functionally graded and multilayered coatings can lead to advantages in performances over comparable single-layer coatings and can combine the attractive properties of different materials in a single protective layer. For example, the graded and multilayered coatings can have a denser structure than that of single-layer coatings [24, 35]; an increase in the adhesion can be obtained between the coating and the substrate [36]; the residual stress can be reduced inside the graded and multilayered coatings [37]; the introduction of a number of interfaces parallel to the substrate surface can act to deflect cracks or provide barriers to dislocation motion, increasing the toughness and hardness [33, 34]. As reported by references, the multilayered CN/TiCN/TiN coatings can obtain good mechanical and tribological properties with respect to single layers [38], and the WS_2/MoS_2 multilayer coating show a significantly improved tribological performance in air compared to the single-layer MoS_2 or WS_2 coating [36]. It is well known that DLC films represent a wide class of amorphous carbon films and hydrogenated amorphous carbon films composed of sp^3 carbon and sp^2 carbon. There are a quite big difference in the composition, structure, properties and the field of application of DLC films that comes from different preparation methods of DLC films. Based on the standards of 'carbon coating' issued by the Association of German Engineers in 2005, DLC films can be divided into seven categories, as follows:

- a-C (amorphous carbon);
- ta-C (tetrahedral amorphous carbon);
- a-C:Me (Me = W, Ti, Mo, Al, etc.) (metal-doped amorphous carbon);
- a-C:H (hydrogenated amorphous carbon);
- ta-C:H (hydrogenated tetrahedral amorphous carbon);
- a-C:H:Me (Me = W, Ti, Mo, Al, etc.) (metal-doped hydrogenated amorphous carbon);
- a-C:H:X (X=Si, O, N, F, B, etc.) (modified hydrogenated amorphous carbon).

A functionally graded and multilayered DLC coating can provide high hardness with low friction surface, which also resisted brittle failure in tests with high contact loads [39]. Consequently, it can be demonstrated that approaches to couple gradient and multilayer architecture with carbon-based coatings should be extremely promising to achieve high levels of performances in a variety of tribological applications.

The main purpose of this chapter is to review recent advances in functionally graded and multilayered carbon-based coatings, mainly focusing on their microstructure and tribological performances as well as some applications, which helps to progressively make their way into the commercial marketplace and meet the ever-increasing performance demands of more severe applications.

4.2 Functionally Graded and Multilayered DLC Coatings with Excellent Tribological Performances

DLC coatings are well known for their excellent wear-protective properties combining high hardness and a very low friction. Nevertheless, those coatings possess an extremely high intrinsic stress, compressive stresses typically reaching value of 10 GPa, which induce a high shear stress on the coating–substrate interface leading usually to delamination of DLC coatings [40–45]. As a result, it limits the thickness of the DLC coatings, that is, coatings in relative high thickness (>2 μm) delaminated due to large compressive stress and the tribological applications in high load. Therefore, a great many of researchers have been trying to solve these limitations of DLC by different approaches [46–48]. The limitations of DLC coatings can be relieved by doping metallic elements, such as metals (W, Ti, Cr, N, Al, Au, Pt and Cu) and light elements (B, Si, N, O or F) have often been used as dopants to modify the hardness, internal stress, adhesion, tribiological properties, biocompatibility and so on [49–55]. In recent years, the use of a functionally graded and multilayered coating is proposed to improve the mechanical and tribological performances. This section will highlight recent developments in tribological DLC coating designs incorporating functional gradients and multilayer where these features are employed to provide exceptional properties.

4.2.1 *Functional Gradient Architecture of DLC Coatings*

Functionally graded DLC coatings refer to a system whose structure and composition gradually varies from one layer to another, that is, no sharp transition exists. The layers have graded interfaces that ensure no abrupt change in structure and composition, and thus this coated system can optimise the stress distribution, reduce microcracking, enhance adhesion strength and allow the operation under severe sliding conditions. Mainly transition metal nitrides and carbides, especially TiN, TiC, TiCN and TiAlN, exhibit many extraordinary properties such as high hardness, low friction, good wear resistance, high melting temperature and high chemical stability, and thereby they are very promising material for use in functional gradient structures together with DLC [48]. Examples of typical gradient architectures are Ti-TiC-DLC and Ti-TiN-TiCN-DLC. Voevodin et al. have designed functionally graded Ti-TiC-DLC coatings based on studies of structural transitions [39]. In this study, structural variations in the Ti–C chemical system prepared by MSPLD were used to achieve a Ti-TiC-DLC transitional architecture. It provided a gradual increase in the hardness from a steel substrate to a DLC layer, and thereby considerably relaxed intrinsic stress and improved the adhesion of the super-hard (60–70 GPa) DLC top coating. Especially, the Ti-TiC-DLC graded coatings maintained a friction coefficient below 0.1 without failures by cracking or delamination typically observed for single-layer DLC coatings by ball-on-disk friction tests with 1.4 and 2.2 GPa contact load. These reveal that the functional gradient design provided a hard DLC coating with a low friction surface and a property of resisted brittle failure in tests with high contact load. Additionally, Choy et al. have designed functionally graded Ti-TiN-TiCN-DLC based on studies of structural transitions in Ti–N thin coating [56]. In their works, several functionally graded DLC coatings were developed to improve the coating adhesion of DLC onto 304 stainless steel and Ti-6Al-4V substrates. The coating systems were investigated including Ti-TiN-TiCN-DLC and Ti-TiC-DLC on titanium alloy and stainless steel substrates, which were fabricated using a magnetron sputter ion plating technique. The scratch test results showed that the functional gradient structure improved the coating adhesion on the substrate. Moreover, the Ti-TiN-TiCN structure was more efficient than Ti-TiC in improving the adhesion of DLC coatings onto both stainless steel and Ti-alloy substrates. The coating adhesion has a strong influence on the wear resistance of the coating. As a result, the wear resistance of graded Ti-TiN-TiCN-DLC was better than Ti-TiC-DLC coating. For all the coating systems, the dry friction coefficient of the DLC coatings against a chrome steel ball was about 0.2. Therefore, Ti-TiN-TiCN-DLC was the coating system that had the best mechanical properties for application such as heart valve prosthesis.

4.2.2 *Multilayer Architecture of DLC Coatings*

Multilayered coatings can achieve benefits in performance over comparable single-layer coatings and can combine the attractive properties of different materials in a single protective layer. The multilayer structures can be achieved by designing and implementing one or more intermediate layers between the coating and the substrate, which can eliminate inertness of DLC and prevent from adhesion failures of coated steel surfaces. In this multilayered structure, transition metal carbides and nitrides are often used as interleaving materials, such as TiC, TiN, CrN and so on. Besides, the introduction of a number of interfaces parallel to the substrate surface can act to deflect cracks or provide barriers to dislocation motion, increasing the toughness and hardness of the coating. For example, an effective multilayer is the introduction of ductile, low elastic modulus layers as metal interlayer into the DLC coating structure to relieve stress and allow crack energy dissipation by plastic deformation in the crack tip. This approach will result in a decreased coating hardness, but the gain in the fracture toughness improvement may be more important in many tribological applications. Some typical works on multilayer architecture by pioneer researchers are presented as follow:

(i) Rincon et al. [57] reported tungsten carbide/DLC multilayer coatings on steel for tribological applications where the coatings are composed of a W-C multilayer base and an upper lubricious DLC layer, which were deposited onto silicon wafers and also onto M2 (AISI) high-speed steel substrates. SEM cross-section morphology of a W-C/DLC multilayer is shown in Fig. 4.1. The DLC upper layer is clearly distinguished and the multilayer stack in which a columnar dense structure is

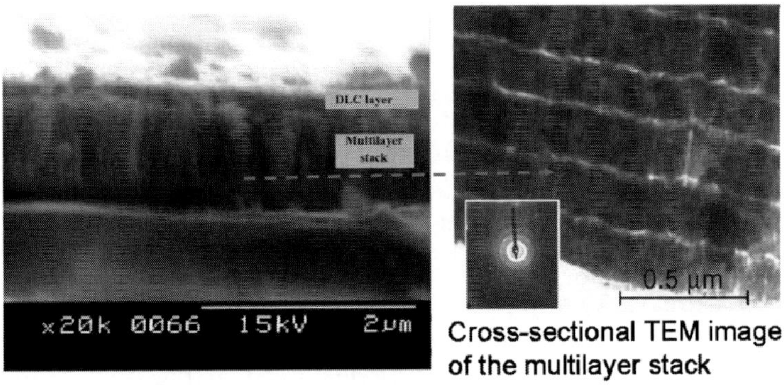

Figure 4.1 Cross-sectional SEM micrograph of a W-C/DLC multilayered coating. Reprinted from Ref. [57], Copyright (2001), with permission from Elsevier.

observed. The multilayer structure cannot be resolved with this technique. Thus, transmission electron microscopy (TEM) cross-section of the multilayer stack with enough resolution is used to clearly differentiate the multilayer structure, and the inset shows the corresponding electron diffraction pattern. The rings correspond to (111) and (222) diffraction planes of a WC_{1-x} phase. The same results have also been obtained from the GIXRD diffraction pattern of the multilayer stack, where exclusively the WC_{1-x} phase appears. Besides, it can be observed that the expected W_2C/WC_{1-x} bilayer structure cannot be distinguished. Instead of this, dark homogenous layers limited by narrow bright stripes are observed. These dark homogenous layers probably correspond to the WC_{1-x} phase and the bright stripes correspond to a carbon-rich interface region. Therefore, the hexagonal W_2C phase does not form while carbon-rich interlayers appear in the multilayer interfaces due to the multilayer deposition process.

Ball-on-disc tests were determined to the friction coefficient of the coatings when sliding against steel and alumina balls. The measured friction coefficient stabilises after a short initial transition period. In the steady state, the friction coefficients are similar for single DLC and multilayer coatings, with a value of 0.12 when sliding against steel balls and 0.08 when sliding against alumina balls. The same ball-on-disc test allows determining the lifetime of the coatings under sliding wear. The test against alumina balls shows that the failure of the coating happens very early as compared to the test against steel balls. The wear scar was observed in the optical microscope after stopping the test at different stages. The images indicated that, when sliding against alumina balls, the coatings are progressively worn, but when sliding against steel balls the coatings suffer no visible wear until failure. In both cases, failures appear to be adhesion failures. Compared to single DLC coatings, the lifetime of tungsten carbide/DLC multilayered coatings are noticeably increased. The ensemble of mechanical tests demonstrates the advantage of the multilayer structures to improve wear resistance of DLC coatings on steel substrates without disturbing their lubricious friction properties.

(ii) Bewilogua et al. [58] also reported about preparation of a-C:H/a-C:H:Si:O and a-C:H/a-C:H:Si multilayer coatings by PACVD. Figure 4.2 shows the fracture cross section SEM images of an a-C:H:Si:O/a-C:H multilayer coating with 36 double layers and a a-C:H:Si:O top layer, prepared by deposition mode (II). Adhesion improving interlayer: a-C:H:Si. Total coating thickness is approximately 3 μm. It is found that the multilayer coatings exhibit lower wear rates and higher hardness values

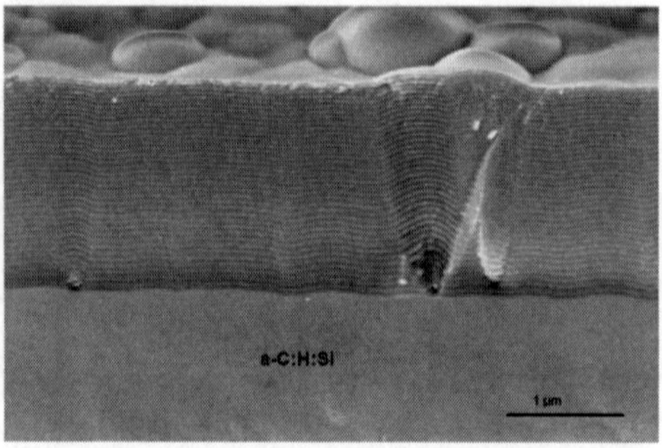

Figure 4.2 Fracture cross section SEM images of an a-C:H:Si:O/a-C:H multilayer coating with 36 double layers and a a-C:H:Si:O top layer, prepared by deposition mode (II). Adhesion improving interlayer: a-C:H:Si. Reprinted from Ref. [58], Copyright (2011), with permission from Elsevier.

than a-C:H:Si:O or a-C:H:Si single layers and lower friction coefficients than pure a-C:H coatings under unlubricated test condition. Utilising duplex processes, combining plasma nitriding pretreatment and a following coating deposition, the adhesion of the multilayer coatings on high-speed and cold-working steel substrates could be considerably improved.

(iii) The multilayered DLC coatings can be deposited via alternating bias during magnetron sputtering, as reported by Li et al. [59]. To combat the high residual stress problem in monolayer DLC coatings, they fabricated multilayered DLC coatings with alternate soft and hard layers via alternating bias during magnetron sputtering. Alternately, biased sputtering deposition provides a combination of hardness, toughness and adhesion strength for the multilayered DLC coating, and thereby can achieve excellent tribological performances.

4.2.3 *Functionally Graded and Multilayered DLC Coatings*

In general, the combination of multilayer and functionally gradient approaches in the design of wear-protective DLC coatings can produce exceptionally tough wear-protective coatings for engineering applications. Figure 4.3a shows a schematic of a multilayer $[Ti/DLC]_n$ coating on a graded load support foundation, where the ductile Ti layers in the multilayer stack were graded at every DLC interface to avoid brittle fracture [60]. A cross-sectional photograph of this coating with 20 [Ti/DLC] pairs is shown in Fig. 4.3b. The ductile Ti layers reduced the composite coating hardness to 20 GPa as compared to a single-layer DLC coating, which has a hardness of about 60 GPa. However due to dramatic improvements in toughness, the multilayer coating design permitted operation during sliding friction at contact pressures as high as 2 GPa without fracture failure compared to 0.6–0.8 GPa for single-layer DLC.

One potential drawback slowing the wide spread use of new coatings was the need for reliable process controls to ensure that the correct compositions, structures, and properties are implemented during growth. However, modern process instrumentation and control technologies are able to meet the challenge and permit successful commercialisation, for example, see Voevodin et al.

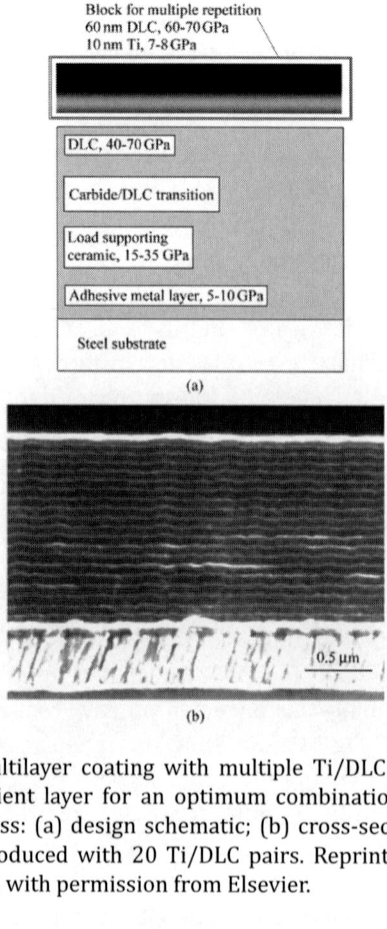

Figure 4.3 A multilayer coating with multiple Ti/DLC pairs on top of a functionally gradient layer for an optimum combination of cohesive and adhesive toughness: (a) design schematic; (b) cross-sectional photograph of the coating produced with 20 Ti/DLC pairs. Reprinted from Ref. [60], Copyright (1997), with permission from Elsevier.

[61] for a review of reactive sputtering control methods during multilayer growth. Thus, functionally graded and multilayered designs will be commonly utilised in the production of modern tribological DLC coatings.

4.3 Multilayered Nanocomposite Carbon-Based Coating with Excellent Tribological Performances

Nowadays, the nanocomposite coatings are intensively attractive to researchers because of their excellent friction and wear responses under severe sliding conditions in extreme environments. In these coatings, the combination of a hard phase (hard carbide or nitride phases) with ductile phases has been assessed as a good approach to achieve advanced properties by appropriate control of chemical composition and microstructure at the nanoscale, as the nature of chemical composition, crystalline size and the amount of each phase determine the final properties of nanocomposite coatings [60–66]. Consequently, the nanocomposite architecture also is employed to fabricate the carbon-based coatings where carbide or nitride nanocrystallites are embedded into amorphous carbon (a-C) matrix, which has been confirmed to be a promising approach to overcome the limitations and extend the range of applications for DLC coatings. Furthermore, some researchers report that the co-introduction of soft aluminum metal in these nanocomposite carbon-based coatings can result in better mechanical and tribological performances. Such as multicomponent nc-TiC/a-C(Al), nc-WC/a-C(Al) carbon-based coatings, which are confirmed to have superior mechanical and tribological properties [67–71].

However, these multicomponent nanocomposite carbon-based coatings still suffer from failure when sliding friction by high applied Hertzian contact pressure. Such as it will give birth to fracture and delamination of coatings to steel substrates by applying Hertzian contact pressures above 1 GPa [72, 73]. Failure analyses reveal that the coating bending and fracture by steel substrate compliance in the contact as well as relatively low adhesion of the coatings to steels are the main factors. Thus, an approach to couple multilayer architecture with nanocomposite carbon-based coatings to achieve higher levels of performance in a variety of tribological applications is considered [74–76]. In this section, the multilayered nc-TiC/a-C(Al), nc-CrC/a-C(Al) and nc-WC/a-C(Al) nanocomposite carbon-based coatings are discussed including their microstructure and tribological performance, as reported by Wang et al. [77]. It is confirmed that these multilayered nanocomposite carbon-based coatings can possess good mechanical and low-frictional properties as well as high load carrying capacity under sliding friction when compared with their corresponding monolayered coatings. The objective is to develop conceptual designs and experimental verification for multilayered nanocomposite carbon-based coatings

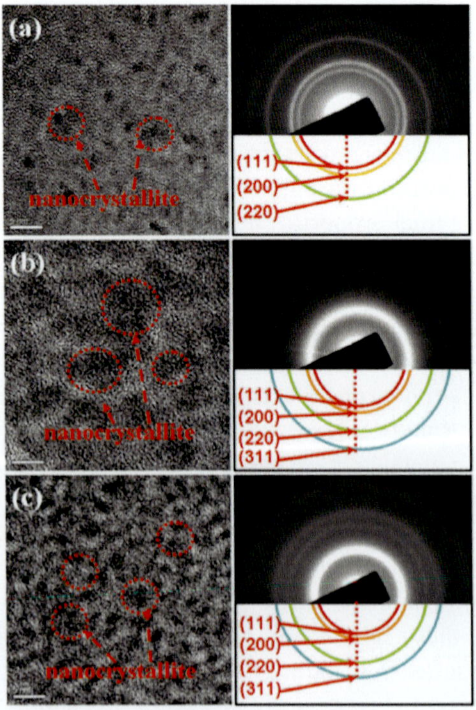

Figure 4.4 HRTEM images and SAED patterns of top-layered nanocomposite carbon-based coatings: (a) nc-TiC/a-C(Al), (b) nc-CrC/a-C(Al) and (c) nc-WC/a-C(Al). Reprinted from Ref. [77], Copyright (2011), with permission from Elsevier.

as wear-protective materials under relatively high applied Hertzian contact pressure.

Figure 4.4 shows the high-resolution transmission electron microscopy (HRTEM) bright-field images and selected area electron diffraction (SAED) patterns of top-layered nanocomposite coatings. HRTEM images reveal that the nc-TiC/a-C(Al), nc-CrC/a-C(Al) and nc-WC/a-C:Al carbon-based coatings have typical nanocrystallites/amorphous microstructure in which carbide nanoparticles (the black dots in the HRTEM images) are uniformly embedded in the amorphous carbon matrix. Besides, HRTEM images present that the size of carbide nanoparticles was about 3–6 nm. The SAED patterns, deriving from the scatter from the crystalline structure of

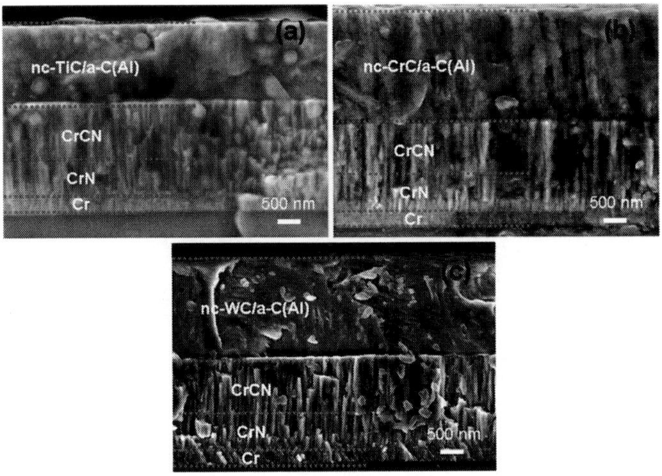

Figure 4.5 Cross-sectional images of multilayered nanocomposite carbon-based coatings. Reprinted from Ref. [77], Copyright (2011), with permission from Elsevier.

the nano-sized metal particles in the amorphous host matrix, display some sharp rings with calculated lattice proving the cubic TiC, CrC and WC_{1-x} phase for these carbide nanoparticles.

The multilayer architecture is Cr, CrN, CrCN and outmost nanocomposite carbon-based layers in turn where the top-layers are nc-TiC/a-C(Al), nc-CrC/a-C(Al) and nc-WC/a-C(Al) coatings, respectively. Figure 4.5 shows the cross-sectional images of multilayered nanocomposite carbon-based coatings by field emission scanning electron microscopy (FESEM). No delaminations are observed in the multilayer configuration, indicating these coatings have good adhesion strength between each layers.

The hardness (H) and Young's modulus (E) of three kinds of multilayered coatings are performed by nanoindentation with a Berkovich indenter. Results show that all coatings exhibit relatively high hardness (H) and Young's modulus (E). The highest hardness (24 GPa) is observed for nc-TiC/a-C(Al) coating, while the highest Young's modulus (217 GPa) is observed for nc-CrC/a-C(Al) coating. It is reported that hardness has long been is regarded as a primary material property affecting wear resistance. However, the elastic strain to failure, which is related to the H/E ratio, is a more suitable parameter for predicting wear resistance [78]. A higher H/E means better wear resistance. The calculated H/E ratio is 0.11, 0.09 and 0.10 corresponding to the multilayered nc-TiC/a-C(Al), nc-CrC/a-C(Al) and nc-WC/a-C:Al carbon-based coatings, respectively. Thus, the multilayered nc-TiC/a-C(Al) should show a lower wear rate. Besides, compressive stresses of −0.5 GPa, −0.7 GPa and −0.7 GPa are calculated for the nc-TiC/a-C(Al), nc-CrC/a-C(Al) and nc-WC/a-C:Al coatings, respectively. Wang et al. [69, 70] have proved that Al doped in the carbon-based coatings can lead to a drastic reduction in internal stress. For the nanocomposite carbon-based coating with Al of nearly 4 μm in total thickness, just −0.5 GPa of internal stress is very low compared to other hard coatings, such as TiN, ZrN and HfN, which could easily have an internal stress of several GPa [79]. Besides, the adhesive behaviour of as-fabricated three kinds of multilayered coatings are evaluated by the scratch tests. The load at which the friction force assumed a sharp increase is defined as the critical load (L_c) of the coatings and is used as a quasi-quantitative criterion to evaluate the adhesion strength of the coatings on the substrate. The critical load is 41 N, 32 N and 35 N corresponding to multilayered nc-TiC/a-C(Al), nc-CrC/a-C(Al) and nc-WC/a-C:Al coatings, respectively. However, the critical load are only 32 N, 25 N and 28 N corresponding to monolayered nc-TiC/a-C(Al), nc-CrC/a-C(Al) and nc-WC/a-C:Al coatings, respectively. Obviously, it is found that these multilayered nanocomposite carbon-based coatings possess superior mechanical properties.

The friction coefficient and wear resistance of multilayered nanocomposite carbon-based coatings are evaluated by ball-on-disc reciprocating tribometer, while their corresponding monolayered coatings are introduced for comparison. Figure 4.6 shows the friction coefficients of the six coatings as a function of sliding distance against steel ball. The results show that the friction coefficient and wear lifetime are affected by the coatings composition and multilayer treatment. For given nanocomposite carbon-based coatings, the obtained results can be classified into two features by monolayer and multilayer treatment.

Comparing the friction coefficient of monolayered and multilayered coatings, it can be found that the friction coefficient of monolayered coatings is relatively lower. For example, the

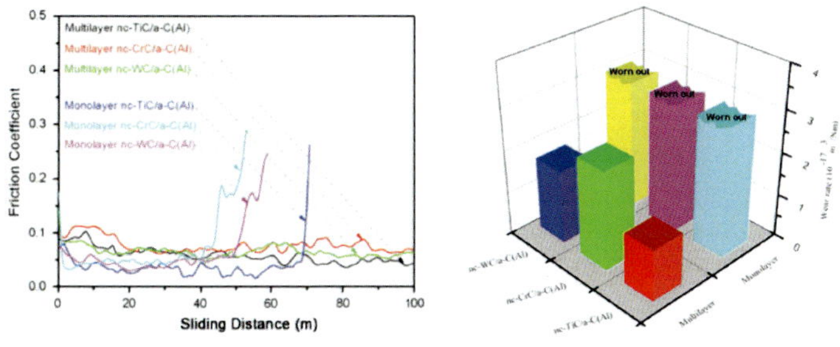

Figure 4.6 Friction coefficient and wear rate for monolayered and multilayered nanocomposite carbon-based coatings against steel ball under ambient air. RH 35 ± 5%. Reprinted from Ref. [77], Copyright (2011), with permission from Elsevier.

monolayered nc-TiC/a-C(Al) coating shows a very low friction coefficient about 0.03 while the corresponding multilayered coating shows an increased friction coefficient about 0.05. Besides, it can be found that the system of nc-TiC/a-C(Al) nanocoposite presents lowest friction coefficient compared to other system, and the system of nc-CrC/a-C(Al) presents the highest friction coefficient. In the other feature, the relatively stable friction coefficient is observed for multilayered coatings. However, an immediate increase in the friction coefficient and abrasion of the substrate are observed for monolayered coatings. Such as monolayered nc-TiC/a-C(Al) coating shows an immediate increase in the friction coefficient at about sliding distance of 70 m. Besides, the values of wear rates are about 1.1×10^{-17} m^3 N-m^{-1}, 1.5×10^{-17} m^3 N-m^{-1} and 1.3×10^{-17} m^3 N-m^{-1} corresponding to multilayered nc-TiC/a-C(Al), nc-CrC/a-C(Al) and nc-WC/a-C(Al) coatings, respectively, while all of monolayered coatings have been wore out after the friction test, shown in Fig. 4.6. These frictional results indicate that the multilayered nanocomposite carbon-based coatings have higher anti-wear and load support capability compared to monolayered coatings.

In order to investigate friction and wear characteristics, SEM is used to observe wear tracks of coatings after friction testing as well as wear scars on steel balls. SEM images show that the wear tracks of monolayered coatings are heavily damaged, if this test is continued, these coatings will be worn out. However,

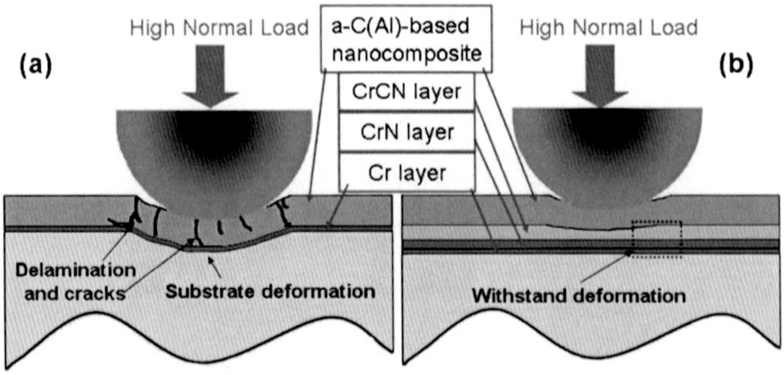

Figure 4.7 Schematic diagram of load support capability: (a) monolayered and (b) multilayered carbon-based coatings. Reprinted from Ref. [77], Copyright (2011), with permission from Elsevier.

the multilayered coatings show relatively smoother and shallower wear tracks after the friction test, indicating relatively mild wear compared to the monolayered coatings. On the other hand, the wear scars on steel balls for multilayered nc-TiC/a-C(Al), nc-CrC/a-C(Al) and nc-WC/a-C(Al) coatings are characterised by SEM, showing that the wear marks on steel balls are covered with transferred materials.

The frictional testing results exhibit that the monolayered nanocomposite carbon-based coatings cannot withstand long-time wear under high applied Hertzian contact pressure. Characteristic failure of monolayer coatings is shown in schematic diagram of Fig. 4.7(a), which is the approximation to a ball-on-disc reciprocating sliding contact. The schematic diagram showed that it is necessary to enhance the hardness of substrate, adhesion strength of substrate to coatings and toughness of carbon-based coatings. Generally, the thicker layer may provide the higher load carrying capacity and deformation resistance of substrate, which is helpful to prevent the coatings from being worn out [80]. But the single-layered thick coatings often have relatively high intrinsic stress and higher brittleness, which is not beneficial to the adhesive strength between coating and substrate. Therefore, a top-layered nanocomposite carbon-based coating with high hardness, low residual stress and good toughness composed of multilayer architecture can be very useful to improve deformation resistance and load support capability and thereby disperse or reduce the friction contacting stress, decrease the brittleness of the top-layer and prevent the cracks from propagating and extending, as shown in Fig. 4.7(b).

Generally, according to the adhesion theory of friction [77], the friction force is the product of the shear strength and the area of welded junctions formed at the real contact area. Thus, the friction coefficient (μ) is given by $\mu = AS/P$, where A was the real contact area, S is the shear strength of the junction and P was the applied normal load. Firstly, the formation of a carbon-rich tribofilm plays an important role in reduction of the friction coefficient under sliding friction [81]. The SEM images of wear scars exhibit that the wear marks on steel balls sliding against multilayered coatings are covered with transferred materials. Raman Analysis reveals that the transferred tribofilms on the frictional contact surface rooted in the graphitisation of top-layered nanocomposite coatings. Thus, these tribofilms on contact surface provide relatively low shear strength (S), which is beneficial to low friction of coatings. Moreover, the real contact area (A) is nearly associated with the deformation resistance of the contact area. The multilayer architecture of nanocomposite carbon-based coatings improves the deformation resistance so as to reduce the real contact area (A), which is also beneficial to low friction. The most important role for the improved deformation resistance is to provide high load-support capability of multilayered nanocomposite carbon-based coatings under ball-on-disk frictional testing. Consequently, the contributions of tribofilm and deformation resistance by multilayer architecture make these multilayered coatings achieve low-friction and anti-wear performances under high applied Hertzian contact pressure.

In summary, three kinds of multilayered nc-TiC/a-C(Al), nc-CrC/a-C(Al), nc-WC/a-C(Al) carbon-based coatings constructed by Cr/CrN/CrCN can possess superior combining mechanical properties, that is, high hardness, low internal stress as well as high adhesion strength and superior anti-wear and long lifetime characteristics under relatively high applied Hertzian contact pressure when compared with the corresponding monolayer coatings. Therefore, the approach to couple multilayer architecture with nanocomposite carbon-based coatings to achieve higher levels of performance is feasible.

4.4 Functionally Graded and Multilayered Carbon-Based Coatings to Improve Tribological Performance of Light-Weight Alloys

Functionally gradient and multilayer architecture can provide excellent mechanical and tribological performance for the carbon-based coatings. Therefore, the applications of these graded and multilayered carbon-based coating on mechanical engineering will contribute to the dramatic energy-saving and a reduction in harmful exhaust emissions to environment. Nowadays, light-weight alloys such as magnesium alloys, aluminum alloys and titanium alloys play an increasing importance as structural materials in the aerospace, automotive and electronic industries due to their excellent properties, especially light weight. However, the poor corrosion resistance and wear resistance of light-weight alloys limit their further applications. In this section, we will introduce these light-weight alloys coated by the graded and multilayered carbon-based coatings to overcome their limitations based on the research works of other and Wang et al.

4.4.1 *Graded and Multilayered Coatings on Magnesium Alloys*

Magnesium alloys as one kind of light-weight alloys are becoming more important in the aerospace, automotive and electronic industries due to its light weight. While the common disadvantages, that is, poor corrosion resistance and wear resistance, limit its further application [82]. DLC, which owns a high hardness and a chemical inertness, has been applied as protective coatings in many areas [83]. Therefore, an attempt to use DLC as a protective coating for magnesium alloys to improve their mechanical and anti-corrosion properties was reported by some researchers [84–87]. Moreover, an approach to couple gradient and multilayer architecture with DLC

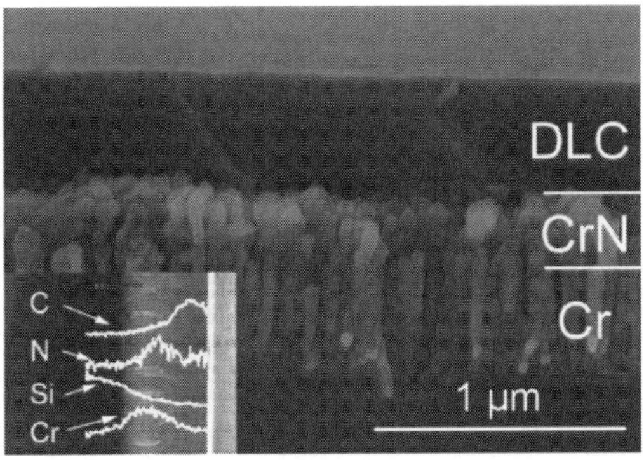

Figure 4.8 Cross section of Cr/CrN/DLC multilayered coating. Reprinted from Ref. [88], Copyright (2010), with permission from Elsevier.

coatings was found to more efficiently improve corrosion resistance and wear resistance of magnesium alloy.

Wu et al. [88] reported a DLC on AZ31 magnesium alloy deposited by ion beam deposition technique. A columnar Cr layer with a (110) preferred texture and a columnar CrN layer with a (111) preferred texture were applied as multilayer architecture. In their works, the influence of interlayers on corrosion resistance of DLC coating on magnesium alloys was discussed. In addition, the effect of bias voltage on the corrosion resistance of CrN/Cr coatings on magnesium alloys was also investigated. Figure 4.8 shows the cross-sectional image of Cr/CrN/DLC graded and multilayered coating. The critical loads of the coated samples were evaluated in the scratch test. With the addition of Cr and CrN interlayers, the critical load of the DLC/AZ31 system became larger. The scratch test is a traditional method to assess the adhesion strength of coating/substrate systems. A loaded diamond stylus is drawn across the coated surface at continuously increasing loads until the coating is stripped from the substrate at a critical load. The adhesion can be thus evaluated by this critical load. The higher the critical load is, the better the adhesion is. Therefore, the result of the scratch test indicates that the addition of Cr and CrN as multilayer architecture is helpful to the improvement of the adhesion between coating and substrate. Besides, Wu et al. also reported the graded and multilayered Al/AlN/DLC coatings can efficiently improve properties of the AZ31 magnesium alloys. In this study, the Al/AlN/DLC coating was successfully deposited on the surface of magnesium alloy by the combination of ion beam deposition and magnetron sputtering. After the formation of the Al/AlN/DLC coating on magnesium alloy, the wear resistance of the magnesium alloy was greatly improved with the increase of the surface hardness and the decrease of the friction coefficient. In addition, the corrosion resistance of the coated magnesium alloy was also significantly improved in the 3.5 wt.% NaCl solution, but the multilayer coating could not prevent the occurrence of pitting corrosion due to the existence of through-thickness holes [89].

Multilayered DLC coatings on magnesium alloys to improvement of corrosion fatigue strength were also reported by Uematsu et al. [90]. Results showed that the laboratory fatigue strength was enhanced by multilayered DLC coatings and increased with increasing film thickness due to suppression of crack initiation on the substrate surface. Whilst the laboratory fatigue strengths of the multilayer DLC-coated specimens were significantly higher than those of the single-layer DLC-coated ones with the equivalent film thicknesses, which could be attributed to reduced size of pre-existing film defects from which cracks were generated. The multilayer 12 μm DLC coating could enhance corrosion fatigue strength up to nearly the same level as the fatigue strength in laboratory, which indicated that the thickening of multilayer DLC coatings could prevent completely the degradation of corrosion fatigue strength in aqueous environments.

4.4.2 *Graded and Multilayered Coatings on Aluminum Alloys*

The characteristic properties of aluminum alloy including high strength stiffness/weight ratio, good machinability, high corrosion resistance and recyclability make it the ideal candidate to replace steel for weight reduction demand in automotive industry [91]. Similar to magnesium alloys, its low hardness, low seizure load and poor tribological performance have prevented its direct usage in moving parts such as cylinder bores, pistons, tappets and bearings [91, 92].

The possible approach to respond to these environmental and tribological requirements is the potential use of high hardness, self-lubricating and wear-resistant coatings on aluminum alloy which usually experiences various severe circumstances such as near-dry and dry friction, oil-lubricated, oil-lack and boundary lubrication conditions. The high hardness, low friction nature, excellent wear and corrosion resistance and high chemical stability of DLC coatings make them good candidates as protective layers for aluminum alloys in engine applications [93]. Among DLC coatings, carbon-based nanocomposite coatings consisting of nanocrystalline ceramic particles embedded in an amorphous hydrocarbon (a-C:H) matrix are able to combine high fracture toughness and wear resistance with a low friction coefficient. An example of these a-C(:H) based nanocomposite coatings is the TiC/a-C:H system [94]. However, the abrupt changes in hardness and thermal expansion coefficient between coating/substrate interface resulted in the poor adhesion strength, low load-carrying ability, easily-deformation and cracking of hard a-C:H coatings during the wear process [95, 96].

Thus, an innovative design involving multilayer and/or duplex diffusion/coating treatment is to improve the adhesion strength and tribological behaviour of aluminum alloys, which includes Al_2O_3/DLC coating [97], Si/SiC/DLC gradient layer [98], TiB_2/DLC multilayers [99] and AlN/Ti/TiN/DLC coatings [100]. It should be noted that when compared with single DLC coatings, the adhesion strength and tribological performance could be improved in certain degree by the multilayer or gradient design within DLC-based coatings. However, several disadvantages still existed including the high surface roughness, inadequate film thickness and relatively higher wear rate for thus multilayered coatings on aluminum alloys.

Work by the author's group presents the possibilities of improving the adhesion of DLC-based coatings on aluminum alloy by fabricating Ti/TiN/Si/(TiC/a-C:H) graded and multilayered coatings using a combined arc ion plating and magnetron sputtering processes. The microstructure, hardness and tribological performance of carbon-based coatings are investigated [101].

The designed Ti/TiN/Si/(TiC/a-C:H) graded and multilayered coatings on aluminum alloy are successfully deposited on aluminum alloys using a combined arc ion plating and magnetron sputtering

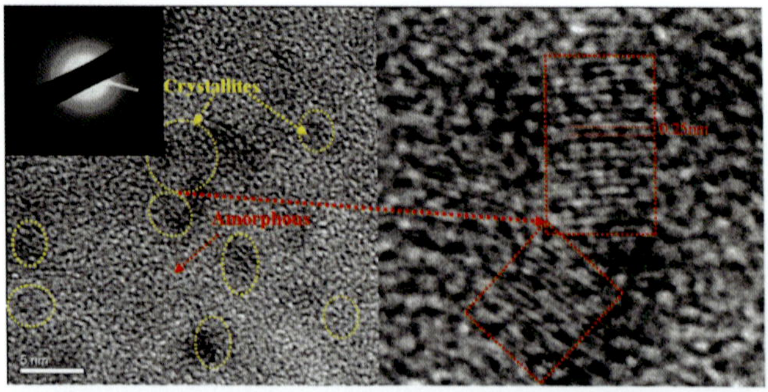

Figure 4.9 HR-TEM micrographs of top TiC/a-C:H layer. Reprinted from Ref. [101], Copyright (2010), with kind permission from Springer Science and Business Media.

process in the same chamber in the present study. The top TiC/a-C:H composite layer is further investigated by high resolution TEM as shown in Fig. 4.9. HRTEM pictures clearly show that small crystalline grains in the range of 3∼5 nm (black area distinguished by the lattice fringe contrast) were embedded in the amorphous (white area) matrix. The electron diffraction patterns of top Ti-DLC layer shows that diffraction rings from TiC (111) and (200) planes were close to each other and located near the transmission spot. The second brightest ring could be from the TiC (220) plane with the interplanar crystal spacing of 0.153 nm. The outer diffraction rings originated from the TiC (311) and (222) planes. The further amplified image of lattice fringes display some nanocrystalline TiC particles with the interplanar crystal spacing of 0.25 nm, which is well in agreement with the typical TiC (111) crystal plane. Therefore, high-resolution TEM presented clearly reveals that the nanocrystalline TiC particles are homogeneously embedded in the a-C:H matrix and the doping of Ti in amorphous carbon matrix greatly promote the formation of an amorphous/crystalline nanocomposite micro-structure (namely, the formation of TiC/a-C:H top layer). Based on the literature, this amorphous/crystalline nanocomposite structure in TiC/a-C:H top layer may be favourable to the high performance of mechanical properties owing to the following concepts [21, 61]: (i) combine TiC nanocrystalline and amorphous carbon matrix to achieve high hardness with good toughness; (ii) maintain TiC nanocrystalline size

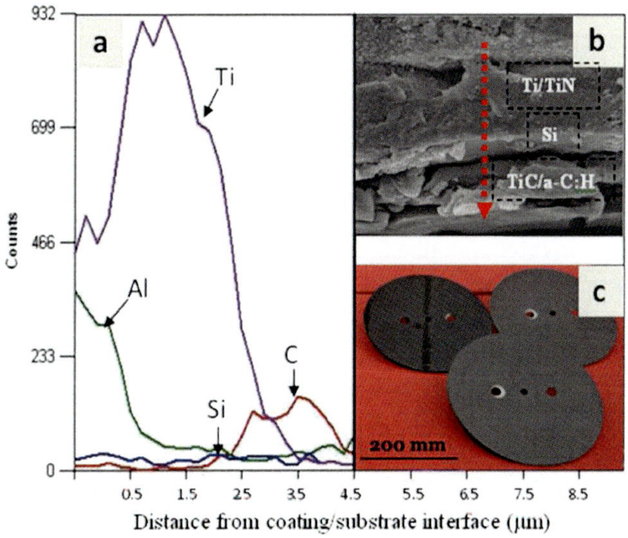

Figure 4.10 (a) Energy-dispersive spectrometry (EDS) depth profile of the Ti/TiN/Si/(TiC/a-C:H) coatings, (b) corresponding SEM micrographs of the fracture cross-sections and (c) large-sized aluminum alloy samples coated by DLC-based multilayer with smooth surface and excellent adhesion with substrate. Reprinted from Ref. [101], Copyright (2010), with kind permission from Springer Science and Business Media.

at several nanometre level to restrict crack and create a large volume of grain boundaries.

The multilayered structure in Ti/TiN/Si/(TiC/a-C:H) coatings was clearly confirmed in Fig. 4.10. EDS depth profiling (Fig. 4.10a) and the SEM micrographs of the fraction cross-sections (Fig. 4.10b) of coatings reflects very clearly the multilayer structure of the four layers (i.e., Ti, TiN, Si and TiC/a-C:H). The whole thickness of multilayered coatings on aluminum alloy substrates was approximately 4 μm. The profile indicates a Ti bonding layer of ~200 nm thick, a TiN supporting layer of ~2300 nm thick, an interface Si layer between the TiN and top DLC layer of less than 150 nm and finally the top TiC/a-C:H layer of ~1500 nm. The photo-micrographs also demonstrate that the supporting Ti/TiN/Si multilayers are very dense, providing the top TiC/a-C:H layer with excellent load support.

The nano-hardness of the corresponding multilayered coating varied with the thickness. The nano-hardness of aluminum alloy substrate is around 2 GPa. The Ti bonding layer with a thickness of ~100 nm is about 4 GPa, a supporting TiN layer is approximately 10 GPa. In addition, the formation of $TiSi_x$ compound can better modify the adhesion between amorphous silicon and TiN or Ti layer [102], consequently an interface Si layer with a hardness of 12 GPa is thus fabricated. Finally, top TiC/a-C:H layer exhibited an highest hardness of approximately 20 GPa. As a result, such better hardness gradient in coatings effectively avoids the bigger hardness transition from Al substrate (very soft) to the upper DLC layer (very hard) and thus relieves the stress during the deposition process. This gradient design in hardness also greatly strengthens the adhesion strength between aluminum alloy and DLC nanocomposite coatings and improves the load carrying capacity.

The adhesion of the graded and multilayered Ti/TiN/Si/(TiC/a-C:H) coatings on the aluminum alloy substrates was studied using a commercial scratch testing system equipped with an acoustic emission signal detecting system. The critical load, that is, the normal load at the coating failure as detected using an acoustic emission detector, was accepted as a quantitative measure to evaluate the adhesion strength. The critical load was further verified by observing the corresponding SEM micrograph of a scratch pattern and the critical load of multilayered coatings on Al alloys substrates are higher than 20 N, which indicated the strong adhesion of coatings on Al alloys. In addition, the thermal shock tests of coatings were performed using the water quench tests, the specimens were heated to 200°C for 20 min in a furnace and then quenched into a container of water with a temperature of 20°C for 2 min before returning to the furnace. This procedure is repeated until the presence of macroscopical cracks or chipping failure of coating. Via thermal shock tests, characteristic of the interfacial bonding strength of multilayered DLC-based coatings on aluminum alloy substrates is strongly influenced by the presence of residual thermal stresses, which is caused by the differential thermal contraction during sudden temperature change. Results show that after 30 thermal cycles the DLC-based multilayered coatings on Al alloy exhibited no delamation or peeling from Al substrates. The higher thermal resistance of multilayered DLC coatings indicated that the multilayered design in composition and hardness can relax thermal stresses and resist the formation and propagation of cracks, therefore interfacial bonding strength between soft Al substrate and hard DLC-based coating can be improved significantly.

The tribological behaviour of Ti/TiN/Si/(TiC/a-C:H) coating, single TiC/a-C:H (DLC) layer and aluminum alloy in dry and oil-lubricated conditions were comparatively investigated as shown in Fig. 4.11a. Under dry friction condition, aluminum alloys exhibit a very high friction coefficient of 0.38, while the multilayered coatings show the lower friction coefficient of 0.16. However, the friction failure of single DLC layer occurs in a very short duration. Under the oil-lubricated conditions, multilayered coatings exhibit a very low and stable friction coefficient of approximately 0.05. Furthermore, the specific wear rate is shown in Fig. 4.11b. When compared with aluminum alloy, the wear rate of the multilayered coatings was more than two orders of magnitude lower in dry wear condition and three orders of magnitude lower in oil-lubricated condition. The morphology of the wear tracks of aluminum alloy substrate shows the larger extent of adhesion wear and severe deformation in the sliding direction under the combined stresses of compression and shear, which results in larger wear rate of aluminum alloy. The morphology of the worn surface of multilayered coatings under dry sliding condition is similar to that of multilayered coatings tested at oil-lubricated condition. The worn surface of multilayered coatings revealed slight wear and rather smooth surface with smaller damaged regions, only some small grooves and few dimples are noted on the worn surface.

For the wear failure of aluminum alloys, adhesion wear becomes critical and predominant [103]. The interesting results obtained in this paper provide important design principles for the fabrication of self-lubricating, corrosion and wear-resistant coatings as protective barrier for aluminum alloy parts. First, control of the structure and composition of the coatings at the micro-/nanoscale is an exciting scientific subject combined with an industrial challenge. For tribological coatings, low internal stress and strong adhesion between 'hard' coatings and 'soft' aluminum substrates is critically important for a long service life [104]. In the present paper, the thick

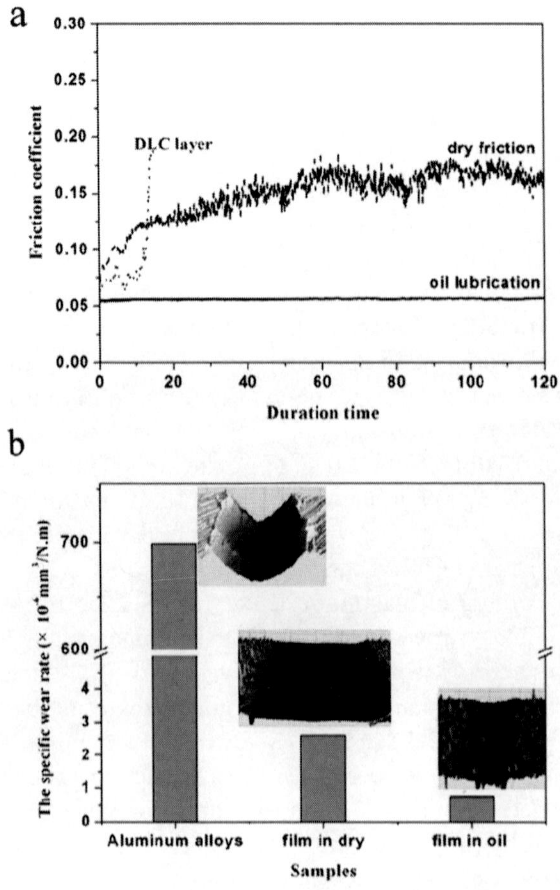

Figure 4.11 The friction coefficient of Ti/TiN/Si/(TiC/a-C:H) multilayered coatings versus the duration time in dry and oil-lubricated conditions (a); the specific wear rate of the Ti/TiN/Si/(TiC/a-C:H) multilayered coatings in dry and oil-lubricated conditions (b), inset pictures are wear track profiles analysed by non-contact 3D surface profiler after the wear test. Reprinted from Ref. [101], Copyright (2010), with kind permission from Springer Science and Business Media.

supporting layers with high strength achieved by the microcontrol of composition and microstructure on each sublayer effectively reduce the internal stress to a minimum [105]. Secondly, the assembly of a self-lubricating DLC top layer with excellent anti-adhesion wear properties is vital to the fabrication of such kinds of high-performance coatings on aluminum alloys. Finally, this illustrates that tribological performance of combined coatings can be further improved by optimising the microstruture and composition without loss of strength or their nanostructural nature.

Graded and multilayered Ti/TiN/Si/(TiC/a-C:H) on aluminum alloy was successfully fabricated on aluminum alloys by a combined arc ion plating and magnetron sputtering process. Graded nanohardness and multilayered structure within DLC-based coatings has shown to significantly improve the load-bearing capacity, anti-wear and self-lubricating ability of Al alloys. The significant improvement in the hardness and tribological performance of these coatings can be attributed to the multilayered microstructure, graded hardness distribution within coatings and the solid lubrication effect of the top DLC layer. Such high-performance carbon-based nanocomposite coatings on aluminum alloys with superior self-lubricating performance and higher wear resistance will be the 'ideal' advanced protective coatings for soft aluminum alloys in a wide range of engineering applications.

4.4.3 *Graded and Multilayered Coatings on Ti-6Al-4V Alloys*

Owing to the combination of light weight, biocompatibility, high specific strength and excellent corrosion resistance, Ti-6Al-4V alloys have been used in fields of aerospace, biomedicine, chemical plant and so on [106, 107]. However, the high friction and wear imposed restrictions on their widely utilisation. A number of surface engineering techniques [108, 109] have been proposed to improve the anti-wear ability of titanium alloys. Recently, one of the most potential methods might be the fabrication of advanced coatings on the titanium alloys to work as the protective and lubricant surfaces of the corresponding substrates. The amorphous carbon (a-C) coating has demonstrated a well performed surface for the titanium alloy in the ambient atmosphere, however, it is rare to find investigations on the improvement of the tribological performance of titanium alloys in aqueous environments.

Among various a-C films, the graphite-like carbon (GLC) film with a high fraction of sp^2 carbon bonding (the graphite-like structure) was proposed. Not only did it possess high chemical stability, low friction coefficient and excellent anti-wear properties, but also it exhibited better anti-wear properties in water than in atmosphere [110, 111], which provided the possibility of compensating for the poor tribological performance of titanium alloys in aqueous environment. However, because of the low hardness of titanium alloys, they were not competent to cover hard GLC films directly. Poor adhesion strength, low load-bearing capacity, apt to deform and delaminate during wear process would be resulted owing to the abrupt changes of thermal expansion coefficient and hardness between coating and substrate interface. An innovative approach to improve the hardness and tribological performance of GLC coatings on the soft titanium alloys was considered to design gradient multilayer coatings and/or duplex approach with implantation and coatings. Wang et al. [101] fabricated gradient Ti/TiN/Si/(TiC/a-C:H) multilayer on aluminum alloy and effectively improved the anti-wear performance of substrate. Wu et al. and Bouzakis et al. [88, 112] investigated that the adhesion strength between substrate and coating was largely improved by Cr/CrN-interlayer. Besides, Yetim et al. [113] demonstrated that duplex treatment with Ti-DLC on nitriding Ti-6Al-4V had low friction coefficient and high wear resistance. Luo et al. [114] reported that multilayer coatings TiAlN/VN which deposited on plasma nitrided steels showed higher adhesion strength and load-bearing capacity as well as better wear resistance compared to coatings deposited on non-nitrided substrates. Investigations on the multilayer or duplex coating on different substrates have largely conducted. Unfortunately, a little information can be available on the tribological behaviours of GLC film with gradient and multilayer coated titanium alloys in aqueous conditions to improve the hardness and wear resistance of substrate.

In this section, two types of GLC coating systems with four gradient interfaces are fabricated on the Ti-6Al-4V alloys, that is, the multilayer coatings of Ti/TiN/GLC and Cr/CrN/GLC and duplex treatments with multilayer coatings on nitrogen-ion implantation pre-treated substrates (N/Cr/CrN/GLC and N/Ti/TiN/GLC). Com-

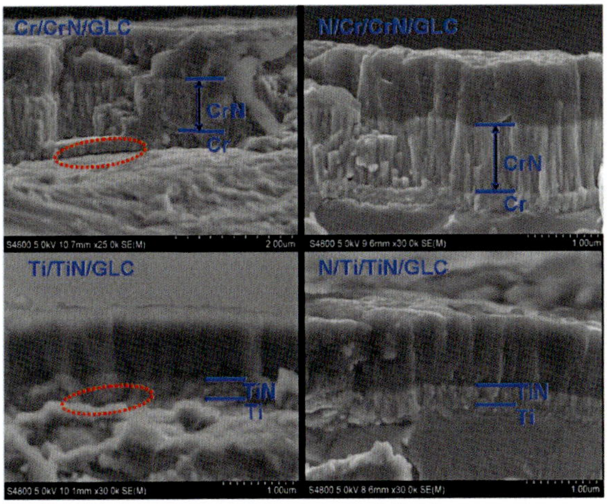

Figure 4.12 The typical cross section fracture of as-deposited coatings on Ti-6Al-4V or nitrogen ion-implanted Ti-6Al-4V substrates. Reprinted from Ref. [115], Copyright (2011), with kind permission from Springer Science and Business Media.

parative studies on the tribological properties of the different coating systems in two aqueous environments of distilled water and sea water were performed. And the influences of different gradient interfaces on the friction and wear behaviours of GLC coatings were discussed [115].

Firstly, the typical fracture sections of coatings with different gradient interlayers on Ti-6Al-4V were investigated, as shown in Fig. 4.12. In each image, three layers can be distinguished clearly and all of them exhibited a continuous structure. The top GLC films were in the same thickness of about 1 μm, but the total thickness of coatings with respect to Ti and Cr gradient layers were about 1.4 μm and 2.0 μm, respectively. It can be inferred that Cr has a higher deposition rate than Ti. Simultaneously, it was clearly observed that both of the duplex coatings were well adherent to the substrates, whereas the coatings free of ion implantation showed micro-gaps between substrates and coatings.

Then, the tribological performances of coatings were evaluated and results showed that N/Cr/CrN/GLC coatings displayed the excellent tribological properties both under low and high applied loads. For excellent tribological behaviour, there must be a strong adhesion strength between the film and the substrate together with an appropriate morphology [116]. Eminent properties, such as high hardness, high adhesion strength, and low shear strength, are also required for a low friction and high anti-wear performance. There are two important interfaces that impact tribological behaviour: the first is that between the top layer and sub-layer, and it is the interface of TiN/GLC or CrN/GLC; the second is that between the substrate and the subsequent layer, and it is the bare surface of the substrate with Ti/Cr, or of the nitrogen ion-implanted substrate with Ti/Cr.

At a low applied load of 3 N, resulting in low contact stress, the tribological behaviour may largely depend on the surface property of the top layer and the first interface. For as-deposited coatings, the same top surfaces of graphites adsorbed water or sea water, resulting in a weakening of the bonds between the basal planes of their hexagonal [117] and, consequently, resulting in low friction coefficients and wear rates of the coatings in the aqueous environment. Therefore, for the different tribological behaviours of coatings with Ti and Cr gradient layer, the first interface (adhesion strength) may play a significant role.

Adhesion strength is related to residual stress, which consists of intrinsic stress and thermal stress. Although intrinsic stress is difficult to measure because it correlates to the microstructure, such as lattice distortion, both of the surfaces in our study were similar to each other, and the intrinsic stress can be assumed to be the same for these two coatings. In contrast, the coefficient of thermal expansion (CTE) multiplied by the temperature difference can be employed to estimate thermal stress. According to previous reports [118–120], the CTEs of CrN, TiN and GLC films are $2.4 \times 10^{-6}\,K^{-1}$, $9.4 \times 10^{-6}\,K^{-1}$ and $2.3 \times 10^{-6}\,K^{-1}$, respectively. The CTE of CrN more closely resembles the GLC film than TiN; namely, the CrN/GLC interface will have a smaller thermal stress and higher adhesion strength than TiN/GLC. Consequently, in terms of GLC deposition on the high CET mismatch interlayer TiN, the high thermal stress contributed to film failure. The poor strength of the TiN/GLC interface contributed to the poor wear behaviour of Ti/TiN/GLC.

However, at an applied load of 9 N, both the first and second interface influence tribological behaviour. Nitrogen ion implantation hardened the surface of the substrate and enhanced the adhesion strength of the second interface. In combination with the better adhesion strength of the first interface of CrN/GLC, the N/Cr/CrN/GLC coating thereby demonstrated the best adhesion strength among all of the coatings. As the higher the adhesion strength, the better tribological performance, the N/Cr/CrN/GLC coating exhibite excellent tribological behaviour under distilled water and sea water conditions. Tribological behaviour was based on the adhesion strength of the coatings, with the N/Cr/CrN/GLC coating demonstrating the best tribiological behaviour, followed sequentially in order of decreasing tribological performance by the N/Ti/TiN/GLC and Cr/CrN/GLC and Ti/TiN/GLCs.

In order to further understand the interface effect responsible for the different friction and wear behaviour of the coatings in an aqueous environment, the schematic diagram of friction and wear systems were simplified in Fig. 4.13, in which the upper GLC film is shown as a black layer, the second TiN or CrN film as a yellow layer, the third Ti or Cr film as a white layer, with the bottom grey layer representing the substrate. As shown in Fig. 4.13e, the surface of the multilayer was smooth, while after the nitrogen ion implantation pre-treatment, the surface of the substrate became rough (the subsurface of substrate depicted as a yellow layer in Fig. 4.13f). Multilayer coatings would duplicate the rough surface of ion-implanted substrates during the deposition process; moreover, the microasperities on the ion-implanted surface can enhance the mechanical interlocking of interfaces and produce a compact and dense interface, and thereby contributing to the improvement of adhesion strength. In contrast to multilayer coatings, the pre-treatment of substrates was also useful for the high adhesion strength of coatings.

At a low applied load of 3 N, scarcely any deformation arose from contact stress. For the smooth surface of coatings free of nitrogen ion implantation (Fig. 4.13a), graphitisation was easily induced during the wear process. Formation of the graphitic tribofilm not only prevented direct contact between the ball and coating, it also allowed easy access of shear between the inner layers, thus decreasing the friction coefficient and protecting the coating from severe wear. For the rough surface of duplex system (Fig. 4.13c),

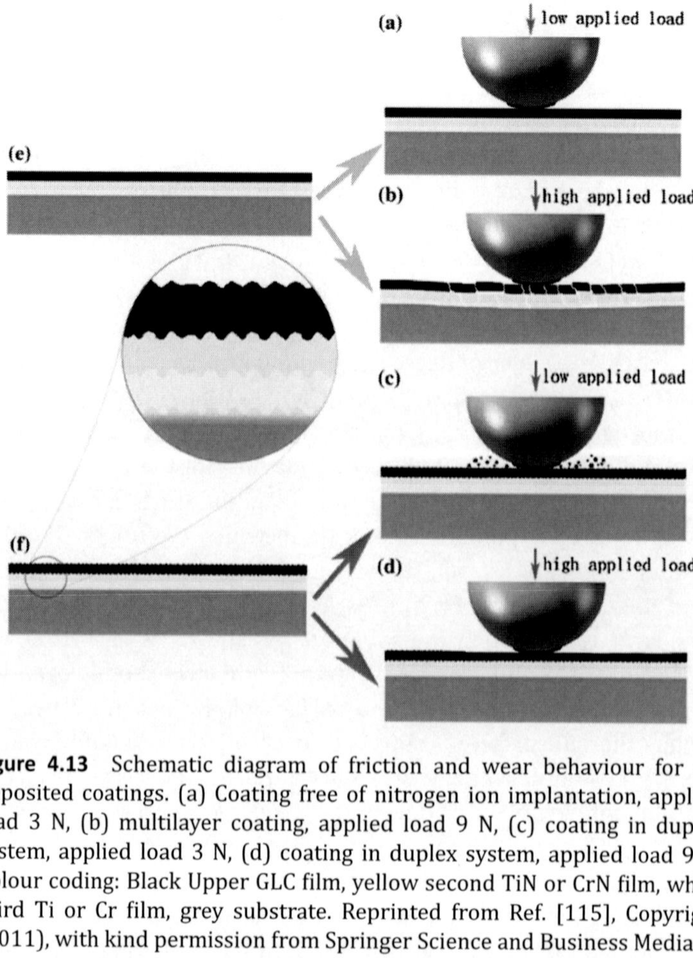

Figure 4.13 Schematic diagram of friction and wear behaviour for as-deposited coatings. (a) Coating free of nitrogen ion implantation, applied load 3 N, (b) multilayer coating, applied load 9 N, (c) coating in duplex system, applied load 3 N, (d) coating in duplex system, applied load 9 N. Colour coding: Black Upper GLC film, yellow second TiN or CrN film, white third Ti or Cr film, grey substrate. Reprinted from Ref. [115], Copyright (2011), with kind permission from Springer Science and Business Media.

the tribofilm lubricated the coating largely, but the asperities on the surface may still act as an abrasive, resulting in a slight increase in the friction coefficient. Thus, we can conclude that under a low applied load, the multilayer coating was more suitable for the better tribological performance. Moreover, taking into account the interface effect, the similarity of CTE between layers is also necessary.

When the applied load increased to 9 N, the deformation caused by contact stress enlarged. For the multilayer coatings (Fig. 4.13b), liquids were able to penetrate into the coatings through longitudinal micro-cracks which originated from film growth. When these liquids reached the surface of substrates, the adhesion strength between films and substrates would be weakened. Together with the corrosion induced by the liquids, the micro-cracks would extend and the damage to coatings accelerate, finally resulting in the delamination of coatings. In contrast to multilayer coatings, the outstanding performance of the duplex systems (Fig. 4.13d) can be attributed to the following. First, the high energy from the friction process could smooth the surfaces and promote the formation of graphitic tribofilm, which would contribute to the low friction coefficients and wear rates under distilled water and sea water conditions. Second, the graded interface layers smoothly facilitated a high level of adhesion and bonding strength between the coating and substrate. Third, hardened substrates formed from N-ion implantation enhanced the second interface and provided strong support to the multilayer coatings by suppressing plastic deformation and restricting crack initiation and propagation. Consequently, for soft substrate materials at high applied load, coatings combining the pre-treatment of substrates, such as by ion implantation, with the appropriate intermediate layers are a prerequisite for excellent tribological performance.

4.5 Summary and Conclusion

This chapter reviews recent progress on functionally graded and multilayered carbon-based coatings, mainly focusing on their microstructure and tribological performances as well as some applications. Designing functional gradient and multilayer architectures offer significant potential for producing higher levels of mechanical and tribological performance carbon-based coatings that can operate across multiple extreme environments. Their development will enable increased utilisation of carbon-based coatings in many types of applications including those in the automotive, tool and aerospace industries where the operational environment is variable and severe. For functionally graded and multilayered carbon-based coatings to become widely accepted, further development and scale-up of deposition processes are required so that engineering components may be coated reliably and inexpensively. To accelerate these coating availability, process controls, life prediction methodologies, design criteria, and nondestructive evaluation tools must also be further developed. The ultimate goal is to progressively make their way into the commercial marketplace and meet the ever-increasing performance demands of more severe applications.

Acknowledgments

The authors are grateful for financial support from the National Natural Science Foundation of China (Grant No. 51322508 and 51105352).

References

1. Y. Drozdov, N. Lukashina, T. Nazarova, *J. Machin. Manuf. Reliability*, 40 (2011), 97.
2. B. Persson, *Sliding Friction: Physical Principles and Applications*. Springer, Berlin: New York, (1998).
3. I. Iliuc, *Math. Comput. Biol. Chem.* 2008, 66.
4. B. Andersson. *Tribol. Ser.*, 8 (1991), 503.
5. G. Zambrano, P. Prieto, F. Perez, C. Rincon, H. Galindo, L. Cota-Araiza, J. Esteve, E. Martinez, *Surf. Coat. Technol.*, 108-109 (1998), 323.
6. J. Sundgren, *Thin Solid Films*, 128 (1985), 21.
7. I. Petrov, L. Hultman, J. Sundgren, J. Greene, *J. Vac. Sci. Technol., A*, 10 (1992), 265.
8. L. Hultmann, J. Sundgren, L. Markert, J. Greene, *J. Vac. Sci. Technol. A*, 7 (1989), 1187.
9. I. Petrov, P. Barna, L. Hultman, J. Greene, *J. Vac. Sci. Technol. A*, 21 (2003), 117.
10. P. Mayrhofer, M. Geier, C. Löcker, L. Chen, *Int. J. Mater. Res.*, 100 (2009), 8.
11. J. Noshiro, S. Watanabe, T. Sakurai, S. Miyake, B. Podgornik, D. Hren, J. Vizintin, *Thin Solid Films*, 476 (2005), 92.
12. H. Wang, B. Xu, J. Liu, D. Zhuang, *Surf. Coat. Technol.*, 201 (2007), 6719.

13. R. Martins, P. Moura, J. Seabra, *Tribol. Int.*, 39 (2006), 1686.
14. A. Vanhulsel, F. Velasco, R. Jacobs, L. Eersels, D. Havermans, E. Roberts, I. Sherrington, M. Anderson, L. Gaillard, *Tribol. Int.*, 40 (2007), 1186.
15. C. Donnet, J. Fontaine, T. Le Mogne, M. Belin, C. Héau, J.P. Terrat, F. Vaux, G. Pont, *Surf. Coat. Technol.*, 120–121 (1999), 548.
16. M. Audronis, P. Kelly, R. Arnell, A. Leyland, A. Matthews, *Surf. Coat. Technol.*, 200 (2005), 1616.
17. R. Gilmore, M. Baker, P. Gibbson,W. Gissler, *Surf. Coat. Technol.*, 105 (1998), 45.
18. Y. Bae, W. Lee, T. Besmann, C. Yust, P. Blau, *Mater. Sci. Eng. A*, 209 (1996), 372.
19. R. Gilmore, M. Baker, P. Gibbson, W. Gissler, M. Stoiber, P. Losbichler, C. Mitterer, *Surf. Coat. Technol.*, 108–109 (1998), 345.
20. A. Voevodin, J. Zabinski, *Wear*, 261 (2006), 1285.
21. A. Voevodin, J. Zabinski, *Compos. Sci. Technol.*, 65 (2005), 741.
22. D. Shtansky, A. Sheveyko, D. Sorokin, L. Lev, B. Mavrin, P. Kiryukhantsev-Korneev, *Surf. Coat. Technol.*, 202 (2008), 5953.
23. H. Wang, B. Xu, J. Liu, D. Zhuang, *Mater. Chem. Phys.*, 91 (2005), 494.
24. P. Si, C. Choi, J. Lee, D. Geng, Z. Zhang, *Mater. Sci. Eng. A*, 443 (2007), 167.
25. H. Tsai, D. Bogy, *J. Vac. Sci. Technol. A*, 5 (1987), 3287.
26. N. Morrison, S. Rodil, A. Ferrari, J. Robertson, W.I. Milne, *Thin Solid Films*, 333 (1999), 71.
27. N. Savvides, B. Window, *J. Vac. Sci. Technol. A*, 3 (1985), 2386.
28. S. Harris, A. Weiner, S. Tung, S. Simko, M. Militello, *Surf. Coat. Technol.*, 62 (1993), 550.
29. N. Paik, *Diamond Relat. Mater.*, 14 (2005), 1556.
30. A. Yan, X. Xiao, I. Kulaots, B. Sheldon, R. Hurt. *Carbon*, 44 (2006), 3116.
31. R. Khan, S. Silva, *Diamond Relat. Mater.*, 10 (2001), 224.
32. E. Gerstner, P. Lukins, D. Mckenzie, D. McCulloch, *Phys. Rev. B*, 54 (1996), 14504.
33. G. Kamath, A. Ehiasarian, Y. Purandare, P. Hovsepian, *Surf. Coat. Technol.*, 205 (2011), 2823.
34. M. Staia, D. Bhat, E. Puchi-Cabrera, J. Bost. *Wear*, 261 (2006), 540.
35. S. Watanabe, J. Noshiro, S. Miyake, *Surf. Coat. Technol.*, 188–189 (2004), 644.
36. S. Watanabe, J. Noshiro, S. Miyake, *Surf. Coat. Technol.*, 183 (2004), 347.

37. E. Kusano, M. Kitagawa, Y. Kuroda, H. Nanto, A. Kinbara, *Thin Solid Films*, 334 (1998), 151.
38. C. Moranta, P. Prietoa, A. Fornb, J. Picasb, E. Elizaldea, J. Sanza, *Surf. Coat. Technol.*, 180–181 (2004), 512.
39. A. Voevodin, M. Capano, S. Laube, M. Donley, J. Zabinski, *Thin Solid Films*, 298 (1997), 107.
40. A. Grill, *Surf. Coat. Technol.*, 94–95 (1997), 507.
41. Y. Pei, D. Galvan, J. De Hosson, A. Cavaleiro, *Surf. Coat. Technol.*, 198 (2005), 44.
42. C. Corbella, E. Pascual, G. Oncins, C. Canal, J. Andújar, E. Bertran, *Thin Solid Films*, 482 (2005), 293.
43. A. Voevodin, J. ONeill, J. Zabinski, *Thin Solid Films*, 342 (1999), 194.
44. J. Damasceno, S. Camargo Jr., M. Cremona, *Thin Solid Films*, 433 (2003), 199.
45. C. Moura e Silva, J. Branco, A. Cavaleiro, *Thin Solid Films*, 515 (2006), 1063.
46. Q. Xu, S. Yu, Y. Kang, *Mech. Res. Commun.*, 26 (1999), 55.
47. A. Voevodin, J. Zabinski, *Diamond Relat. Mater.*, 7 (1998), 463.
48. J. Bulır, M. Novotny, M. Jelınek, T. Kocourek, V. Studnicka, *Surf. Coat. Technol.*, 200 (2005), 708.
49. W. Wu, J. Ting, *Thin Solid Films*, 420–421 (2002), 166.
50. L. Huang, Electrochem. *Commun.*, 8 (2006), 262.
51. J. Musil, *Surf. Coat. Technol.*, 125 (2000), 322.
52. K. Bewilogua, C.V. Cooper, C. Specht, J. Schroder, R. Wittorf, M. Grischke, *Surf. Coat. Technol.*, 127 (2000), 223.
53. W. Wu, J. Ting, *Carbon*, 44 (2006), 1210.
54. G. Zhang, P. Yan, P. Wang, Y. Chen, J. Zhang, L. Wang, J. Zhang, *Surf. Coat. Technol.*, 202 (2008), 2684.
55. W. Hsieh, P.-S. Shih, C.-C. Lin, C.-Y. Wang, H.-C. Shih, *Surf. Coat. Technol.*, 200 (2006), 3175.
56. K. Choy, E. Felix, *Mater. Sci. Eng. A*, 278 (2000), 162.
57. C. Rincon, G. Zambrano, A. Carvajal, P. Prieto, H. Galindo, E. Martınez, A. Lousa, J. Esteve, *Surf. Coat. Technol.*, 148 (2001), 277.
58. K. Bewilogua, I. Bialuch, H. Ruske, K. Weigel, *Surf. Coat. Technol.*, 206 (2011), 623.
59. F. Li, S. Zhang, J. Kong, Y. Zhang, W. Zhang, *Thin Solid Films*, 519 (2011), 4910.

60. A.Voevodin, S. Walck, J. Zabinski, *Wear*, 1997, 203–204: 516.
61. A. Voevodin, P. Stevenson, J. Schneider, A. Matthews, *Vacuum*, 46 (1995), 723.
62. M. Murakawa, T. Komori, S. Takeuchi, K. Miyoshi, *Surf. Coat. Technol.*, 120–121 (1999), 646.
63. D. Galvan, Y. Pei, J. Hosson, *Surf. Coat. Technol.*, 201 (2006), 590.
64. R. Hauert, J. Pastcheider, *Adv. Eng. Mater.*, 2 (2000), 247.
65. S. Veprek, S. Reiprich, *Thin Solid Films*, 265 (1995), 64.
66. M. Abad, M.A. Muñoz-Márquez, S. El Mrabet, A. Justo, J.C. Sánchez-López, *Surf. Coat. Technol.*, 204 (2010), 3490.
67. S. Dub, Y. Pauleau, F. Thiery, *Surf. Coat. Technol.*, 180–181 (2004), 551.
68. B. Tay, Y. Cheng, X. Ding, S. Lau, X. Shi, G. You, D. Sheeja, *Diamond Relat. Mater.*, 10 (2001), 1082.
69. G. Zhang, P. Yan, P. Wang, Y. Chen, J. Zhang, *J. Phys. D: Appl. Phys.*, 40 (2007), 6748.
70. S. Zhou, L. Wang, S. Wang, Q. Xue, *Appl. Surf. Sci.*, 257 (2011), 6971.
71. S. Zhou, L. Wang, Q. Xue, *Tribol. Lett.*, 43 (2011), 329.
72. H. Miki, T. Takeno, T. Takagi, *Thin Solid Films*, 516 (2008), 5414.
73. P. Hovsepian, A. Ehiasarian, A. Deeming, C. Schimpf, *Vacuum*, 82 (2008), 1312.
74. W. Zhang, A. Tanaka, B. Xu, Y. Koga, *Diamond Relat. Mater.*, 14 (2005), 1361.
75. M. Delplancke, O. Monteiro, *Surf. Coat. Technol.*, 108–109 (1998), 484.
76. S. Zhang, X. Bui, Y. Fu, D. Butler, H. Du, *Diamond Relat. Mater.*, 13 (2004), 867.
77. S. Zhou, L. Wang, Q. Xue, *Surf. Coat. Technol.*, 206 (2011), 387.
78. A. Leyland, A. Matthews, *Wear*, 246 (2000), 1.
79. A. J. Perry, V. Valvoda, D. Rafaja, *Thin Solid Films*, 214 (1992), 167.
80. D. Wang, K. Kato, *Wear*, 252 (2002), 210.
81. Y. Liu, A. Erdemir, E.I. Meletis, *Surf. Coat. Technol.*, 82 (1996), 48.
82. J. Gray, B. Luan, *J. Alloys Compd.*, 336 (2002), 88.
83. J. Robertson, *Mat. Sci. Eng. R*, 37 (2002), 129.
84. N. Yamauchi, N. Ueda, A. Okamoto, *Surf. Coat. Technol.* 2007, 201, 4913.
85. N. Yamauchi, K. Demizu, N. Ueda, *Thin Solid Films*, 506–507 (2006), 378.

86. N. Yamauchi, K. Demizu, N. Ueda, *Surf. Coat. Technol.*, 193 (2005), 277.
87. J. Choi, S. Nakao, J. Kim, *Diam. Relat. Mater.*, 16 (2007), 1361.
88. G. Wu, L. Sun, W. Dai, L. Song, A. Wang, *Surf. Coat. Technol.*, 2010, 204, 2193.
89. G. Wu, W. Dai, H. Zheng, A. Wang, *Surf. Coat. Technol.*, 205 (2010), 2067.
90. Y. Uematsu, T. Kakiuchi, T. Teratani, Y. Harada, K. Tokaji, *Surf. Coat. Technol.*, 205 (2011), 2778.
91. W. Miller, L. Zhuang, J. Bottema, A. Wittebrood, P. De Smet, A. Haszler, A. Vieregge, *Mater. Sci. Eng. A* 2000, 280(1), 37.
92. S. Kim, T.H. Kim, K.-T. Rie, *Surf. Coat. Technol.*, 131 (2000), 121.
93. M. Kano, *Tribol. Int.*, 39 (2006), 1682.
94. Y. Pei, D. Galvan, C. Strondl, *J. Eur. Ceram. Soc.*, 26 (2006), 565.
95. J. Ouyang, S. Sasaki, *Wear* 2009, 266(1-2), 96.
96. J. Sanchez-Lopez, D. Martınez-Martınez, A. Fernandez, *Surf. Coat. Technol.*, 202 (2008), 4011.
97. X. Nie, A. Wilson, A. Leyland, A. Matthews, *Surf. Coat. Technol.*, 121 (2000), 506.
98. C. Schwarz, J. Heeg, M. Rosenberg, *Diam. Relat. Mater.*, 17 (2008), 1685.
99. J. Rao, R. Cruz, K. Lawson, *Diam. Relat. Mater.*, 14 (2007), 1805.
100. J. Liao, L.F. Xia, M.R. Sun, W.M. Liu, Q.J. Xue, *Surf. Coat. Technol.*, 183 (2004), 157.
101. L. Wang, S. Wan, S. Wang, R. Wood, Q. Xue, *Tribol. Lett.*, 38 (2010), 155.
102. K. Ezoe, T. Yamamoto, K. Ishii, S. Matsumoto, *Thin Solid Films*, 369 (2000), 244.
103. H. Torabian, J.P. Pathak, S.N. Tiwari, *Wear*, 172 (1994), 49.
104. L. Xia, Z.H.Yan, J.X. Liao, *Wear*, 257 (2004), 599.
105. L. Wang, J.Y. Zhang, L.T. Hu, Q.J. Xue, *Nanotechnology*, 17 (2006), 4614.
106. M. Long, H.J. Rack, *Biomaterials*, 19 (1998), 1621.
107. B. Yilbas, A.Z. Sahin, A.Z. Al-Garni, S.A.M. Said, Z. Ahmed, B.J. Abdulaleem, M. Sami, *Surf. Coat. Technol.*, 80 (1996), 287.
108. T. Bell, P.H. Morton, A. Bloyce, *Mater. Sci. Eng. A*, 184 (1994), 73.
109. G. Silva, M. Ueda, C. Otani, C.B. Mello, C.M. Lepoenski, *Surf. Coat. Technol.*, 2010, 204, 3018.
110. V. Kulikovsky, K. Metlo, A. Kurdyumo, P. Bohac, L. Jastrabik, *Diam. Relat. Mater.*, 11 (2002), 1467.

111. Y. Wang, L.P. Wang, S.C. Wang, G.A. Zhang, R.J.K. Wood, Q.J. Xue, *Tribol. Lett.*, 40 (2010), 301.

112. K. Bouzakis, S. Makrimallakis, G. Katirtzaglou, G. Skordaris, S. Gerardis, E. Bouzakis, T. Leyendecker, S. Bolz, W. Koelker, *Surf. Coat. Technol.*, 205 (2010), 1564.

113. A. Yetim, A. Celik, A. Alsaran, *Surf. Coat. Technol.*, 205 (2010), 320.

114. Q. Luo, P.Eh. Hovsepian, D.B. Lewis, W. Munz, Y.N. Kok, J. Cockrem, M. Bolton, A. Farinotti, *Surf. Coat. Technol.*, 193 (2005), 39.

115. X. Guan, Z. Lu, L. Wang, *Tribol. Lett.*, 44 (2011), 315.

116. O. Glozman, G. Halperin, I. Etsion, A. Berner, D. Shectman, G.H. Lee, A. Hoffman, *Diam. Relat. Mater.*, 8 (1999), 859.

117. R. Savage, *J. Appl. Phys.*, 19 (1948), 1.

118. M. Pellizzari, A. Molinari, G. Straffelini, *Surf. Coat. Technol.*, 142–144 (2001), 1109.

119. S. Tsuchiya, T. Oshika, F. Tsushima, A. Nishiyama, *Mater. Res. Soc. Jpn.*, 29 (2004), 425.

120. X. Ding, B.K. Tay, H.S. Tan, S.P. Lan, W.Y. Cheung, S.P. Wong, *Surf. Coat. Technol.*, 138 (2001), 301.

Chapter 5

High-Temperature Coatings

Ben D. Beake[a] and German S. Fox-Rabinovich[b]

[a]*Micro Materials Ltd, Willow House, Yale Business Village, Ellice Way, Wrexham LL13 7YL, UK*
[b]*Department of Mechanical Engineering, McMaster University, 1280 Main Street West, Hamilton, Ontario L8S 4L7, Canada*
ben@micromaterials.co.uk

Nomenclature and coating compositions

PVD	physical vapour deposition
CVD	chemical vapour deposition
FAD	filtered arc deposition
ISO14577	international standard for instrumented indentation
TiAlN	$x = 0.5$ in $Ti_xAl_{1-x}N$
AlCrN	$x = 0.7$ in $Al_xCr_{1-x}N$
AlTiN	$x = 0.67$ in $Al_xTi_{1-x}N$
TiAlCrN	$Ti_{0.2}Al_{0.65}Cr_{0.1}N$ and $Ti_{0.1}Al_{0.70}Cr_{0.2}N$ TiAlCrSiYN/TiAlCrN = $Ti_{0.2}Al_{0.55}Cr_{0.2}Si_{0.03}Y_{0.02}N/Ti_{0.25}Al_{0.65}Cr_{0.1}N$ and $Ti_{0.2}Al_{0.6}Cr_{0.2}Si_{0.03}Y_{0.02}N/Ti_{0.25}Al_{0.65}Cr_{0.1}N$.

Multifunctional Materials for Tribological Applications
Edited by Robert J. K. Wood

ISBN 978-981-4463-90-4 (Hardcover), 978-981-4463-91-1 (eBook)
www.panstanford.com

5.1 Introduction

Hard and wear-resistant coatings have been successfully applied to cutting tool inserts to prolong tool life in machining applications [1, 2]. They are increasingly used under more challenging conditions such as high-speed cutting to increase productivity, cutting without coolant for environmental reasons and/or machining hard-to-cut materials such as advanced Ti alloys or Ni-based superalloys due to their increasing usage in aerospace applications [3, 4]. To achieve long tool life under these conditions coatings need to be multifunctional and display several interlinked characteristics to minimise wear. A mechanical property approach to minimising wear based on the ratio of hardness of a material and its elastic modulus described by Leyland and Matthews has been effective in general tribological applications such as sliding or abrasion [5, 6].

However, as the severity of the mechanical contact increases, the coated components may operate above the elastic limit (stresses can be over 1 GPa [1, 7]), therefore plasticity and the ability to dissipate the energy of friction can become important. Different machining operations require coatings to have a different balance of mechanical properties. Figure 5.1 shows the variation in relative wear intensity in two different contact situations, heavily loaded dry sliding and turning, for three TiN PVD coatings on high-speed steel with different hardness and plasticity index [8]. The plasticity index is a non-dimensional parameter which is a measure of the relative proportions of elastic and plastic deformations occurring in an indentation contact. It is defined as

$$\text{plasticity index} = \text{plastic work/total work} \tag{5.1}$$

The different properties were obtained by altering the nitrogen pressure during deposition. At low nitrogen pressure, the TiN hardness was 35 GPa. At high nitrogen pressure, the axial texture (111) increased, hardness decreased to 25 GPa and plasticity increased. This higher plasticity index improved the cutting behaviour in turning where adhesive wear dominates, due to the greater ability to dissipate energy. The data in Fig. 5.1 clearly show that microstructural design to reduce wear in one situation is not optimum for another. In sliding the requirement for high plasticity

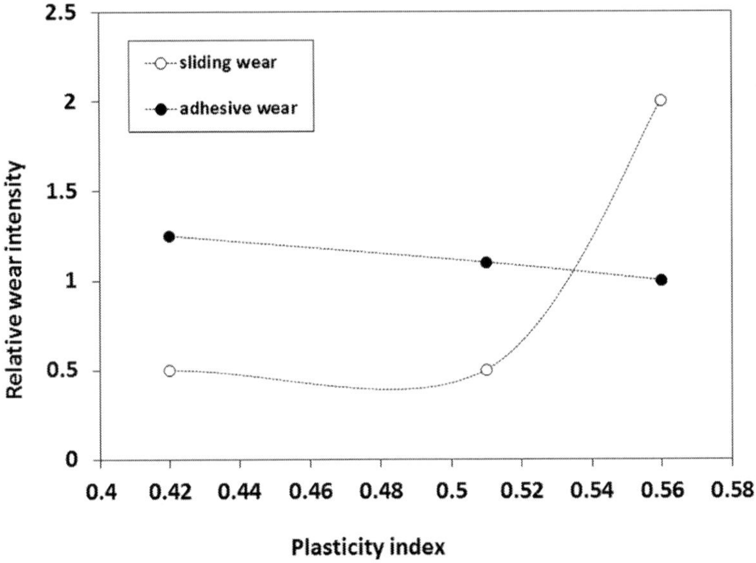

Figure 5.1 Influence of the plasticity index on relative wear intensity of TiN in sliding and turning (after [8]).

is lessened, compared to that for load support and consequently coatings with higher hardness typically improve wear resistance under these conditions. Plasticity in contact is governed by the roughness and elastoplastic properties of the contacting surfaces. In an indentation contact, plasticity is directly correlated with the ratio of hardness (H) and elastic modulus (E), H/E_r. The hardness and elastic modulus of coatings are routinely measured by nanoindentation. The load-displacement curve obtained in a nanoindentation test is essentially a mechanical fingerprint for a material. In addition to H, E and H/E_r, the indentation curve provides the elastic and plastic work done during indentation and the plasticity index, and another very useful parameter, H^3/E_r^2, which is a measure of the resistance to plastic deformation.

For applications involving high-temperature contact, the mechanical properties at the operating temperature are more relevant than those measured at room temperature [9]. High-temperature mechanical properties should be measured by testing at temperature rather than simply retesting thermally annealed samples at

25°C after cooling, which is less effective since it can only provide information on coating stability not on changes to mechanical properties due to enhancement of dislocation movement and creep processes occurring at operating temperatures. Until relatively recently nanoindentation instruments were not sufficiently advanced to perform accurate depth-sensing high-temperature nanoindentation testing. However, recent advances in commercial nanomechanics instrumentation (NanoTest system) have enabled tests to be performed at temperatures up to 750°C [10]. It has become possible to directly test coating mechanical properties (H, H/E_r, H^3/E_r^2) at temperatures close to/matching those in high-speed machining and other mechanical contact situations where high friction results in elevated temperature in the contact zone.

The configuration for high-temperature nanoindentation is shown schematically in Fig. 5.2. The NanoTest uses a horizontal loading approach that has the advantage that the displacement measuring electronics are well away from the hot zone. A thermal shield between the loading head and the hot stage avoids radiant heating of the displacement sensing system. The hot stage has a highly effective thermally insulating ceramic block with porosity to ensure that it does not increase the compliance of the instrument.

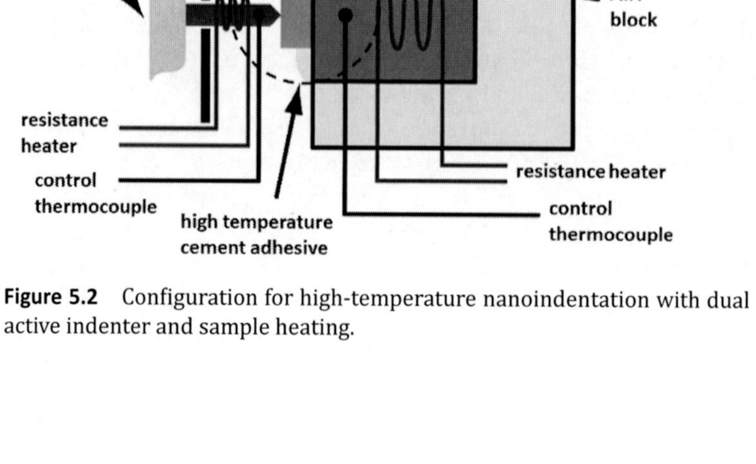

Figure 5.2 Configuration for high-temperature nanoindentation with dual active indenter and sample heating.

To minimise/remove any thermal gradient during indentation, the indenter and the sample are heated separately to achieve an isothermal contact. This approach produces a highly localised hot zone whilst the rest of the instrument remains at room temperature; for example, when the stage is operating at 500°C, the increase in temperature of the plate that supports the block is less than 1°C. Wherever possible, results of high-temperature nanoindentation measurements have been compared to more restricted test methods such as non-depth sensing hardness testing or modulus measurement by acoustic methods with good agreement.

The importance of testing at high temperature is illustrated by Fig. 5.3, showing the variation in hardness over the temperature range 25–500°C for two different commercial TiAlN coatings deposited on cemented carbide substrates. Although one coating has higher hardness when conventionally measured at room temperature, its hardness decreases more rapidly at high temperature. Differences such as these directly influence coating stability during cutting and therefore tool life. The coatings are thermally insulating so that heat flow to the substrate is impeded. More heat remains in the chip and dissipates via chip removal. Nevertheless, with

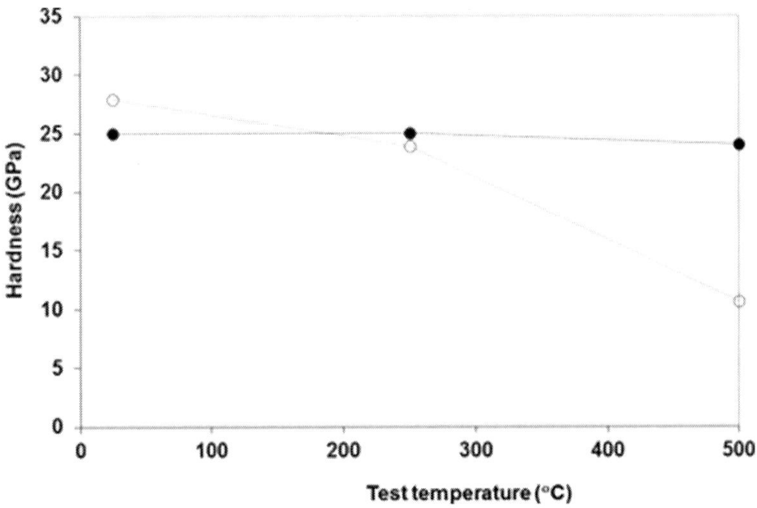

Figure 5.3 Variation in hardness with temperature for different TiAlN coatings.

high temperatures at the cutting edge, hardness and yield stress can decrease and this thermal softening can influence tool life. Simply optimising coating deposition conditions to achieve high room-temperature hardness is likely to be insufficient since in high-speed machining the contact temperature can reach 900°C or more. It might at first appear surprising that coatings can reliably be used to increase cutting tool life since their hardness at high temperature decreases dramatically. However, more aggressive cutting conditions can even increase tool life through the formation of protective and lubricious tribo-oxides [11]. The high-temperature mechanical properties of the coatings are critical in providing a stable low-wear environment for these films to form and the system to display the adaptive behaviour that minimises wear. Differences in the composition and mechanical properties of the coatings at operating temperatures both affect the formation and stability of these protective and lubricating tribofilms. By correlating high-temperature nanomechanical data (H, H/E_r, plasticity index, H^3/E_r^2) with the results of cutting tests, it is becoming possible to replace trial-and-error coating development involving expensive field trials with knowledge-based surface engineering to develop improved coatings that can display the multifunctional behaviour necessary to show low wear under the tribologically extreme conditions of high-speed cutting of hard-to-cut materials.

In this chapter, we describe the technique of high-temperature nanoindentation and review coating machining and evaluation studies where high-temperature nanoindentation testing and its predecessor hot microhardness have aided the understanding of the causes of differences in tool life between coatings. High-temperature tribological testing is also discussed. We highlight where high Al-fraction nitride coatings display adaptive and multifunctional behaviour to improve tool life. The studies described concern coatings deposited on cemented carbide substrate that has higher high-temperature hardness than high-speed steel [3, 7] and is more suited to extreme applications. The performance of state-of-the-art monolayer and multilayer coatings are contrasted and the reasons for improved behaviour shown by multilayer coatings are discussed.

5.2 Nanoindentation, *H/E*$_\mathrm{r}$ and Plasticity Index

Nanoindentation (or depth sensing indentation) is a popular method for measuring H and E of coatings and bulk samples. Full details of the technique are given in references [12, 13]. One direct output is E_r, the reduced indentation modulus (also referred to as E'), however, the plane strain modulus, $E^* = E/(1-\nu^2)$, can also be quoted (e.g., in ISO14577 [14]) and it is therefore essential when presenting nanoindentation data to specify clearly which modulus is being reported (E_r, E^* or E). Conversion of the directly measured reduced modulus to the Elastic (Young's) modulus of the sample requires that Poisson ratio be known or can be reliably estimated. For tool coatings $\nu = 0.2$–0.25. As an example, for a sample with $\nu = 0.24$, an E_r of 310 GPa is equivalent to $E = 400$ GPa when a diamond indenter is used. For a given indenter geometry, the load–displacement curve recorded in a nanoindentation test is a fingerprint for a material. Plasticity in indentation is directly correlated to H/E_r as illustrated for a AlTiN coating in Fig. 5.4.

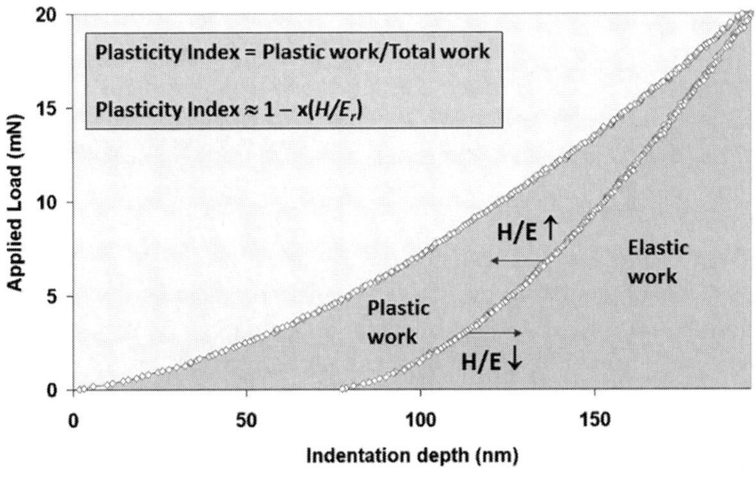

Figure 5.4 Plastic and elastic work done during a 20 mN indentation on an AlTiN coating.

Provided H/E_r is not greater than about 0.1, a good approximation to experimental data is given by the relationship

$$\text{plasticity index} = 1 - x(H/E_r) \qquad (5.2)$$

Finite element analysis gives $x \sim 5$ [15] but values around 6–7 were found for coatings and metallic bulk materials, with only fused silica and K7 glass showing lower x values [16]. Other researchers also have re-investigated this relationship and reported that $x \sim 5$ for materials that sink-in (glasses) and $x \sim 7$ for materials that exhibit pile-up (metals) [17]. The differences in the proportionality constant x between metals and glasses such as fused silica could also be influenced by differences in radial displacement on unloading that causes the size of the residual indent to be underestimated by calculation of the projected contact area from the contact depth (an underlying assumption in standard contact mechanics is that the projected contact area under load and on full unloading are equal) which is more noticeable for fused silica and other factors such as Poisson ratio and variation in the exponent of the force removal curve [12, 18]. Nevertheless, the plasticity index has the advantage that it does not depend on pile-up or sink-in effects so in principle could be used to provide a measure of hardness if the modulus is known [18].

H/E_r is correlated with how elastic the contact is in other related mechanical contact situations such as scratching/sliding [5, 6] or impact/erosion [18]. Contact is more likely to be elastic on higher H/E_r materials. Higher values result in higher critical loads for the onset of yield (non-elastic deformation) in indentation or scratch testing. Figure 5.5 compares the critical load for yield of 0.4–1.1 μm TiSiN coatings on Si with the H/E_r of the coatings determined from nanoindentation at 1 mN [20]. The trend is clear with the non-linearity possibly also reflecting a small influence of increasing thickness providing additional load support, as the coatings with higher H/E_r tended to be slightly thicker in this study [20].

In machining of hardened steels a correlation has been established between the plasticity index from room temperature nanoindentation measurements and the tool life [16]. Coatings with higher Al-fraction and higher plasticity such as AlCrN, AlTiN and TiAlCrN all show longer tool life in end milling than TiAlN which

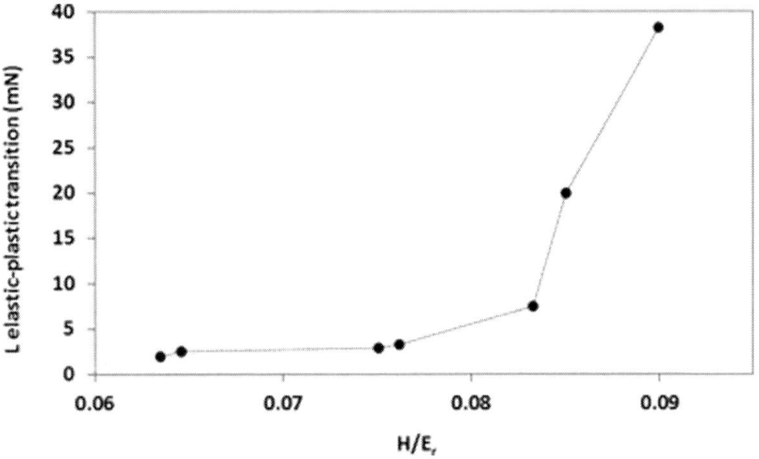

Figure 5.5 Correlation between the critical load for non-elastic deformation in nano-scratch testing of TiSiN coatings on Si and their H/E_r determined from nanoindentation.

has higher hardness but lower plasticity. The relative ranking may alter with the cutting operation and in continuous turning, TiAlN showed longer tool life than TiAlCrN. Similarly, annealing AlTiN at 700°C or more decreases its plasticity resulting in longer tool life in continuous turning than as-deposited AlTiN [21]. The different cutting conditions have differing relative requirements in terms of coating hardness (load support) and plasticity (energy dissipation). Plasticity can be considered as an indirect measure of the toughness of the coating, since coatings with higher plasticity can more easily relieve accumulated strain by plastic deformation rather than brittle fracture [22, 23]. Despite the encouragement from these correlations, depending on the materials being machined, the room temperature plasticity index is not on its own the key factor as coatings need to be multifunctional. Elevated temperature properties are more relevant to cutting tool life, since as shown in Fig. 5.2, relative rankings can switch with increasing temperatures. Similarly increasing the Al-fraction has additional benefits beyond plasticity. High Al-fraction coatings such as AlCrN, TiAlCrN and AlTiN are multifunctional and can display adaptive behaviour-forming protective tribofilms to decrease tool wear.

5.3 Elevated Temperature Mechanical and Tribological Testing

5.3.1 *High-Temperature Tribology*

Tribological tests are commonly performed at room temperature. However, at high sliding speed, these wear tests are effectively high-temperature wear tests due to the undefined increase in the temperature in the contact zone from frictional heating. An alternative approach is to perform elevated temperature tribological tests (e.g., with a ball-on-disc high-temperature tribometer) where the temperature is controlled and friction and wear rate are studied. The tribological behaviour of coatings of transition metal nitrides and transition metal dichalogenides (TMDs) has been extensively studied [24–27]. The coatings can display adaptive behaviour through the formation of tribofilms. Increasing temperature and/or applied load can reduce the wear rate by promoting the formation and growth of these complex tribofilms. At yet higher temperatures, thermal and oxidative stability can be limited and wear tends to increase [26]. Coatings of TMDs such as MoS_2 can act as solid lubricants displaying moderate hardness and low friction and form transfer films that result in low-wear rates. Kubart and co-workers showed that the friction coefficient of MoS_2 sliding against steel decreases from around 0.12 at 25°C to 0.04 at 100°C, although it increased at higher temperatures due to oxidation and the operating temperatures were limited to 350°C [24]. The wear resistance of Cr/CrN multilayers was found to be better than CrN monolayers at high temperature (600–800°C), although the room temperature mechanical properties of the coatings were virtually identical [27]. In both cases, a low-wear rate was due to the formation of a well-adhered tribolayer composed of oxidised coating material (Cr_2O_3). Increased oxidative stability (or tribo-oxidative stability) does not necessarily confer lower wear rate as it limits the formation of the tribofilms required for low wear in sliding contact. Polcar and Cavaleiro have studied the high-temperature tribological behaviour of CrAlSiN coatings with varying Cr/Al ratios (CrAlSiN and AlCrSiN) and a reference CrAlN [26, 27]. X-ray diffraction (XRD) revealed good thermal stability and oxidation resistance to 800°C for all the coatings and sliding against alumina resulted in low coating wear to 300°C. At 400°C, the CrAlSiN showed high wear but the higher Al-fraction AlCrSiN showed the presence of a well-adhered tribolayer but with only limited oxygen incorporation. Up to 500°C, the tribolayer on the CrAlN showed considerable oxygen incorporation due to tribo-oxidation. At 600°C, coatings showed high wear and adhesive failures. Poor adhesion at high temperature can also be detected by high-temperature scratch testing, which although relatively uncommon can be particularly revealing. For example, Allsopp and Hutchings showed [28] that critical loads of AlTiN and TiCN coatings at 350°C were considerably higher than at room temperature though TiAlN did not show the same effect. Since the ranking between coatings changed between room and high-temperature tests, they cautioned that the room temperature data may be misleading [28]. The same trends in the performance of AlTiN and TiAlN coatings have been observed in elevated temperature micro-scratch testing (NanoTest system). Table 5.1 shows the variation in critical load (L_{c2}) in micro-scratch testing for different commercial coatings on WC-Co. Increasing the Al:Ti ratio has a large effect on the high-temperature behaviour. The higher critical load on AlCrN is related to its improved high-temperature hardness and thermal stability [29].

5.3.2 *Measuring High-Temperature Mechanical Properties by Nanoindentation*

Accurate high-temperature nanoindentation requires that (i) tests can be performed without thermal drift or other instrumental

Table 5.1 Variation in micro-scratch test critical load with test temperature for commercial nitride coatings on WC-Co

	Critical load at 25°C/N	Critical load at 500°C/N
TiAlN	4.2 ± 0.5	3.7 ± 0.2
AlTiN *	4.5 ± 0.1	5.8 ± 0.7
AlCrN	5.7 ± 0.2	6.0 ± 0.4

Note: Micro-scratch probe = 25 μm end radius spheroconical diamond.
*Previously unpublished data.

factors influencing the raw data, (ii) the elastic contact mechanics at room temperature can successfully be applied to high-temperature measurements where appreciable time-dependent non-elastic deformation more commonly occurs, (iii) oxidative wear of tip and sample does not complicate the testing, and (iv) reliable results to be obtained on reference materials with known elastic properties at the temperatures of interest. In room temperature nanoindentation testing, fused silica is a common reference material as it is relatively inexpensive, highly polished, mechanically homogeneous and has relatively little time-dependent behaviour due to its high melting point ($\sim 1650°C$). Typically its mechanical properties are given as $H \sim 9.2$ GPa, $E = (72 \pm 1)$ GPa and Poisson ratio $= 0.17$. In practice, actual values may vary slightly from different suppliers. For example, a recent certified reference sample purchased from NPL had $E =$ 73 GPa and Poisson ratio $= 0.16$.

Fused silica is a common reference material in room temperature nanoindentation testing. It is relatively inexpensive, highly polished, mechanically homogeneous and has relatively little time-dependent behaviour due to its high melting point ($\sim 1650°C$). Typically its mechanical properties are quoted as $H \sim 9.2$ GPa, $E = (72 \pm 1)$ GPa and Poisson ratio $= 0.17$. In practice actual values may vary slightly from different suppliers. For example, a recent certified reference sample purchased from NPL had $E = 73$ GPa and Poisson ratio $= 0.16$.

Despite being highly popular, its mechanical behaviour is quite complex. For example, (i) its near surface properties may be reduced by hydration, (ii) it is an anomalous glass whose elastic modulus increases with temperature, (iii) contact area determination can be influenced by radial dilation [12, 18], (iv) densification occurs under the indenter, (v) cracking occurs at high load, dependent on the acuity of the indenter. Nevertheless, it has potential advantages for a candidate reference material for high-temperature nanoindentation testing. These include (i) it shows relatively little creep at elevated temperature, since $T/T_m = 0.5$ at 750°C; (ii) it is mechanically homogeneous and its elastic properties have been well characterised by other techniques such as resonance ultrasound; (iii) it is not particularly thermally conductive.

This last property may be something of a double-edged sword. With its low thermal conductivity fused silica is less susceptible to thermal drift and so can be fairly reliably tested with a strong thermal gradient between indenter tip and sample. However, measurements on fused silica do not necessarily prove thermal drift was minimised so that measurements could be made on any conductive materials. Data on fused silica have been reported relying on thermal equilibrium achieved by contact between a heated sample and a non-actively heated indenter for a long equilibration time (e.g., 75 min) prior to starting the indentation loading ramp [30–31]. The limits to the applicability of this approach have been investigated in experimental and finite element studies [32]. They conclude that it is more applicable to relatively shallow indentation depths and moderate test temperatures. Even if a local thermal equilibrium is achieved prior to indentation, as the indentation proceeds, it will involve deformation of the colder non-equilibrated material below the surface resulting in drift, which progressively influences the indentation curve as the indentation depth increases.

In many applications such as evaluating coatings for high-speed machining, it is necessary to be able to perform high-temperature indentations to greater depths, to be free of the influence of surface roughness and to ensure full plasticity so that meaningful hardness measurements can be made. Currently, the most popular approach is to minimise/remove the thermal gradient by heating the indenter and the sample separately to achieve an isothermal contact. This dual heating approach has been popular in high-temperature microhardness testing. The first report on a commercial nanoindentation system (NanoTest) was by Smith and Zheng in 2000 [33]. Subsequently, this approach has been adopted in many high-temperature nanoindentation publications, including the vast majority of those at over 300°C. The configuration for this has been shown in Fig. 5.2.

The methodology to obtain a stable environment using the dual heating approach [34] involves three steps for the indenter heating control side:

(i) the indenter is heated to the target temperature by feedback control

(ii) once the target temperature is reached, the controller measures the average power supplied to maintain the target temperature
(iii) the indenter is then supplied with this constant power during indentation

This provides more stable temperature control than using proportional-integral-derivative control for both sample and indenter. Nanoindentation testing with this dual heating approach has recently been compared to tests performed with the same instrument using only sample heating [32]. Data on fused silica were not particularly sensitive to whether or not the indenter was heated. FEA showed that for a high-conductivity sample such as gold, without the additional indenter heating only a small region of the sample reaches thermal equilibrium and a steep thermal gradient is created resulting in heat flow during indentation. Consequently, indentation curves on gold at 300°C acquired with or without additional indenter heating were quite different. With dual heating, the curves appeared similar to those at room temperature but with more creep. In contrast, when the experiments were repeated without indenter heating, significant 'negative creep' occurred due to drift and it proved impossible to test at higher temperatures. Tang and Zhang have performed an extensive evaluation of the high-temperature capabilities of a different nanoindenter that does not use additional indenter heating using Si wafers as test samples [35]. Although the drift rate prior to indentation could be minimised by an extended equilibration time (>30 min in pre-contact at 2 μN force) once the indentation starts the drift rate was found to increase dramatically (e.g., to $> 2\ \mathrm{nm\ s^{-1}}$ at 156°C) due to contact with colder material resulting in highly anomalous curves that were impossible to analyse meaningfully.

The international standard for nanoindentation, ISO14577, comments that care should be taken that any time-dependent deformation during unloading does not influence the accuracy of modulus measurements determined from the unloading curve. In general, significant creep can occur when the nanoindentation test is performed at high homologous temperature ($T/T_{\mathrm{m}} > 0.6$). For the hard ceramic coatings used on cutting tools creep is typically

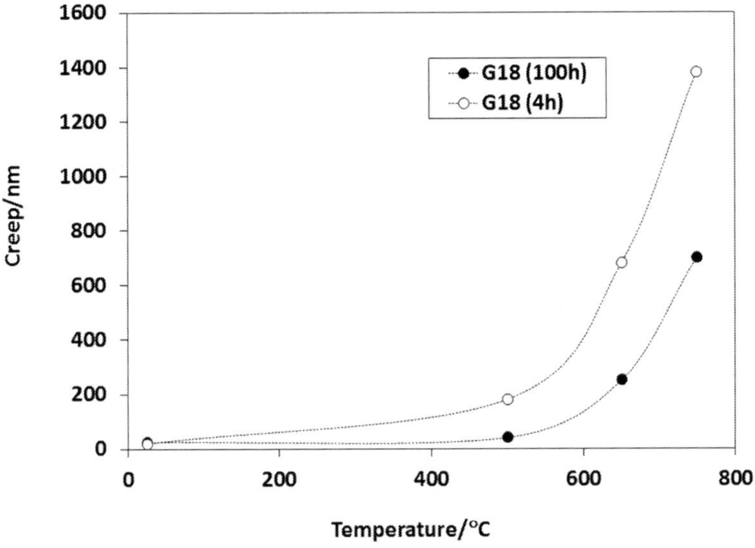

Figure 5.6 Influence of prior aging for 4 hr or 100 hr at 800°C on subsequent creep of a glass-ceramic in high-temperature nanoindentation.

small, even at 600°C. Nevertheless, it can be beneficial to use slower loading rates and/or longer hold time at peak load to ensure accuracy [21, 36]. On other samples, such as solders [37], polymers [38] or CMSX-4 superalloys [39], considerable creep at high temperature has been reported. Under these conditions, creep compliance or analytical data corrections to improve the accuracy of the elastic modulus measurement may be considered [39].

Oxidation of diamond indenters will occur in air above ~500°C, so for higher temperature work other indenter materials such as cBN or sapphire are commonly used. Figure 5.6 shows increasing creep with temperature during indentation of aged glass-ceramic seal materials tested with a cBN indenter [10]. Prior ageing for 100 hr at 800°C decreased the susceptibility to creep of the glass-ceramic compared with aging for 4 hr. A sapphire indenter was used to measure the elevated temperature behaviour of the top coat of E-beam PVD yttira-stabilised zirconia thermal barrier coatings (TBCs). Mean hardness decreased from (5.8 ± 1.0) GPa at 25°C to

Table 5.2 Novel elevated temperature nano- and micro-mechanical test techniques

Technique	Application	Temperature (°C)
Micro-friction	Ag–Bi alloys [42]	270
Nano-friction	Nitride coatings [82]	750
Micro-scratch and wear	AlCrN, TiAlN, AlTiN [21, 29]	500, 600
Nano-impact	TiAlN and AlTiN [43]	500
Micro-pillar compression	Si, Ni-based superalloys [41, 44, 45]	400, 665
Cantilever bending	W and Si beams	600, 700
SPM imaging	W and Si beams and surfaces	700

(2.9 ± 0.5) GPa at 750°C [40]. The magnitude of the decrease in hardness at temperature was consistent with hot microhardness measurements on similar TBCs. Another approach is to test in vacuum [41].

5.3.3 *Novel Nanomechanical Techniques Performed at Elevated Temperature*

There are many situations that require high-temperature measurements since dislocation mobility (and therefore hardness) is temperature sensitive and indentation creep studies can provide quantitative measurements of stress exponents and activation energies for creep processes. In addition to nanoindentation, several other nano- and micro-mechanical test techniques can be performed at elevated temperature. Table 5.2 illustrates the range of techniques and maximum test temperatures to date.

Smith and co-workers used a WC-Co probe to study nanoscale friction on a range of TiN-based coatings at 25, 400 and 750°C [82]. The tests were performed at very low contact pressure so that the ploughing contribution to the total measured friction, usually the dominant term at higher contact pressure, was minimised and the interfacial friction could be measured directly. A maximum in friction was observed at 400°C (Fig. 5.7). The decrease in friction at 750°C was associated with formation of lubricating surface oxides and oxidation-associated surface roughening.

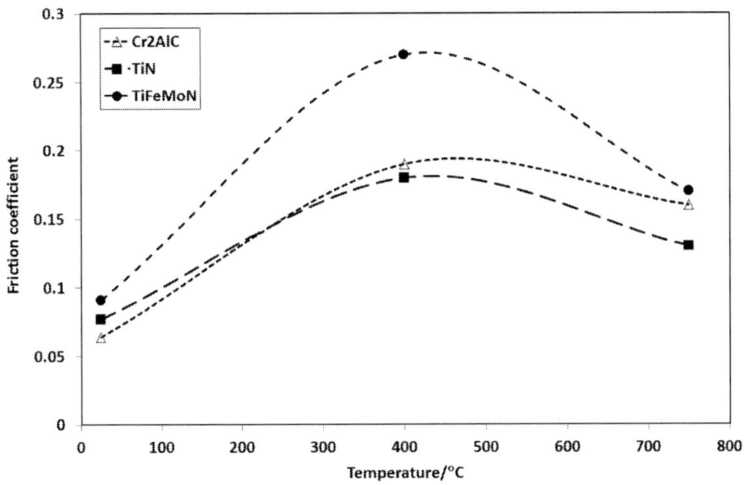

Figure 5.7 Variation in friction with temperature.

5.4 Review of Coating Machining Studies

Flank wear of coated tools controls tool life, and cutting performance deteriorates rapidly when flank wear reaches 0.3 mm so the time or number of cuts to this point is commonly taken as a measure of the tool life. Using FEM Erkens and co-workers have shown that the maximum von Mises stress appears close to the rank where the coating fractures first occur [46]. It has been realised that, although for convenience, coating hardness is commonly measured at room temperature, the high-temperature hardness can be a more relevant predictor of actual performance when the application involves significant heat generation in mechanical contact [8]. High-temperature hardness has historically been evaluated by hot micro-hardness testing. To be tested by hot micro-hardness, it was necessary to produce much thicker coatings ($\sim$10 μm) than is commonly used in PVD applications. Since the coating microstructure can vary strongly with increasing coating thickness, typically coarsening [47], it is not possible to directly relate these results to the cutting tests. However, with the advent of high-temperature nanoindentation, it has become possible to test directly on the actual coated inserts used in the cutting tests.

Different cutting applications require different combinations of high hardness combined with a suitable reserve of plasticity depending on the severity of the mechanical contact. That is, a combination of high coating hardness with some toughness is required and the relative importance of these factors differs with the cutting conditions. For demanding operations such as face-and-end milling, intensive adhesive wear and oxidation can occur during the cycling impact and a reserve of plasticity to dissipate energy generated during friction is necessary. The fatigue properties of the surface engineered layers are also of extreme importance for the tool life during cutting [1]. A typical failure mechanism of cutting tools during face milling is thermal fatigue, whereas during end milling it is low cycling fatigue. This failure mechanism is important for end milling of Ti alloys when intensive adhesive interaction between work piece material and cutting tool surfaces take place. The adhered material tears off during further cutting resulting in intensive adhesive-fatigue wear of tool accompanied by intensive chipping of its cutting edges.

Microstructural changes to alter plasticity can potentially be accompanied by other factors during tool operation, such as adhesion, friction, oxidative stability, efficiency of chip formation, elevated temperature mechanical properties and tribofilm formation. It has been highly challenging to determine the relative importance of these factors to tool life. To improve our understanding of the factors in-depth studies have fully characterised the properties of a relatively small number of advanced coatings and investigated the correlation between these properties and tool life under the same cutting conditions. Several of these studies are now reviewed with the aim of combining knowledge from them to move towards possible coating design rules for different cutting applications.

Unless otherwise stated, the composition of the PVD coatings discussed is as given in the nomenclature section of this chapter. With enhanced oxidation resistance compared to TiN, TiAlN has proved a highly popular PVD coating for a range of applications so in several of the reported studies the performance of newly developed adaptive coatings have been compared directly to it. The room temperature hardness of the coatings was > 20 GPa and they were deposited on WC-Co to a thickness of 3.0–3.5 μm. The materials

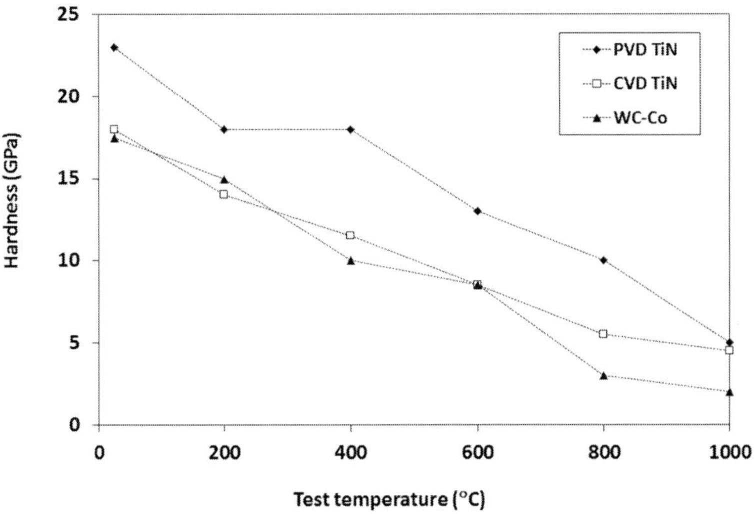

Figure 5.8 Variation in hardness vs. temperature for PVD and CVD TiN coatings and uncoated cemented carbide substrate.

being machined included H13 hardened steel, 4340 hardened steel, 1040 structural steel, and the Ni- and Ti-based aerospace alloys Inconel DA 718, Waspaloy, powder metallurgical ME16 superalloy and Ti6Al4V.

5.4.1 *CVD versus PVD TiN Coatings, Comparison to Alumina-Based Ceramics*

Quinto and co-workers investigated the temperature dependence of the microhardness of a range of 8–15 μm thick PVD and CVD stoichiometric nitride coatings on cemented carbide, including TiN and HfN [9]. Although the hardness of the PVD coatings was much higher than their CVD counterparts at room temperature, it decreased more rapidly so that values of all the coatings were similarly low by 1000°C. Figure 5.8 illustrates their data on titanium nitride after conversion to GPa, for a Magnetron Sputtered PVD TiN coating, a CVD TiN and uncoated cemented carbide substrate. Even though the hardness of both coatings decreases with increasing temperature, they nevertheless retain higher hardness than the cemented carbide and can provide improved load support protecting the underlying WC-Co in high-temperature contact.

Differences in high-temperature mechanical properties between PVD and CVD coatings were correlated with microstructure. The CVD coatings possessed larger, defect free, thermally equilibrated grains whilst the PVD coatings were finer grain size with non-equilibrated microstructure that resulted in a more rapid thermal recovery due to its stored energy (i.e., relaxation of high compressive residual stresses), together with grain boundary sliding [48].

The decrease in high-temperature hardness of the PVD and CVD coatings can be compared to the performance of ceramics, which also show pronounced reduction in hot hardness. Alpert and co-workers used hot microhardness with a 10 N load to investigate the decrease in hardness of a range of alumina-based ceramics over the temperature range 20–1000°C [49]. Alpert and co-workers reported that hardness decreased steadily for all the materials studied, consistent with thermally activated slip, with the hardness of single crystal sapphire decreasing from 19 GPa at room temperature to under 5 GPa at 1000°C. At low temperatures, the hardness of the sapphire and pure polycrystalline aluminas was greater than debased (with some partially crystallised alumino-silicate glass as second phase) polycrystalline aluminas, but at higher temperatures the hardness values converged, indicating the indentation behaviour was controlled by slip in the hard Al_2O_3 component.

5.4.2 *TiAlN versus TiCN*

Jinhal and co-workers later investigated the decrease in micro-hardness with temperature for 8 μm PVD TiAlN, TiCN and TiN coatings [50]. Throughout the temperature range the TiN coating showed the lowest hardness and poorest machining performance in continuous turning of medium carbon steel, ductile cast iron and Inconel 718. TiCN retained higher hardness than TiAlN from room temperature until 600°C. Between 600°C and 800°C, its hardness decreased more rapidly than TiAlN so that at 800°C and above the highest hardness was observed on TiAlN. This, along with its lower thermal conductivity, is a key factor in the superior tool life of TiAlN over TiCN in continuous turning of inconel and carbon steel. The relative improvement of TiAlN over TiCN was more noticeable at higher cutting speeds where the contact temperature is greater [50]. Direct measurements of mechanical properties such as H/E_r at the contact temperature can provide an improved route to coating selection to minimise wear. More recently, 3.5 μm TiAlN and TiCN coatings on WC-Co have been tested by nanoindentation. Their mechanical properties varied strongly with temperature as illustrated for H^3/E_r^2 in Fig. 5.9. The data in Fig. 5.9 were acquired in an environmental chamber using a cubic Boron Nitride indenter. Despite having lower hardness, H/E_r and H^3/E_r^2 at room temperature than TiCN, the TiAlN coating showed better stability at higher temperatures. The coatings show the same ranking switch with temperature although it occurs lower than was reported in the hot hardness testing of the thicker coatings. Bouzakis and co-workers have shown that the microstructure of PVD coatings coarsens with increased coating thickness [47]. The faster drop off in hardness with temperature might be expected for these thinner PVD coatings with their finer grain size, than for the thicker coatings previously studied by hot microhardness testing. Similarly, Polcar and co-workers have reported that the oxidation resistance of 1.7 μm TiCN decreased above 300°C and in wear tests the coating was more susceptible to fracture and delamination above this temperature [51].

The deposition technique and post-treatment can influence the coating microstructure and compressive stresses. Post-deposition micro-blasting of TiAlN induces compressive stresses which can be beneficial in reducing wear although the results are highly dependent on pressure and the grain size and geometry of the abrasive materials since the induced compressive stresses can result in increased brittleness [59]. PVD TiAlN coatings deposited by filtered arc deposition (FAD) have a nanocrystalline microstructure with lower grain size than conventional PVD TiN [52]. In turning 1040 structural steel, the conventional TiAlN coating displays much longer tool life than nanocrystalline TiAlN at low cutting speed, but as the cutting speed increases, its performance deteriorates while the performance of the nanocrystalline coating is maintained to 450 m/min [52]. The FAD TiAlN coatings display adaptive behaviour in high-speed machining forming tribofilms predominantly composed

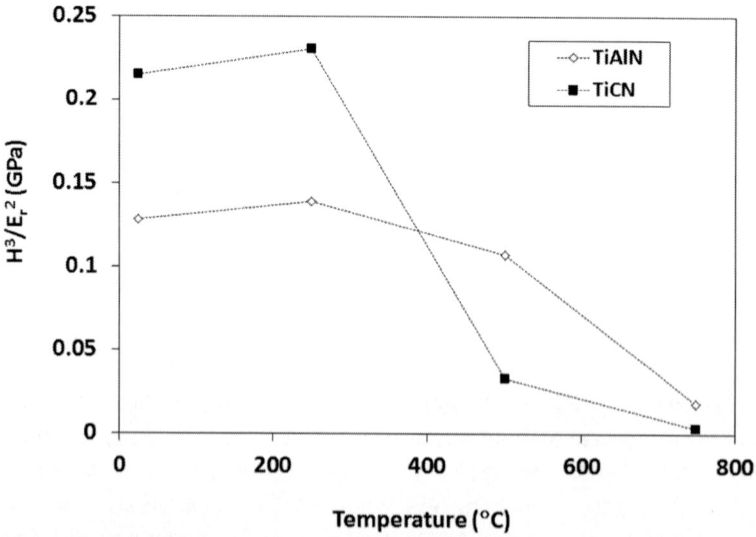

Figure 5.9 Variation in H^3/E_r^2 versus temperature for TiAlN and TiCN coatings.

of protective alumina whereas for the conventional TiAlN tribofilms were predominantly non-protective rutile.

5.4.3 *High Al-Fraction Coatings Displaying Adaptive Behaviour*

5.4.3.1 TiAlCrN, AlCrN and AlTiN

Tribological adaptability in high-speed machining has also been observed for TiAlCrN which has superior high-temperature hardness [54] and oxidation resistance [53]. The latter property is a result of the formation of tribofilms containing protective alumina-based and lubricious chromia (CrO_x) phases. The TiAlCrN coating has been described as synergistically alloyed as the alumina and CrO_x oxide layers work in synergy and enhance the surface protection against intensive adhesive interaction under the severe friction conditions of high temperature and stress in dry end milling of hardened steels. The coating displays adaptive behavior since when stable tribofilms are formed the wear process achieves greater stability

Table 5.3 Nanomechanical test data and machining performance of TiAlN and AlCrN coatings

	H at 500°C (GPa)	H/E_r at 500°C	Impact final depth (μm)[a]	Relative cutting life[b]
AlCrN	20.5	0.040	1.0	5
TiAlN	15.4	0.042	1.8	1

[a]Depth after 45 impacts in a 180 s test using a cube corner diamond impact probe at 15 mN.
[b]Length of cut in end milling of annealed 1040 structural steel, normalised to the cut length for TiAlN.

without intensive adhesive surface damage over a longer period of time (i.e., there is a shift to a milder wear mode). TiAlN has higher hardness (30 GPa) than AlCrN (25 GPa) at room temperature though in cutting tests, it can show lower tool life [29, 36]. A range of nano and micro-mechanical tests, including nanoindentation and micro-scratch testing at room and elevated temperature and nano-impact, have been performed to investigate the influence of mechanical factors on this behaviour. Table 5.3 summarises the results of some of these tests.

TiAlN softens more rapidly than AlCrN at elevated temperature and its lack of stability results in a higher plasticity index at temperature. At 500°C the hardness of AlCrN is ~25% lower than at room temperature, but on re-cooling back to 25°C the hardness was virtually the same as before heating, suggesting thermal activation of dislocation motion as the cause of the lower hardness at temperature.

Nano-impact testing (a fast repetitive nanoindentation technique) was developed to test coating properties at higher strain rates (~10^3 s^{-1}) than conventional nanoindentation, and to investigate surface fatigue and fracture due to repetitive contact with the aim of using results to optimise the coating properties for improved durability in interrupted contact situations [55–57].

This high strain rate contact allows much closer simulation of the performance of coatings systems under highly loaded intermittent contact (such as interrupted cutting and erosive wear) and the evolution of wear under these conditions. The repetitive contacts are true impact events where the probe leaves the surface between each subsequent impact which occurs at the same location on the sample surface. The progression of surface damage is followed using the displacement-sensing capability of the instrumentation. Practice has shown a high degree of correlation between results of nano-impact tests and the performance of coated systems operating in extreme intermittent contact environments. Good agreement between the evolution of tool wear in cutting of 1040 structural steel and the progression of coating damage on repetitive nano-impact testing has been reported for AlCrN and TiAlN [29, 36]. With its lower plasticity, the TiAlN fractures more extensively in the nano-impact test and in the cutting test resulting in greater wear.

AlTiN exhibits longer tool life than TiAlN in end milling of hardened steel and aerospace alloy Ti6Al4V, and in face milling of structural steel [43] (Table 5.4). Increasing the Al content to 67% is beneficial for end and face milling applications despite the drop in hardness that results. Increasing the Al content results in the formation of more protective Alumina tribofilms. Similar results have been reported by several groups, for example Hörling and co-workers have noted that $Ti_{0.34}Al_{0.66}N$ performs marginally better in face milling of low carbon steel than $Ti_{0.5}Al_{0.5}N$ [58] and Erkens et al. reported significantly longer tool life in high-speed milling of hardened steel for a dense $Ti_{0.33}Al_{0.67}N$ coating (deposited by a highly ionized plasma and described as a supernitride, nano-crystalline coating), compared to a $Ti_{0.46}Al_{0.54}N$ [46].

Table 5.4 Correlation between fracture probability in the nano-impact test and tool life in face and end milling

Test temperature/°C	Coating	Plasticity index[a]	Fracture probability[b]	Relative tool life in end milling[c]	Relative tool life in face milling[d]
25	AlTiN	0.52	0.4		
25	TiAlN	0.48	0.8		
500	AlTiN	0.73	<0.2	1.2	1.9
500	TiAlN	0.70	0.5	1	1

[a]From nanoindentation.
[b]Defined as probability of fracture occurring within the duration of the impact test.
[c]End milling of 4340 steel.
[d]Face milling of 1040 steel.

To investigate whether differences in coating brittleness were also important, elevated temperature nano-impact tests have been performed for TiAlN- and AlTiN-coated WC-Co cutting tool inserts. Although the room-temperature plasticity is similar, the nano-impact test showed significantly more brittle behaviour for TiAlN.

At 500°C, fracture probability data on TiAlN and AlTiN coated WC-Co inserts correlate with the plasticity index and cutting data [43]. The fracture probability decreased markedly, consistent with the higher plasticity shown by nanoindentation. TiAlN fractures at 500°C, consistent with more fracturing and unstable wear in interrupted cutting applications that generate significant heat (milling) as has been observed in cutting tool tests. The lower brittleness of the AlTiN coating helps provide more favourable, low wear conditions for effective tribofilm formation. This protects the cutting tool surface against the chipping that is typical for cutting operations with intensive adhesive interaction with workpiece materials such as machining of Ti-based alloys.

TiAlCrN performs well when machining hardened steels to due to factors such as its elevated temperature mechanical properties and oxidation resistance [60, 61]. In contrast, when machining strong aerospace materials such as Ti6Al4V or the Ni-based superalloy Waspaloy, increased plasticity is required, even if at the cost of hot hardness. Related coatings such as AlTiN can outperform TiAlCrN in these applications. Staia and co-workers have also recently reported that AlTiN coatings soften markedly over the same temperature range [49]. The AlTiN surface is able to dissipate frictional energy by plastic deformation instead of crack formation and this results in longer tool life under the intense adhesive-fatigue interactions with the workpiece.

The machining conditions have been simulated by nano-impact testing. Although initial nano-impact resistance is similar for both coatings, on continued impact, the TiAlCrN coating fractures dramatically resulting in much larger final impact depth, consistent with its poorer performance in machining Ti6Al4V and Waspaloy [60, 61].

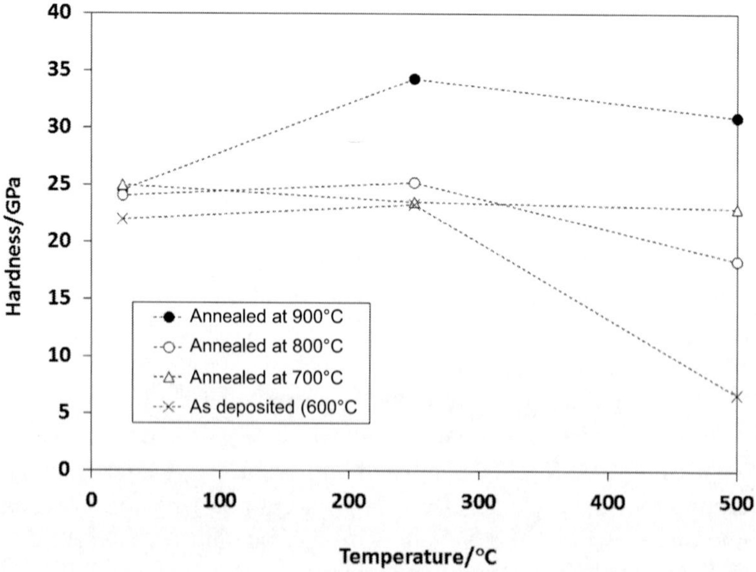

Figure 5.10 Influence of annealing at 700–900°C for 2 hr in vacuum on the high-temperature hardness of AlTiN.

5.4.3.2 Influence of annealing on AlTiN

Annealing AlTiN at 700°C or more has a clear effect on its durability in different machining operations [21, 62]. A partial phase transformation occurs during annealing and the solid solution of Al in TiN (i.e., the as-deposited coating) partially transforms into a beneficial nanocomposite including nano-scale domains of c-AlN in a c-(Ti,Al)N matrix.

Figure 5.10 shows the variation in high-temperature hardness with annealing temperature. The hardness of as-deposited AlTiN increases slightly at 250°C before decreasing at 500°C. When testing another high Al-fraction nitride coating, $Ti_{0.35}Al_{0.65}N$, Bouzakis and co-workers reported a complex dependence of hardness versus temperature with a pronounced increase at 150–200°C, which was interpreted as dislocation pile up due to strain hardening [63]. Annealing AlTiN at 700–900°C has a dramatic effect on its high-temperature mechanical properties. The hardness of the AlTiN coating annealed at 900°C is higher at 500°C than at room temperature.

Table 5.5 Influence of annealing on relative tool life in different cutting conditions

Annealing temperature (°C)	High-speed turning of 1040 steel	Interrupted turning of 4340 steel	End milling of H13 steel
700	1.8	1.3	1.6
800	2.0	1.2	1.3
900		0.8	

Note: The tool life data are relative to the as-deposited AlTiN which is assigned a value of 1 for each cutting condition.

Hörling and co-workers have suggested that the retention of high hardness during annealing of $Ti_{0.34}Al_{0.66}N$ is due to competing mechanisms: softening by residual stress relaxation being balanced by hardening from formation of a coherent nanocomposite structure of cubic TiN and AlN domains by spinodal decomposition [58]. The nanocomposite structure stabilises the high-temperature hardness, potentially due to higher resistance to dislocations at the c-AlTiN/c-AlN phase interfaces. The increase in grain size may be an additional factor in the general increase in high-temperature stability.

The influence of these changes on the relative tool life in continuous high-speed turning of 1040 structural steel, interrupted turning of 4340 steel and end milling of H13 hardened steel are shown in Table 5.5.

Annealing at 700°C improves tool life in all these conditions compared to the as-deposited coating (temperature during deposition was approximately 600°C). The success of increasing the temperature of annealing to 900°C depended strongly on the cutting conditions. The longer tool life in high-speed turning of 1040 structural steel after annealing AlTiN at 900°C has been attributed to the combination of high temperature at the tool/workpiece interface and relatively low stresses at the cutting zone [21], which nanoindentation shows is much greater after annealing at 900°C. In contrast, in interrupted cutting conditions toughness and plasticity appear more critical. In end, milling high hot hardness (and yield stress [7]) should be combined with improved plasticity (and high-temperature fracture strength [7]) for longer tool life and after annealing at 900°C, the AlTiN coating does not display an optimum balance of these properties.

Annealing at 700°C results in the AlTiN coating with an optimum combination of properties for both interrupted cutting tests. Mechanisms of nanograin and compressive stress toughening have been suggested. It has a combination of properties that provide a more stable environment for tribofilms to form and grow in interrupted cutting conditions than the coating after annealing at higher temperatures, which display lower plasticity.

5.4.3.3 TiAlCrSiYN/TiAlCrN multilayers

Multifunctional TiAlCrSiYN-based coatings have shown potential in dry ultra-high-speed machining of hardened tool steels and hard-to-cut aerospace materials [4, 64, 65]. Si and Y additions are thought to work synergistically in these coatings. Si is believed to refine grain size, and Yttrium to prevent grain size coarsening at elevated temperatures, segregating to grain boundaries to block diffusion paths, and to increase phase stability.

Figure 5.11 shows tool life in machining hardened tool steel at 400–700 m/min for a TiAlCrSiYN/TiAlCrN multilayer coating with 55% Al in the Si+Y layers and monolayer TiAlCrSiYN. The TiAlCrSiYN monolayer coating dramatically outperforms the

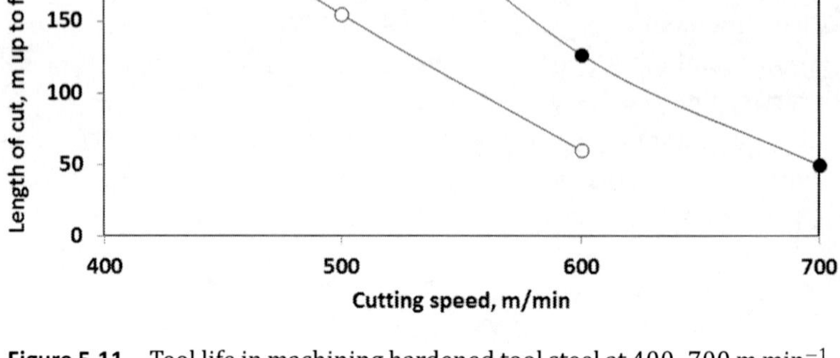

Figure 5.11 Tool life in machining hardened tool steel at 400–700 m min^{-1}.

TiAlCrN monolayer, which cannot sustain speeds over 300 m min^{-1} (length of cut = 50 m at 300 m min^{-1}). The TiAlCrSiYN/TiAlCrN multilayer coating shows better high-speed machining performance significantly outperforming the TiAlCrSiYN monolayer coating over the range of cutting speeds of 500–700 m min^{-1}. The TiAlCrSiYN/TiAlCrN multilayer has only slightly higher hardness than either of the monolayers at room temperature. At high-temperature differences become much more marked.

It showed a considerably smaller reduction in the mechanical properties H, H/E_r, H^3/E_r^2 with increasing temperature than either of the TiAlCrSiYN and TiAlCrN monolayer coatings and it retained a high hardness of around 20 GPa at 600°C. During machining, formation of protective tribofilms form on the surface of TiAlCrSiYN-based coatings results in beneficial heat-redistribution in the cutting zone. The tribofilms act as a thermal barrier creating a strong thermal gradient. The presence/absence of non-beneficial AlN_{hex} in worn surfaces may be correlated with tool life as an indicator of the temperature of the coating/tribofilm interface, and the effectiveness of the tribofilms as a thermal barrier. XRD data suggest for the TiAlCrSiYN/TiAlCrN multilayer, the temperature decreases from ~1100–1200°C at the outer surface to ~600°C at the tribofilm/coating interface [65].

The TiAlCrSiYN/TiAlCrN multilayer has only slightly higher hardness than either of the monolayers at room temperature. At high temperature, differences become much more marked. The nanomechanical properties of the underlying coating layer provide a stable low-wear environment for the tribofilms to form and regenerate so it can survive high temperatures under operation (600°C). The high wear resistance of the multilayer TiAlCrSiYN/TiAlCrN coating under extreme operating conditions is due to this combination of characteristics. TiAlCrSiYN and TiAlCrN monolayer coatings have a less effective combination of adaptability and mechanical characteristics and therefore lower tool life. Both monolayer coatings have better high-temperature mechanical properties than AlTiN or TiAlN studied previously by similar methods. The TiAlCrSiYN/TiAlCrN multilayer has a good combination of properties for high-speed machining of hardened steel. Although it has rather limited plasticity, this is offset by having a dense multilayered structure capable of crack deflection under severe conditions [64].

Multilayer coatings have been prepared with the % Al in the Si + Y containing layers increased from 55% or 60% Al. This higher % Al multilayer coating has noticeably lower hardness (24.5 GPa at room temperature) with greater plasticity. These different properties result in different multilayer compositions being optimum for different Ni-based aerospace superalloys [4]. For turning the thermally resistant aerospace alloy ME16 the multilayer TiAlCrSiYN/TiAlCrN coating with 55% Al in the Si+Y layers performs better due to its improved oxidation resistance and high-temperature hardness [4]. However, for turning of Inconel DA 718 alloy plasticity appears a more dominant factor than maximising elevated temperature H/E and H^3/E_r^2 as the multilayer with 60% Al in the Si + Y layers performs slightly better.

5.5 Knowledge-Based Design Rules for Improving Coating Durability

Room- and high-temperature nanoindentation test data on a range of advanced monolayer multicomponent nitride coatings on cemented carbide substrate have been correlated with the tool life results from cutting tests. Based on this correlation, it has been possible to investigate the optimum combinations of hardness and plasticity (or H/E_r) that can result in longer tool life in different cutting conditions.

In addition to the mechanical properties, 'knowledge-based design rules' for improving coating durability under the extreme frictional conditions in cutting also need to consider (i) the influence of the substrate, (ii) coating architecture—for example, monolayer or multilayer—and composition, (iii) cutting conditions, (iv) the material being machined, and (v) tribofilm composition, mechanical and thermal barrier properties.

Several of these factors are interlinked and the coatings need to be multifunctional to minimise wear during cutting. The choice of substrate influences the strain accumulation under load. The coating composition influences the composition of the tribofilm layer formed. The coating microstructure influences the deformation behaviour.

The severity of the cutting conditions and the material being machined both influence the relative importance of factors such as energy dissipation (plasticity) and high-temperature hardness, which can influence the effectiveness of tribofilm formation and the ability of the coatings to display adaptive, multifunctional behavior.

5.5.1 *Substrate Properties*

The mechanical properties of the substrate influence strain accumulation under load. Knotek and co-workers investigated the fatigue strength of CrN and TiZrCN PVD coatings deposited on high-speed steel, cemented carbide and alumina [66]. They reported markedly greater fatigue strength when the coatings were deposited on high-speed steel, due to stress relief by plastic deformation. Yoon and co-workers studied impact wear of TiAlN and TiN coatings on AISI D2 steel and WC-Co [67]. Mechanical properties were determined at a depth around one-fourth of the coating thickness so contain considerable substrate influence. The more ductile substrate minimises accumulation of elastic strain at high load. On D2 steel, the coating system with higher (composite) H^3/E^2, TiAlN, showed both lower impact wear volume and a longer number of cycles-to-fracture. For TiAlN on cemented carbide the behaviour was completely different. With continued impacts the it showed pronounced brittle cracking so that the impact wear volume was increased dramatically. However, it should be noted that these studies were performed at room temperature and temperature may also play a role. High-speed steel softens markedly with temperature, particularly around 600°C, and this softening has been shown to cause extreme coating cracking as the brittle PVD coatings cannot accommodate this resulting in tool failure [7]. Very high H^3/E^2 may also create adhesion problems. Ronkainen et al. [68] considered that differences in substrate stress relaxation could explain why hard ta-C perform more poorly on hard WC-Co and alumina substrates than a-C:H coatings with low internal stress. For the more highly stressed ta-C coatings moderate loading could result in adhesion failure. These problems may be mitigated with a duplex or graded deposition to improve adhesion [69].

5.5.2 *Continuous Cutting*

The cutting temperatures in continuous high-speed turning operations are very high (>1000°C) and consequently high temperature hardness is paramount. Coatings with excellent hot hardness, such as AlTiN after annealing at 900°C, and the multilayer TiAlCrSiYN/TiAlCrN coating with 55% Al in the Si + Y layers show longer tool life in high-speed turning of hardened steel and the Ni-based superalloy ME 16. The requirement for plasticity is not as important.

5.5.3 *Interrupted Cutting*

In contrast, interrupted cutting conditions can alter the relative importance of high temperature hardness (and hence yield stress) and plasticity (and hence fracture resistance). Monolayer coatings with higher plasticity than TiAlN, such as TiAlCrN, AlTiN and TiAlN perform better in interrupted cutting of structural steels, Ti6Al4V and Ni-based superalloys. Similar data have been reported by other groups. For milling, Hörling and co-workers have shown that AlTiN exhibits longer tool life than TiAlN [58] and Zeng and co-workers have shown that both AlCrN and AlTiN out-perform TiAlN [70–71]. An optimum plasticity index determined from room temperature data for end/face milling has been estimated as 0.53.

In addition to the plasticity index determined from the quasi-static indentation test, a more direct measure of dynamic toughness is provided by nano-impact tests that directly show enhanced cracking for coatings that crack more in the cutting tests. Nano-impact tests also correlate well with solid particle erosion tests [82]. Chen et al. investigated the role of changing coating microstructure and mechanical properties due to thermal aging at 1100–1500°C on the repetitive impact and erosion behaviour of 150 μm thick thermal barrier coatings (TBCs) for aero-engines, reporting good agreement between the tests [82]. It is generally thought that coating hardness is more important when cutting, whilst coating toughness is critical when requiring high erosion resistance under crack-dominated conditions. This classification is not ideal for semi-brittle nitride coatings that are susceptible to solid particle erosion involving cracking (fracture) and cutting (flow) [66]. Increasing the particle impingement angle promotes erosive impact damage and more brittle rather than ductile dominated damage.

Coatings therefore can display optimum mechanical properties for erosion resistance at intermediate values of hardness and other mechanical property indices such as H^3/E_r^2 [72]. For example, 6 μm TiAlN coatings on Ti6Al4V designed for aerospace applications showed significantly poorer erosion performance at very high H^3/E_r^2 [73]. Erosion resistance shows a much stronger dependence on the mechanical properties at 90° impact (more severe) than 30° (less severe). Although there have been many reports of improved erosion and impact resistance by increasing H/E or H^3/E_r^2 [74], we caution against this as a general conclusion since coatings with very high H^3/E_r^2 can be very brittle unless microstructurally designed to mitigate this. In the nano-impact test, the transition to more severe test conditions with increasingly crack-dominated behaviour is achieved by increasing the impact force. Typically coatings with higher H^3/E_r^2 perform better at lower force due to enhanced load support and crack initiation threshold but less well as the impact force is increased. At higher force, the low plasticity and toughness in the higher H^3/E_r^2 coatings can result in significantly greater cracking compared to coatings with higher plasticity and lower H^3/E_r^2. This behaviour has been reported for TiAlSiN coatings on high-speed steel [74] and TiFeN coatings on silicon [23].

The extent of correlation between mechanical properties (specifically plasticity index and nano-impact fracture resistance) at room temperature and tool life implies either (i) that any trend in mechanical properties with temperature stays broadly similar and/or (ii) plasticity (toughness) is significantly more important than hardness/load support in the application—that is, cracking is life-limiting. The mechanical properties provide a stable environment for tribofilm formation and coatings with higher Al percentage form more protective alumina and less non-protective rutile phases in the tribofilms. Since the dynamics of the tribofilm formation occur at elevated temperature, elevated temperature properties are likely

Table 5.6 Temperature dependence of H/E_r in advanced PVD coatings

	H/E_r at 25°C	H/E_r at 500°C
TiAlCrSiYN/TiAlCrN	0.096	0.072
TiAlCrSiYN	0.093	0.061*
TiAlCrN	0.084	0.06*
AlTiN (annealed at 900°C)	0.088	0.082
AlTiN (annealed at 800°C)	0.086	0.055
AlTiN (annealed at 700°C)	0.084	0.056
AlTiN (as deposited)	0.075	0.032
AlCrN	0.065	0.04
TiAlN	0.079	0.042

*Data at 450°C.

to be more important. These vary considerably for the coatings studied. For example, Table 5.6 shows marked changes in H/E_r between room temperature and 500°C.

There is a general trend towards decreasing H/E_r with increasing temperature due to the decrease in the hardness of the PVD coatings together with little change in their elastic modulus. The decrease in hardness is due to relaxation of high compressive residual stresses, grain boundary sliding and thermally activated dislocations, such as dislocation glide [76–77]. For the monolayer coatings, optimum values of hardness and plasticity at 450–500°C to achieve longer tool life in interrupted cutting conditions such as end milling were found to be $H \sim 20$ GPa, $H/E_r \sim 0.06$ and plasticity index ~ 0.6 [16, 36]. This reflects the requirement for both high hot hardness/yield stress and toughness/fracture strength in this application [7]. Coatings with values far outside of these typically have significantly lower tool life under these conditions.

For high-speed machining (ball nose end milling) of hardened steels, neither coating hardness, the plasticity index or nano-impact behaviour are necessarily the best predictor of tool life when these properties are determined at room temperature. The high temperature properties are of more interest as they influence the environment where the tribofilms form. The mechanical properties are a critical factor in providing a stable environment for the system to display adaptive behaviour. Under these conditions, TiAlCrN with its metal-covalent character, high hot hardness and low plasticity outperforms AlTiN with its metallic bonds, low hot hardness and higher plasticity [60, 61].

5.5.4 *Multilayer Advantages*

Different types of coating multilayer architecture have been designed. These include systems such as (i) the 'hard–hard' TiAlCrSiYN/TiAlCrN multilayer described above, (ii) other nanolaminates which can show significant hardness enhancements ('superhardness') over the rule of mixtures and (iii) hard–soft (brittle/ductile) systems. Pronounced hardness enhancements are due to dislocation hindering at interfaces due to differences in shear moduli and by coherency strain in lattice-mistmatched materials [76, 78]. Multilayer design in the hard–soft system is for load support from hard layers in combination with shear in the ductile layers. Matthews and co-workers performed modelling studies of the deformation of alternating hard–soft (or high E/low E) layers multilayer coatings [79]. They found that the main shear deformation occurred in the low modulus layers enabling the more brittle layers to effectively slide over each other without experiencing as high bending stresses as a monolayer of the same total thickness composed of the same (hard) material. The multilayers could then deflect with the substrate under load without cracking and survive heavily loaded tribo-contacts. Castanho and Vieira have reported that ductile interlayers of Ti, Al in TiAlN/M multilayers on M2 steel improved the scratch adhesion whilst Cu interlayers reduced it [80]. Multilayers comprising superlattice TiN/CrN alternating with pure Ti have been deposited on Ti6Al4V as candidate coatings for erosion protection [81]. Yang and co-workers have reported poor performance—extensive cracking and localised interfacial debonding—in erosion and scratch testing due to the differences in properties, primarily weak adhesion between (i) the brittle superlattice layers and the ductile Ti and (ii) between the coating and the substrate [81].

The TiAlCrSiYN/TiAlCrN multilayers have improved performance in ultra-high-speed machining of hardened steels and high-speed machining of Ni-based superalloys compared to the state-of-the-art high Al-fraction monolayer coatings. The microstructure and mechanical properties of the multilayer are important factors in its ability to perform under the extreme frictional conditions. Compared to the monolayer coatings, it has very high H^3/E_r^2 which gives high resistance to plastic deformation and crack formation. The nanolaminated structure enables crack deflection along interfaces providing an additional route to relieving strain accumulation in heavy loaded cutting. A similar toughening effect in scratch testing was observed by Yau and co-workers for nc-TiAlN/a-Si_3N_4 nanolaminate coatings on high-speed steel [77]. In contrast, the microstructure of the monolayer coatings means they cannot readily combine high H^3/E_r^2 with a mechanism that avoids large scale fracture under increasingly severe conditions and the cutting speed range is more limited.

5.6 Summary

To achieve long tool life in the tribologically extreme conditions in high-speed machining of hard-to-cut materials, PVD coatings need to be multifunctional and display several interlinked characteristics to minimise wear. Different cutting conditions require coatings to have a different balance of mechanical properties. For highly load tribo-contact designing, coatings for minimising strain accumulation has become a successful strategy. Factors such as the influence of the substrate, the coating composition and architecture—for example, monolayer or multilayer, the material being machined and the cutting conditions should all be considered when attempting to optimize coating mechanical properties.

Coating design to minimize high-temperature wear and fracture provides a stable environment for tribofilms to form that can act as a thermal barrier and protect the underlying coating. The coating composition influences the composition of the tribofilm layer formed with high Al-fraction coatings forming more protective alumina tribofilms. The coating microstructure and the resultant mechanical properties influence the deformation behaviour. The severity of the cutting conditions and the material being machined both influence the relative importance of factors such as energy dissipation (plasticity) and high-temperature hardness.

Recently developed multifunctional and adaptive high Al-fraction multilayer coatings, with a combination of mechanical properties and microstructure that can minimise crack formation and dissipate energy by crack deflection along interfaces, have shown advantages over the state of the art monolayer coatings that result in improved durability.

References

1. P. Wright, E.M. Trent, *Metal Cutting*, 4th Edition (Butterworth-Heinmann, Boston, USA) (2000).
2. G.S. Fox-Rabinovich, A.I. Kovalev, J.L. Endrino, S.C. Veldhuis, Shuster, L.S., I.S. Gershman, Chapter 9: Self-organization during friction: Advanced surface engineered materials and systems design, (eds. G.S. Fox-Rabinovich, G. Totten), *Surface-Engineered Tool Materials for High-Performance Machining*, Taylor and Francis Books/CRC Press LLC: USA (2007), pp. 231–296.
3. E.O. Ezugwu, J. Bonney, Y. Yamane, An overview of the machininability of aeroengine alloys, *J. Mater. Proc. Technol.*, 134 (2003), 233.
4. G.S. Fox-Rabinovich, B.D. Beake, K. Yamamoto, M.H. Aguirre, S.C. Veldhuis, G. Dosbaeva, A. Elfizy, A. Biksa, L.S. Shuster, A.Y. Rashkovskiy, Structure, properties and wear performance of nano-multilayered TiAlCrSiYN/TiAlCrN coatings during machining of Ni-based aerospace superalloys, *Surf. Coat. Technol.*, 204 (2010), 3698.
5. A. Leyland, A. Matthews, On the significance of the H/E ratio in wear control: a nanocomposite coating approach to optimised tribological behavior. *Wear*, 246 (2000), 1.
6. A. Leyland, A. Matthews, Design criteria for wear-resistant nanostructured and glassy-metal coatings. *Surf. Coat. Technol.*, 177–178 (2004), 317.
7. S. Hogmark, M. Olsson, Wear mechanisms of HSS cutting tools, 1st International High Speed Steel Forum Conference, Feb 2-3, Aachen, 2005. (2005)
8. D.T. Quinto, G.J. Wolfe, P.C. Jinhal, High temperature microhardness of hard coatings produced by physical vapour deposition, *Thin Solid Films* 153 (1987), 19.
9. G.S. Fox-Rabinovich, S.C. Veldhuis, V.N. Scvortsov, L. Sh. Shuster, G.K. Dosbaeva, M.S. Migranov, Elastic and plastic work of indentation as

a characteristic of wear behaviour for cutting tools with nitride PVD coatings, *Thin Solid Films*, 469–470 (2004), 505.

10. J. Milhans, D.S. Li, M. Khaleel, X. Sun, M.S. Al-Haik, A. Harris, H. Garmenstani, Mechanical properties of solid oxide fuel cell glass-ceramic seal at high temperatures, *J. Power Sources*, 196 (2011), 5599.
11. G.S. Fox-Rabinovich, K. Yamamoto, A.I. Kovalev, Chapter 10: Self-organization during friction: Advanced surface engineered materials and systems design, (eds. G.S. Fox-Rabinovich, G. Totten) *Synergistic Alloying of Self-adaptive Wear-resistant Coatings*, Taylor and Francis Books/CRC Press LLC: USA, (2007), pp. 297–334.
12. T. Chudoba, Chapter 6: Nanostructured coatings (eds. A. Cavaleiro, J.Th.M. De Hosson), *Measurement of Hardness and Young's Modulus of Coatings by Nanoindentation,* Springer: New York (2006), pp. 216–260.
13. A.J. Fischer-Cripps, *Nanoindentation*, 3rd Edition, Springer: USA (2011).
14. ISO 14577: Metallic Materials—Instrumented Indentation Test for Hardness and Materials Parameters, Part 4 - Test Method for metallic and non-metallic coatings.
15. Y.-T. Cheng, C.-M. Cheng, Relationships between hardness, elastic modulus, and the work of indentation. *Appl. Phys. Lett.*, 73 (1998), 614.
16. B.D. Beake, G.S. Fox-Rabinovich, S.C. Veldhuis, S.R. Goodes, Coating optimisation for high-speed machining with advanced nanomechanical test methods *Surf. Coat. Technol.*, 203 (2009), 1919.
17. Y. Choi, H.-S. Lee, D. Kwon, Analysis of sharp-tip-indentation load-depth curve for contact area determination taking into account pile-up and sink-in effects. *J. Mater. Res.*, 19 (2004), 3307.
18. T. Chudoba, N.M. Jennett, Higher accuracy analysis of instrumented indentation data obtained with pointed indenters, *J. Phys. D: Appl. Phys.*, 41 (2008), 215407.
19. S.J. Bull, Using work of indentation to predict erosion behaviour in bulk materials and coatings *J. Phys. D: Appl. Phys.*, 39 (2006), 1624.
20. B.D. Beake, V.M. Vishnyakov, R. Valizadeh, J.S. Colligon, Influence of mechanical properties on the nanoscratch behaviour of hard nanocomposite TiN/Si_3N_4 coatings on Si. *J. Phys. D: Appl. Phys.*, 39 (2006), 1392.
21. G.S. Fox-Rabinovich, J.L. Endrino, B.D. Beake, A.I. Kovalev, S.C. Veldhuis, L. Ning, F. Fotaine, A. Gray, Impact of annealing on the microstructure, properties, and cutting performance of AlTiN coating, *Surf. Coat. Technol.*, 201 (2006), 3524.

22. S. Zhang, H.L. Wang, S.-E. Ong, D. Sun, X.-L. Bui, Hard yet tough nanocomposite coatings—present status and future trends. *Plasma Process. Polym.*, 4 (2007), 219.

23. B.D. Beake, V.M. Vishnyakov, J.S. Colligon, Nano-impact testing of TiFeN and TiFeMoN films for dynamic toughness evaluation, *J. Phys. D: Appl. Phys.*, 44 (2011), 085301.

24. T. Kubart, T. Polcar, L. Kopecky, R. Novak, D. Novakova, Temperature dependence of tribological properties of MoS_2 and $MoSe_2$ coatings, *Surf. Coat. Technol.*, 193 (2005), 230–233.

25. T. Polcar, R. Martinez, T. Vitu, L. Kopecky, R. Rodriguez, A. Cavaleiro, High temperature tribology of CrN and multi-layered Cr/CrN coatings, *Surf. Coat. Technol.*, 203 (2009), 3254–3259.

26. T. Polcar, A. Cavaleiro, High-temperature tribological properties of CrAlN, CrAlSiN and AlCrSiN coatings, *Surf. Coat. Technol.*, 206 (2011), 1244–1251.

27. T. Polcar, A. Cavaleiro, Structure and tribological properties of AlCrTiN coatings at elevated temperature, *Surf. Coat. Technol.*, 205 (2011), S107–S110.

28. D.N. Allsopp, I.M. Hutchings, Micro-scale abrasion and scratch response of PVD coatings at elevated temperatures, *Wear*, 251 (2001), 1308–1314.

29. G.S. Fox-Rabinovich, B.D. Beake, S.C. Veldhuis, J.L. Endrino, R. Parkinson, L.S. Shuster, M.S. Migranov, Impact of mechanical properties measured at room and elevated temperatures on wear resistance of cutting tools with TiAlN and AlCrN coatings, *Surf. Coat. Technol.*, 200 (2006), 5738.

30. C.A. Schuh, C.E. Packard, A.C. Lund, Nanoindentation and contact-mode imaging at high temperatures, *J. Mater. Res.*, 21 (2006), 725–736.

31. J.C. Trenkle, C.E. Packard, C.A. Schuh, Hot nanoindentation in inert environments, *Rev. Sci. Instr.*, 81 (2010), 073901.

32. N.M. Everitt, M.J. Davies, J.F. Smith, High temperature nanoindentation—the importance of isothermal contact, *Philos. Mag.*, 91 (2011), 1221.

33. J.F. Smith, S. Zheng, High temperature nanoscale mechanical property measurements, *Surf. Eng.*, 6 (2000), 143.

34. UK Patent GB2478134B and Int Pat App WO2011/104529.

35. F. Tang, L.C. Zhang, On the reliability of nanoindentation on Si wafer at elevated temperatures, *Adv. Mater. Res.*, 325 (2011), 684.

36. B.D. Beake, S.R. Goodes, J.F. Smith, G.S. Fox-Rabinovich, S.C. Veldhuis, Handbook of Nanostructured Thin Films and Coatings, Mechanical

Properties (ed. S. Zhang), Chapter 6 *Using nanomechanics to optimise coatings for cutting tools* CRC Press (2010), pp. 205–244.

37. Y.C. Liu, J.W.R. Teo, S.K. Tung, K.H. Lam, High temperature creep and hardness of eutectic 80Au/20Sn solder, *J. Alloys Compd.*, 448 (2008), 340.
38. A. Gray, B.D. Beake, Elevated temperature nanoindentation and viscoelastic behaviour of poly(ethylene terephthalate) films, *J. NanoSci. Nanotech.*, 7 (2007), 2530–2533.
39. A. Sawant, S. Tin, High temperature nanoindentation of a Re-bearing single crystal Ni-base superalloy, *Scr. Mater.*, 58 (2008), 275.
40. J.R. Nicholls, et al., conference presentation at ICMCTF San Diego, USA. (2003).
41. S.K. Korte, R.J. Stearn, J.M. Wheeler, W.J. Clegg, High temperature microcompression and nanoindentation in vacuum, *J. Mater. Res.*, 27 (2012), 167.
42. B. He, G. Ghosh, Y.-W. Chung, Q. Wang, Effect of Melting and Microstructure on the Microscale Friction of Silver–Bismuth Alloys, *Tribol. Lett.*, DOI 10.1007/s11249-010-9606-4 (2010)
43. B.D. Beake, J.F. Smith, A. Gray, G.S. Fox-Rabinovich, S.C. Veldhuis, J.L. Endrino, Investigating the correlation between nano-impact fracture resistance and hardness/modulus ratio from nanoindentation at 25-500°C and the fracture resistance and lifetime of cutting tools with $Ti_{1-x}Al_xN$ (x=0.5 and 0.67) PVD coatings in milling operations *Surf. Coat. Technol.*, 201 (2007), 4585.
44. S.K. Korte, W.J. Clegg, Micropillar compression of ceramics at elevated temperature, *Scr. Mater.*, 60 (2009), 807.
45. S.K. Korte, J.S. Banard, R.J. Stearn, W.J. Clegg, Deformation of silicon, insights from microcompression testing at 25–500°C, *Int. J. Plasticity*, 27 (2011), 1853.
46. G. Erkens, R. Cremer, T. Hamoudi, K.-D. Bouzakis, et al., Properties and performance of high aluminium containing (Ti, Al)N based supernitride coatings in innovative cutting applications, *Surf. Coat. Technol.*, 177–178 (2004), 734.
47. K.-D. Bouzakis, S. Hadjiyiannis, G. Skordaris, I. Mirisidis, N. Michailidis, K. Efstathiou, et al., The effect of coating thickness, mechanical strength and hardness properties on the milling performance of PVD coated cemented carbide inserts, *Surf. Coat. Technol.*, 177–178 (2004), 657.
48. M.H. Staia, M. D'Alessandria, D.T. Quinto, F. Roudet, M. Marsal Astort, High-temperature tribological characterisation of commercial TiAlN coatings, *J. Phys.: Condens. Matter.*, 18 (2006), S1727.

49. C.P. Alpert, H.M. Chan, S.J. Bennison, B.R. Lawn, Temperature dependence of hardness of alumina-based ceramics, *J. Am. Ceram. Soc.,* 71 (1988), C-371.

50. P.C. Jinhal, A.T. Santhanam, U. Schleinkofer, A.F. Shuster, Performance of PVD TiN, TiCN and TiAlN coated cemented carbide tools in machining, *Int. J. Refract. Metals. & Hard Mater.,* 17 (1999), 163.

51. T. Polcar, R. Novak, P. Siroky, The tribological characteristics of TiCN coating at elevated temperatures, *Wear,* 260 (2006), 40.

52. G.S. Fox-Rabinovich, G.C. Weatherley, A.I. Dodonov, A.I. Kovalev, L.S. Shuster, et al., Nano-crystalline filtered arc deposited (FAD) TiAlN PVD coatings for high-speed machining applications, *Surf. Coat. Technol.,* 177–178 (2004), 800.

53. G.S. Fox-Rabinovich, K. Yamamoto, S.C. Veldhuis, A.I. Kovalev, G.K. Dosbaeva, Tribological adaptability of TiAlCrN PVD coatings under high performance dry machining conditions *Surf. Coat. Technol.,* 200 (2005), 1804.

54. G.S. Fox-Rabinovich, S.C. Veldhuis, G.K. Dosbaeva, K. Yamamoto, A.I. Kovalev, D.L. Wainstein, I.S. Gershman, L.S. Shuster, B.D. Beake, Nanocrystalline coating design for extreme applications based on the concept of complex adaptive behaviour, *J. Appl. Phys.,* 103 (2008), 083510.

55. B.D. Beake, S.R. Goodes, J.F. Smith, Micro-impact testing: a new technique for evaluating fracture toughness, adhesion, erosive wear resistance and dynamic hardness, *Surf. Eng.*, 17 (2001), 187.

56. B.D. Beake, J.F. Smith, Nano-impact testing—an effective tool for assessing the resistance of advanced wear-resistant coatings to fatigue failure and delamination, *Surf. Coat. Technol.,* 188–189C (2004), 594.

57. B.D. Beake, Evaluation of the fracture resistance of DLC coatings on tool steel under dynamic loading, *Surf. Coat. Technol.*, 198 (2005), 90.

58. A. Hörling, L. Hultman, M. Odén, J. Sjölén, L. Karlsson, Mechanical properties and machining performance of $Ti_{1-x}Al_xN$ coated cutting tools, *Surf. Coat. Technol.*, 191 (2005), 384.

59. K.-D. Bouzakis, G. Skordaris, E. Bouzakis, A. Tsouknidas, S. Makrimallakis, et al. Optimisation of wet micro-blasting on PVD films with various grain materials for improving the coated tools' cutting performance, *CIRP Ann. Manuf. Technol.*, 60 (2011) 587–590.

60. A.I. Kovalev, D.L. Wainstein, A.Y. Rashkovskiy, G.S. Fox-Rabinovich, K. Yamamoto, S.C. Veldhuis, M. Aguirre, B.D. Beake, Impact of Al and Cr

alloying in TiN-based PVD coatings on cutting performance during machining of hard to cut materials, *Vacuum*, 84 (2010), 184.

61. G.S. Fox-Rabinovich, S.C. Veldhuis, K. Yamamoto, M.H. Aguirre, A. Kovalev, D.L. Wainstein, B.D. Beake, J.L. Endrino, Design and performance of AlTiN and TiAlCrN PVD coatings for machining of hard to cut materials, *Surf. Coat. Technol.*, 204 (2009), 489.
62. G.S. Fox-Rabinovich, J.L. Endrino, B.D. Beake, M.H. Aguirre, S.C. Veldhuis, D.T. Quinto, C.E. Bauer, A.I. Kovalev, A. Gray, Effect of annealing below 900°C on structure, properties and tool life of an AlTiN coating under various cutting conditions, *Surf. Coat. Technol.*, 202 (2008), 2985.
63. K.-D. Bouzakis, M. Pappa, E. Bouzakis, G. Skordaris, S. Gerardis, Correlation between PVD coating strength properties and impact resistance at ambient and elevated temperatures, *Surf. Coat. Technol.*, 205 (2010), 1481.
64. Fox-Rabinovich, K. Yamamoto, B.D. Beake, A.I. Kovalev, M.H. Aguirre, S.C. Veldhuis, G. Dosbaeva, D.L. Wainstein, A. Biksa, A.Y. Rashkovskiy, Emergent behavior of nano-multilayered coatings during dry high speed machining of hardened tool steels, *Surf. Coat. Technol.*, 204 (2010), 3425.
65. B.D. Beake, G.S. Fox-Rabinovich, Y. Losset, K. Yamamoto, M.H. Agguire, S.C. Veldhuis, et al., Why can TiAlCrSiYN-based adaptive coatings deliver exceptional performance under extreme frictional conditions? *Faraday Discuss.*, 156 (2012), 267.
66. O. Knotek, B. Bosserhoff, A. Schrey, T. Leyendecker, et al., A new technique for testing the impact load of thin films: the coating impact test, *Surf. Coat. Technol.*, 54/55, 102.
67. S.Y. Yoon, S.-Y. Yoon, W.-S. Chung, K.H. Kim, Impact-wear behaviors of TiN and Ti-Al-N coatings on AISI D2 steel and WC-Co substrates, *Surf. Coat. Technol.*, 177–178 (2004), 645.
68. H. Ronkainen, J. Koskinen, S. Varjus, K. Holmberg, Load-carrying capacity evaluation of coating/substrate systems for hydrogen-free and hydrogenated diamond-like carbon films, *Tribol. Lett.*, 6 (1999), 63.
69. S. Guruvenket, D. Li, J.E. Klemberg-Sapieha, L. Martinu, J. Szpunar, Mechanical and tribological properties of duplex treated TiN, nc-TiN/a-SiN_x and nc-TiCN/a-SiCN coatings deposited on 410 low alloy stainless steel, *Surf. Coat. Technol.*, 203 (2009), 2905.
70. X.Z. Ding, X.T. Zeng, Y.C. Liu, Fang, F.Z., Lim, G.C. SimTech Technical Report 7 (2006), 10.

71. X. Zeng, C. Liang, Y.C. Liu, L.-S. Tan, *SimTech Tech. Rep.* PT/99/003/ST.
72. Q. Yang, D.Y. Seo, L.R. Zhao, X.T. Zeng, Erosion resistance performance of magnetron sputtering deposited TiAlN, *Surf. Coat. Technol.*, 188–189 (2004), 168.
73. S. Hassani, M. Bielawski, W. Beres, L. Martinu, M. Balazinski, J.E. Klemberg-Sapieha, Predictive tools for the design of erosion resistant coatings, *Surf. Coat. Technol.*, 203 (2008), 204.
74. J. Chen, R. Ji, R.H.U. Khan, X. Li, B.D. Beake, H. Dong, Effects of mechanical properties and layer structure on the cyclic loading of TiN-based coatings, *Surf. Coat. Technol.*, 206 (2011), 522.
75. L. Hultman, *Nanostructured coatings,* (eds. A. Cavaleiro, J.Th.M. De Hosson), Chapter 13, *Synthesis, Structure, and Properties of Superhard Superlattice Coatings,* Springer: New York (2006), pp. 539–554.
76. L. Hultman, C. Mitterer, *Nanostructured coatings,* (eds. A. Cavaleiro, J.Th.M. De Hosson), Chapter 11, *Thermal stability of advanced nanostructured wear-resistant coatings,* Springer: New York (2006), pp. 464–510.
77. B.-S. Yau, J.-L. Huang, H.-H. Lu, P. Sajgalik, Investigation of nanocrystal-$(Ti_{1-x}Al_x)N_y$/amorphous-Si_3N_4 nanolaminate films, *Surf. Coat. Technol.*, 194 (2005), 119.
78. A. Matthews, R. Jones, S. Dowey, Modelling the deformation behaviour of multilayer coatings, *Tribol. Lett.*, 11 (2001), 103.
79. J.M. Castanho, M.T. Vieira, Effect of ductile interlayers in mechanical behaviour of TiAlN thin coatings, *J. Mater. Proc. Technol.*, 143–144 (2003), 352–357.
80. Q. Yang, D.Y. Seo, L.R. Zhao, Multilayer coatings with alternate pure Ti and TiN/CrN superlattice, *Surf. Coat. Technol.*, 177–178 (2004), 204.
81. J. Chen, B.D. Beake, R. Wellman, J.R. Nicholls, and H. Dong, An investigation into the correlation between nano-impact resistance and erosion performance of EB-PVD thermal barrier coatings on thermal ageing, *Surf. Coat. Technol.*, 206 (2012), 4992–4998.
82. J.F. Smith, V.M. Vishnyakov, M.I. Davies, and B.D. Beake, Nanoscale friction measurements up to 750°C, *Tribol. Lett.* 49(2013), 455.

Chapter 6

Multifunctional Requirements for Surfaces Subjected to Tribocorrosion

Robert J. K. Wood
The national Centre for Advanced Tribology at Southampton (nCATS), Engineering Sciences, University of Southampton, Southampton, SO17 1BJ, UK
r.wood@soton.ac.uk

6.1 Introduction and Definitions

The term 'tribo' has origins in the Greek word 'tribos' (rubbing), which led to the sciences of lubrication, friction and wear being called 'tribology' while the term corrosion has its origins in medieval Latin word 'corrodere', which mean to gnaw through with 'cor' meaning intensive force and 'rodere' to gnaw [1]. Therefore, the term tribocorrosion refers to the surface degradation mechanisms when mechanical wear and chemical/electrochemical processes interact with each other. Therefore, the subject marries two sets of stimuli together that of mechanical and electrochemical processes driven at the tribological interface. There is a growing understanding that not all tribological surfaces are appropriate for operating in corrosive environments and that there are few design guides for selecting

Multifunctional Materials for Tribological Applications
Edited by Robert J. K. Wood

ISBN 978-981-4463-90-4 (Hardcover), 978-981-4463-91-1 (eBook)
www.panstanford.com

surfaces for resistance to tribocorrosion. The materials and coatings used in such applications need to be multifunctional providing low wear, corrosion and friction as well as being anti-fouling or decorative in some circumstances.

The subject of tribocorrosion includes the interaction of corrosion and erosion (solids, liquid flow and droplet impingements or cavitation bubbles), abrasion, adhesion, fretting and fatigue wear processes. Tribocorrosion is often linked to the synergy resulting from the coupling of mechanical and environmental effects. Many of these mechano-electrochemical interactions are time dependent and non-linear. Tribocorrosion research aims to address the need to select or design new surfaces for future equipment as well as minimise the operating costs and extend the life of existing machinery and medical devices. For example, passive alloys (i.e., most engineering alloys) rely on a 1–10 nm thick surface film for their protection from aggressive and corrosive environments. This film forms by reaction with the environment but wear or aggressive fluid flows can lead to the local rupture or complete removal of the film and permanent deformation of the substrate. This can lead to areas of the substrate being exposed to the aggressive environment and unless repassivation (repair or self-healing) mechanisms reform the passive film to inhibit corrosion processes, accelerated anodic dissolution will occur within these sites. Should surfaces also be required to act as a bearing surface (i.e., plain bearings), then low friction under sliding is needed and this is also likely to be dominated by the surface films formed similar to other tribology systems where surface tribolayers dictate bearing performance. The coupling of mechanical and environmental effects can also create surfaces of specific reaction layers on materials which could inhibit corrosion and/or wear. Examples of this are self-lubricating and/or self-healing surface layers [2]. Perhaps unlike other tribology systems, corrosion adds complications as surface films formed under static corrosion can take several months to fully develop and passivate; so tribocorrosion response for mechanical systems used only intermittently are likely to be very different to those used continuously.

On the positive side, tribocorrosion phenomena can be used as a manufacturing process such as in chemical–mechanical polishing of silicon wafers. Enhanced material removal for chemical mechanical polishing processes has been achieved using electrochemical techniques [3].

Unpredictable interactions can result between wear and corrosion when the surfaces in contact have complex, multiple phase microstructures that can lead to micro-galvanic activity and selective phase corrosion (a localised attack) as well as three-body wear modes. Examples of such surfaces include composites or surfaces that undergo compositional changes induced by tribological interactions. For instance, the presence of carbides in a metallic surface, typically formed for improved wear resistance, establishes a micro-galvanic corrosion cell as the carbide is likely to be cathodic with respect to the surrounding metallic matrix [4]. This can result in preferential anodic dissolution of the metallic matrix close to or at the matrix/carbide interface and thereby accelerate carbide removal from surfaces and reduce the anti-wear properties of the surface.

Erosion is one of several wear modes involved in tribocorrosion. Solid particle erosion is a process by which discrete small solid particles, with inertia, strike the surface of a material causing damage or material loss to its surface. This is often accompanied by corrosion due to the environment and is termed erosion–corrosion. A major environmental factor with significant influence on erosion–corrosion rates is that of flow velocity but it should be set in context of the overall flow field as other parameters such as wall shear stress, wall surface roughness, turbulent flow intensity and mass transport coefficient (this determines the rate of movement of reactant species to reaction sites and thus can relate to corrosion wall wastage rates). A critical flow velocity is often quoted to represent a transition from flow-induced corrosion to enhanced mechanical-corrosion interactive erosion–corrosion processes and used to indicate the resistance of the passive and protective films to mechanical breakdown [5].

6.2 Particle–Surface Interactions

Corrosion is typically an electrochemical process involving electron transfer at the solid/liquid interface needs reactants at the surface and it will produce corrosion products. When a corrosion rate is partially or wholly controlled by mass transfer of reactants to or products from the surface, then local conditions under erosion may well influence the mass transfer kinetics (measured by the mass transfer coefficient k_m). Under such conditions, the corrosion will be controlled by the mass transfer (typically diffusion processes) and the driving concentration gradient (relative concentrations of active species near surface compared to free stream concentrations) [6]. Both the mass transfer and concentration gradient will be affected by solid particle impingement that influence the local fluid flow field and increase surface roughness. Erosion will also increase the surface area wetted by the corrosive electrolyte and could establish micro-galvanic cells on the surface with damaged areas being anodic to the passive and (cathodic) unaffected areas. Erosion will also increase the dislocation density in affected surface areas, which can lead to potential differences between eroded and less eroded areas being established and the formation of anodic and cathodic sites generating a micro-galvanic effect enhancing material loss from the surface. It is also likely that the corrosion kinetics will be altered on eroded areas.

6.3 Depassivation and Repassivation Kinetics

As suggested above, depassivation of film surface is associated with the flow velocity. Figure 6.1 shows the trend for such a passive system as a function of flow velocity and its vulnerability to mechanical processes, which can accelerate corrosion loss rates due to depassivation.

6.3.1 *Depassivation*

Depassivation in erosion–corrosion processes is assumed to be instantaneous and associated with the mechanical removal or stripping of the passive surface layers through particle, cavitation bubble or liquid droplet impact. Figure 6.2 illustrates the level of damage to passive films as a function solid particle impingement angle. Mechanical removal and/or rupture of the passive layer

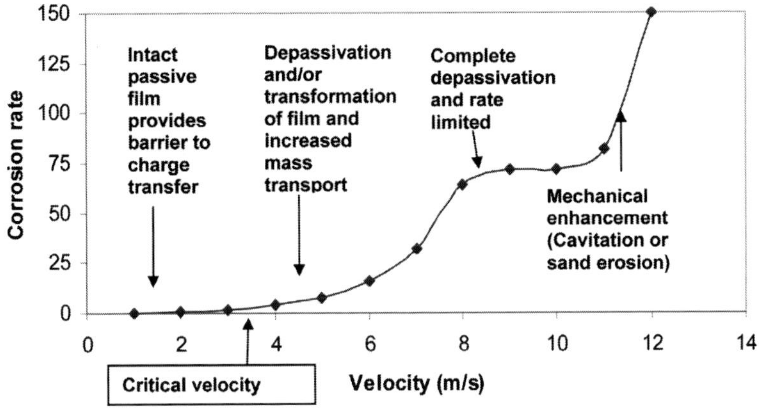

Figure 6.1 Illustrates possible corrosion trends (with arbitrary units) for a passive metallic surface [7].

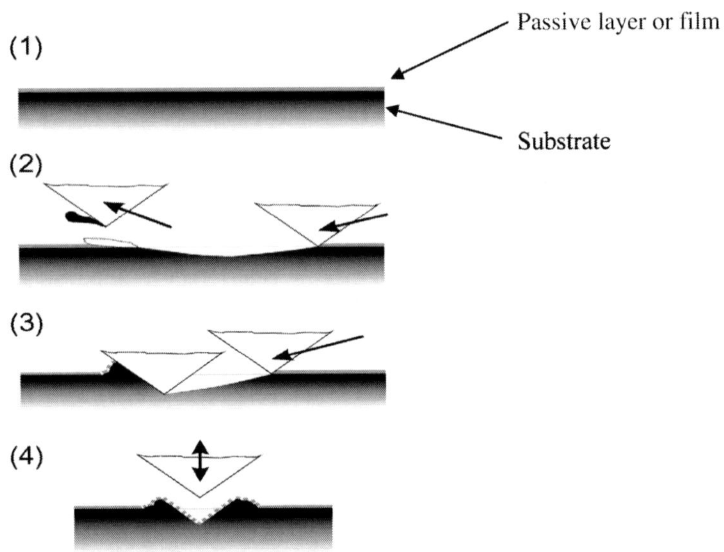

Figure 6.2 Schematic showing that the level of passive layer damage is dependent on particle impact angle [7].

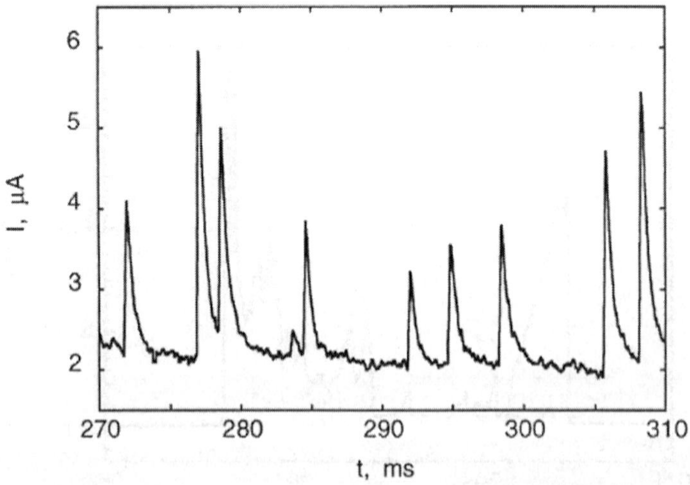

Figure 6.3 Current transients in erosion–corrosion due to depassivation/repassivation [8].

enables charge transfer to proceed at varying rates and parent metal dissolution is likely. 'Recovery' or repassivation aspects of the passive film are therefore important but are very system dependent.

Figure 6.3 shows the current response over time for individual solid particle impacts on the naturally passivating system of stainless steel. Some impacts will only result in partial passive layer removal or cracking, which will influence repassivation kinetics and possibly the composition and thickness of the regrown layer.

Clearly over large surface areas, depassivation is localised and thus not all the film area will be ruptured or removed. Clearly, a key aspect for materials and coatings is to not only to resist the mechanical aspects of erosion that includes plastic deformation at high strain rates but also to repassivate quickly thereby limiting dissolution of material and limiting corrosion loss.

Figure 6.4 is an attempt to map surface responses to the whole range of erosion and corrosion mechanisms from erosion dominated through erosion–corrosion, corrosion–erosion zones to corrosion dominated mechanisms. To identify what zone a particular application may be operating under is difficult without some data on erosion and corrosion performance of the surface in

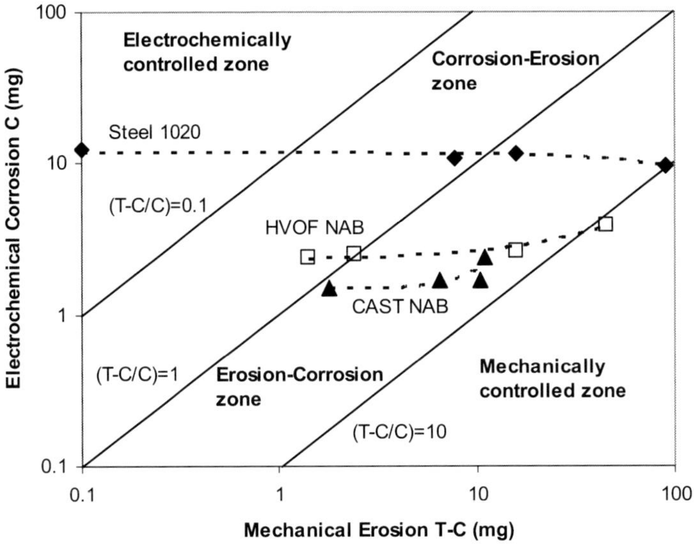

Figure 6.4 Mapping electrochemical material loss against mechanical erosion rates for a non-passivating surface carbon steel (AISI 1020) along with two potentially passivating surfaces of NAB, one that has been thermally sprayed by high-velocity oxy-fuel deposition as a coating on carbon steel and the other that has been cast. The results have been obtained by jet impingement erosion–corrosion tests. Reprinted from Ref. [4], Copyright (2008), with permission from Elsevier.

question. However, the sand particle energy is directly related to the likely mechanical erosion damage rates given by $T-C$, where T is the total loss from wear-corrosion and C is the corrosion loss. This will define a position on the $T-C$ axis of Fig. 6.4 and the rate of depassivation as is likely to be linked to the kinetic energy of the sand particle E_k. Surface repassivation, on the other hand, is material-electrolyte-flow field specific and this will dictate the corrosion rate C. Three systems are plotted on this mechanical versus corrosion map. Two are passivating systems based on cast and high velocity oxygen fuel (HVOF) nickel aluminium bronze (NAB) surfaces in 3.5% NaCl solution and have a positive slope (increase with increasing mechanical erosion effects). This increase is related to repassivation kinetics being unable to repassivate the eroded surface under increasing erosion. The third is carbon steel in saline solution, which is not passivating and has a slope near to zero or slightly negative. Here, the corrosion products formed on the surface are non-adherent and easily removed and the trend is dominated by erosion rate while the corrosion rate remains relatively constant. These results were obtained using a free jet impingement erosion–corrosion rig.

6.3.2 *Repassivation Rate*

Exploring the repassivation rate of materials and coatings for tribocorrosion systems is therefore vital. However, little work has been reported and limited test facilities are able to measure these. An attempt has been made by the author and a visiting scientist, Maggy Sun, from Quindío University of Technology in China. A set of reciprocating wear tests rubbing a 6 mm diameter alumina ball against cast NAB in 3.5% NaCl solution at room temperature has been conducted with a stroke length of 21.5 mm between 1–12 Hz frequency and 5–35 N applied load. The electrochemical response is presented in Fig. 6.5. The test comprised of a 1 hr static immersion period with the pin loaded onto the test surface without reciprocation, which was followed by a 1 hr wear-corrosion test. As seen, the level of increased current is dependent on a combination of load and speed. Figure 6.6 shows currents from a set of experiments conducted at 35 N but with varying frequencies. Clearly, the corrosion current increases with increasing frequency but this increase is non-linear. A transition from relatively low current between 1–5 Hz and higher currents above 10 Hz is seen. The interesting trend is seen at 8 Hz where initially the current is low (40 μA) but rises steadily throughout the wear test period to 80 μA. This indicates a transition between predominantly passivated to depassivated conditions in the wearing contact. This frequency coincides with the expected repassivation time for this material of between 150–200 ms. It is also unclear whether trends predicted for depassivated surfaces hold here. Equation (6.1) predicts wear current for non-passivating surfaces and shows a dependence on load times the square root of frequency [9]. However, such trends are neither seen here nor between wear volume or friction as a function of frequency. Hence, temporal effects of repassivation are

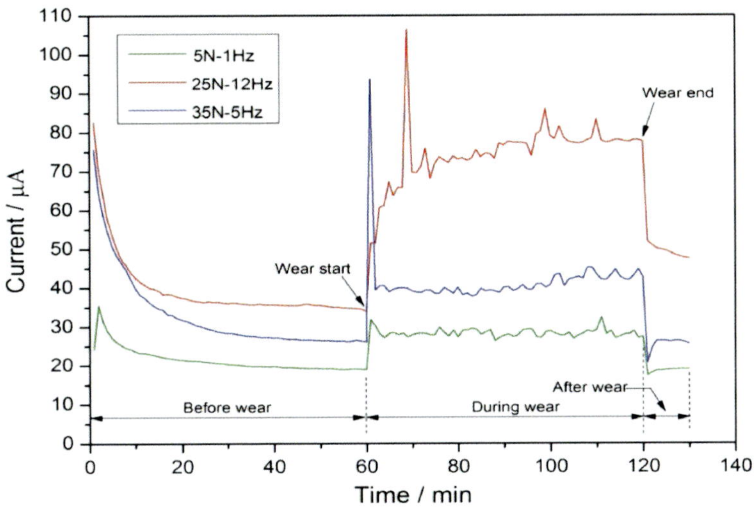

Figure 6.5 Sliding wear-corrosion of NAB in 3.5% NaCl solution, Stroke 21.5 mm, 6 mm diameter alumina ball, Reference electrode: Ag/AgCl, Counter Electrode graphite, Static 60 min, wear-corrosion 60 min and 10 min post wear-corrosion [10].

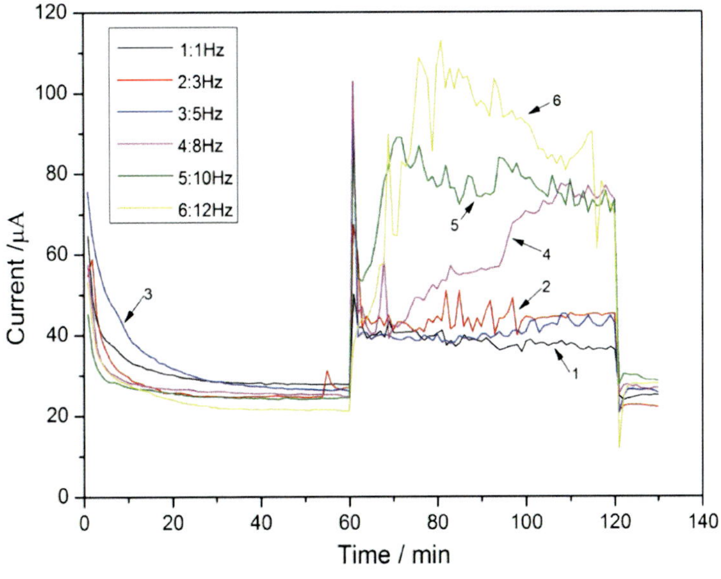

Figure 6.6 Sliding wear-corrosion of NAB in 3.5% NaCl solution at 35 N and at room temperature [10].

clearly important but wear mechanisms and friction levels are also varying to complicate analysis but illustrate the complex nature of tribocorrosion of sliding couples. It must be noted that the passive film mentioned here is not fully developed as on NAB this takes up to 3 months in static (i.e., no flow) seawater conditions.

This work shows the importance of a full appraisal of materials to be used in tribo systems, where corrosion is a factor. It also requires a good understanding of the contact conditions.

$$I_r = A_r i_r = K_w l f \sqrt{\frac{F_n}{H}} \int_0^{1/f} i_a dt \qquad (6.1)$$

where
wear accelerated current $= I_r$
real repassivation area $= A_r$
real repassivation current density $= i_r$
sliding distance $= l$
sliding frequency $= f$
normal load $= F_n$
hardness of test material $= H$
proportionally constant $= K_w$
anodic current density $= i_a$

6.4 Models and Mapping

Erosion damage by solid particle impact or cavitation bubble collapse to an oxide or passive film will reveal the underlying nascent (highly reactive and passive film free) surface inducing a higher activity (higher corrosion current), for a limited duration, than for the intact oxide surface. Bozzini et al. [11] employed a simple approximate model using a 'recovering target' concept. This erosion–corrosion model has the advantage in that it can be applied to both passivating and actively corroding conditions. The impacting particles are assumed to be rigid mono-dispersed spheres of radius r_p. The particle impact process is assumed to be Poissonian with parameter λ (impacts per square metre per second). It is assumed that each impact gives rise to an alteration of the corrosion rate through a localised change in corrosion current density for a period of time, relating to a recovery to the unaffected state. The effective corrosion current density, i_{corr} (nA cm^{-1}), at a given electrode potential (typically the corrosion potential) can be related to the mechanically affected corrosion component of the synergistic damage through a coefficient, f_a, (such that $0 \leq f_a \leq 1$) expressing the fraction of the corroding surface, which is affected by the erosive action of impinging particles, by Eq. (6.2):

$$i_{corr} = f_a i_a + (1 - f_a)\, i_u \tag{6.2}$$

where the subscripts 'a' and 'u' stand for 'affected' and 'unaffected', respectively. The current densities i_a and i_u are characteristic for the corroding material in the absence and in the presence of the erosive action and can be measured separately by means of suitable experiments. In general, the coefficient f_a can be defined in Eq. (6.3):

$$f_a = \left(\frac{\text{no. of impacts}}{\text{control area}}\right) \times \left(\frac{\text{damaged area}}{\text{impact}}\right) \times \text{recovery time} = \lambda A_a \tau \tag{6.3}$$

where A_a = affected surface area, τ = passive recovery time

Equation (6.2) indicates an accelerating element to corrosion caused by film damage or removal and is one element of the complex interactions between erosion and corrosion that needs to be understood.

These interactions can be defined as follows. The total damage under wear-corrosion, T, can be represented as

$$T = E + C + S \tag{6.4}$$

Where E is the pure erosion material loss, C is the solids free flow corrosion rate [possibly derived from the corrosion current density i in Eq. (6.2)]. The synergistic effect (interactive term), S, is referred to as ΔE_c or $(\Delta C_e + \Delta E_c)$ depending on the literature source where ΔE_c is the enhanced erosion loss due to corrosion and ΔC_e is the enhanced corrosion due to erosion. The term ΔC_e can be partially expressed by Eq. (6.2) but additional terms are required relating to the effect of the erodent deforming the surface leading to increased corrosion activity. The S terms and how they should be measured

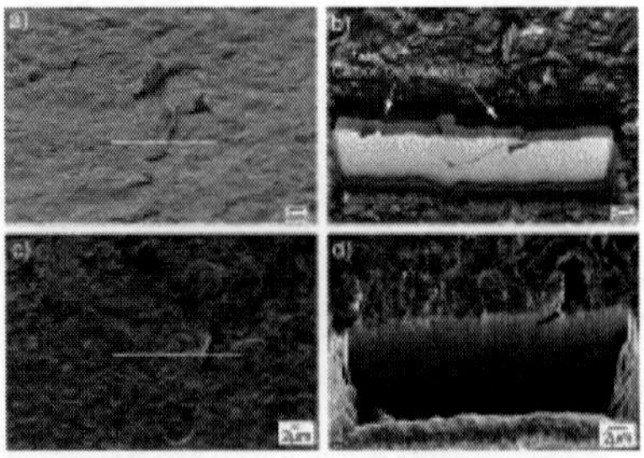

Figure 6.7 Secondary electron images (a and c) of erosion–corrosion surfaces and corresponding location (white line) of the focused ion beam (FIB) cross-sections shown in (b) and (d). Reprinted from Ref. [14], Copyright (2013), with permission from Elsevier.

are given by the ASTM G119-93 standard, which is a useful guide to evaluate synergy [12]. It is also important to note that the synergistic term S can be positive or negative. Bozzini et al. [11] shows that annealed carbon steel has a more active corrosion potential before it was work hardened resulting in $i_u > i_a$ and by using Eq. (6.2) shows a reduction (i.e., a negative synergy) in overall corrosion rate with erosion present.

It has also been shown that synergistic effects, which result in the damage due to separate corrosion and erosion processes are normally greater than the sum of the individual damage processes and can accelerate material removal significantly (i.e., $T = 50 \times (C + E)$ for grey cast iron), see Wood and Hutton and Fig. 6.1 [13].

However, the ability to predict or map erosion–corrosion relies on including the correct mechanisms and material responses (i.e., phase transformations).

The use of modern advanced surface analysis techniques such as FIB and tunnelling electron microscopy (TEM) allows detailed sub-surface characterisation of eroded samples at the nanoscale [14]. Slurry pot erosion and erosion–corrosion testing of cylindrical samples of 316 have been conducted in abrasive silica slurry at 7 m s^{-1} velocity for 1 hr Conventional scanning electron microscopy (SEM) showed both erosion and erosion–corrosion surfaces exhibited similar surface features including impact crater, particle embedment and crater lip formation. FIB cross-sectioning of similar areas from both surfaces revealed a nanograin surface layer as well as sub-surface cracks, attributed to flow of material from the edge of impact crater lips. Site-specific FIB–TEM lamella samples were created from cross-sections and showed how subtle differences existed as a result of corrosion activity. Selected area diffraction patterns revealed how particle impact was responsible for the formation of martensite from the austenite microstructure. The presence of a corrosive NaCl solution resulted in dissolution of the martensitic phase and a lamellar microstructure. The decrease in the volume fraction of martensite was believed to be responsible for reducing the work hardening rate and increasing elongation to failure, resulting in longer lip formation and ultimately material removal. This work shows how subtle nanoscale microstructural changes can influence erosion–corrosion rates and why such surface damage accumulation mechanisms need to be taken into account when developing theoretical models to predict material loss.

6.5 Negative Synergy: A Design Requirement for Multifunctional Tribocorrosion Surfaces

6.5.1 *Abrasion–Corrosion*

More complex and sometimes negative synergistic interactions between wear and corrosion can occur. An example is given below of abrasion–corrosion on a WC/Co/Cr composite surface deposited by thermal spray and often used in the offshore industry [4]. Here, preferential corrosion at the carbide/binder interface resulted in a 20% reduction in abrasion–corrosion wear rate compared to purely abrasive conditions. The physical mechanism of this negative synergy is thought to relate to the preferential corrosion of metallic tungsten at the carbide/binder interface, resulting in the hard carbide and wear-resistant phase being loosened and making the contact a more compliant surface under abrasion conditions reducing contact stresses that in turn reduce material loss, see Figs. 6.8 and 6.9.

In situ electrochemical measurements made during abrasion-corrosion tests are useful as they record the kinetics of the reactions occurring and as well inform on the degree of depassivation and

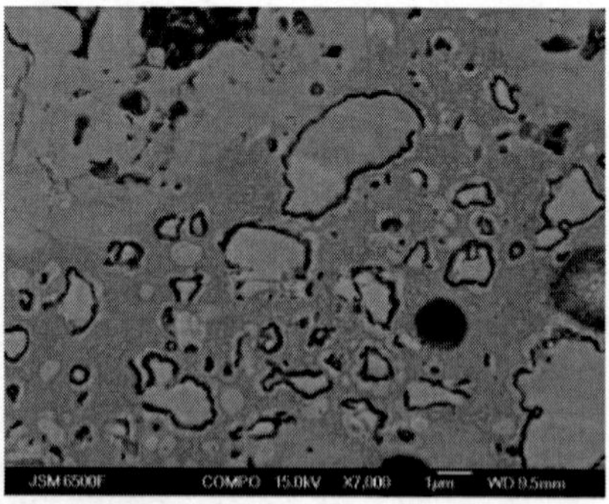

Figure 6.8 Selective corrosion of W and W_2C phases (which are formed due to decarburisation of WC) around the periphery of the carbide grain when exposed to an alkaline environment of pH 11 for 168 hr [15].

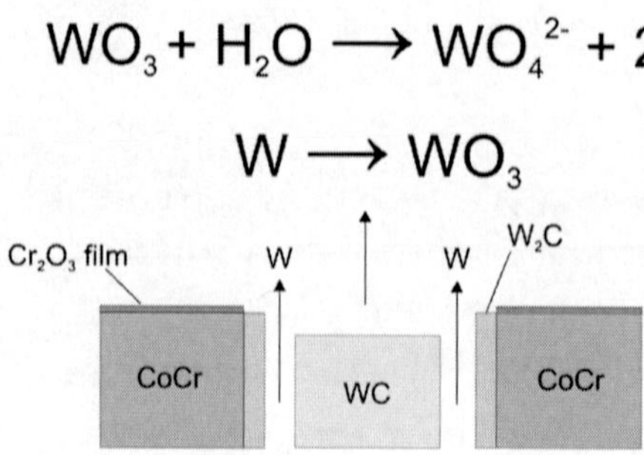

Figure 6.9 Schematic showing the selective corrosion of W and W_2C phases in a WC-10Co-4Cr coating [15].

repassivation. Figure 6.9 is a typical example of such measurements made from a micro-abrasion tester where a rotating ball is covered in corrosive and abrasive slurry and rotates against a test sample creating a wear dimple or scar. The measurements shows the open circuit potential (OCP) and current recorded during repeat tests under micro-abrasion at pH 11 on a thermally sprayed tungsten carbide coating. The potential drops by about 60 mV over a 250 s period once ball rotation and abrasion commences whilst the current instantly increases from low freely corroding levels to 2 μA. Both these trends show that depassivation is occurring in the wear scar area. On stopping rotation of the ball, and thus abrasion, at

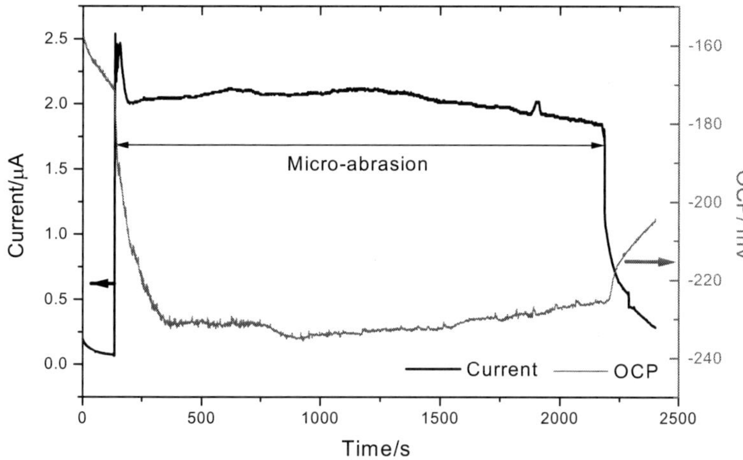

Figure 6.10 OCP and current as a function of time during micro-abrasion of a 250 μm thick D-Gun WC-10Co-4Cr coating on stainless steel 316 for a sliding distance of 90 m using 0.16 volume fraction slurry of 4.5 μm SiC abrasives suspended in NaOH solution of pH 11. The applied load was 0.2 N and a Zirconia ball of 25.5 mm diameter was rotated to give a sliding velocity of 0.05 m s^{-1} [15].

2200 s, the potential recovers in the electropositive direction and the current decreases towards the initial freely corroding levels, indicating repassivation of the wear scar.

6.6 Multifunctional Tribocorrosion Resistant Surfaces for Use within the Body

6.6.1 *Metal-on-Metal Hip Joints*

Researchers are becoming increasingly interested in metal-on-metal joint implants which have just been banned by the NHS in the UK with studies being focused on their wear and corrosion (ion release) behaviour. The corrosion, wear and wear-corrosion behaviour for three materials (high-carbon CoCrMo, low-carbon CoCrMo and UNS S31603) have been discussed by Yan et al. [16]. In the steady-state regime, 20–30% of the material degradation was attributed to corrosion-related damage. High-carbon CoCrMo showed excellent corrosion, wear and corrosion-wear resistance and therefore delivered the best overall performance in terms of a lower wear rate, a lower friction coefficient and a higher resistance to corrosion. The proteins in the joint fluids are thought to play a key role.

Sinnett-Jones et al. [17] investigated the synergistic effects of corrosion and wear of surgical grade cast F-75 cobalt-chromium-molybdenum (CoCrMo) alloy. Micro-abrasion-corrosion identified depassivation and repassivation processes and showed strong synergistic effects ranging from negative to positive. The synergistic levels appear to be dependent on the integrity of the passive films and the repassivation kinetics.

The role of protein in the tribocorrosion of metal implants is not well understood but initial studies show conflicting trends with corrosion either being enhanced or reduced when protein is present and adsorbed onto the articulating surfaces. Sun [18] has studied abrasion–corrosion of cast CoCrMo in saline [0.9% NaCl and phosphate buffer solution (PBS)] and protein containing solutions of bovine serum (BS). In situ current noise results, see Fig. 6.11, show reduced current levels when proteins are present.

Using the same test geometry, but removing the abrasives, so the alumina ball abrades the CoCrMo surface under sliding wear mode, again shows that the presence of protein alters and reduces the wear mechanism and rate as seen in Fig. 6.12.

6.6.2 *Dental Materials*

Titanium (Ti) has been widely used for implant applications due its excellent biocompatibility and high corrosion resistance. However, Ti has poor tribological properties with low wear resistance. For example, dental implants are often subjected both to a corrosive environment and cyclic micro-movements at the implant/bone interface, becoming part of a tribocorrosion system. Tribocorrosion is characterised by the migration of wear debris and corrosion products to peri-implant sites, inducing inflammatory responses and implant loosening [19]. Further, the lack of osseointegration is also one of the most frequent causes of implants failure [20]. Hence, recent work has focused on modifying the surface features of Ti in order to produce bio-multifunctional surfaces able to display improved tribocorrosion and biological performances.

Commercially pure (cp)-Ti grade 2 were treated by plasma electrolytic oxidation (PEO). A commercial electrolyte containing NaF and a Ca- and P-based solution (Cβ) were used for PEO treatments. Reciprocating sliding tests were performed in artificial saliva at 37°C. The tribocorrosion behaviour was analysed by OCP measurements. Biological assays were conducted in order to assess the best tribocorrosion resistant surfaces and their interactions with MG63 bone cells. The osteoblastic cells morphology and viability were evaluated for 1, 3 and 7 days of culture. The results show PEO treatments greatly improved the tribocorrosion performance of Ti and the electrolyte composition appears to play a crucial effect by decreasing the tendency of corrosion and mechanical damage. PEO generated with Cβ solution surface shows significantly improved tribocorrosion and biological performance of cp Ti and are thus very attractive candidates for implant applications.

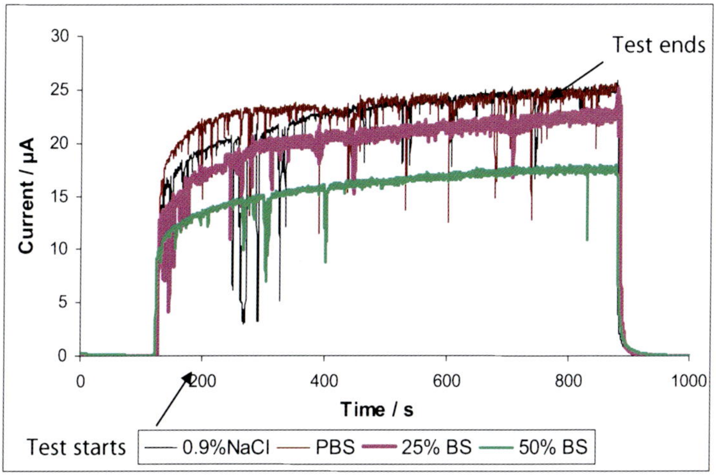

Figure 6.11 Electrochemical current noise obtained in microabrasion-corrosion test on as-cast CoCrCo (1.0 g cm^{-3} SiC) in 0.9% NaCl, PBS solutions, 25% and 50% BS at 37°C. Reprinted from Ref. [17], Copyright (2005), with permission from Elsevier.

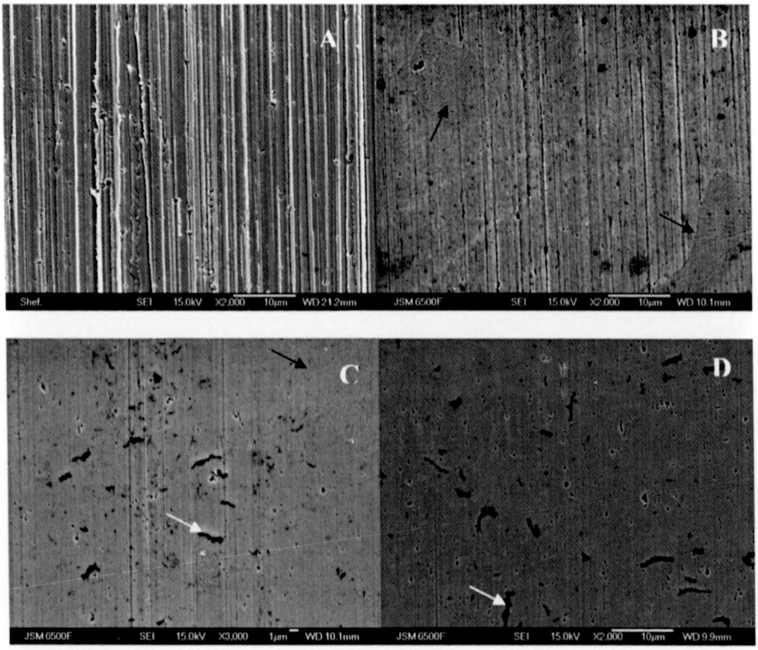

Figure 6.12 SEM micrographs of wear scars on as-cast CoCrMo under pure sliding conditions in (a) 0.9% NaCl, (b) PBS, (c) 25% BS, (d) 50% BS at 37°C. Black arrows show a carbide standing proud of the surface. The white arrows show possible protein and wear debris on the worn surfaces. Polished wear scars were produced in proteinaceous solutions indicating proteins form a lubrication film, which protect the surface. Reprinted from Ref. [18], Copyright (2009), with permission from Elsevier.

6.7 Decorative Coatings

Tribocorrosion of decorative single-layer zirconium oxynitride ZrO_xN_y films (between 3–6 μm thick) have been studied recently [21]. Such films have potential use on watches, writing instruments and spectacle frames. The presence of oxygen allows the film properties to be tailored between those of metallic nitrides, MN_y and those of the corresponding insulating oxides MO_x. Altering the metallic/covalent and ionic bonding characteristics allows the mechanical, electrical, chemical and optical properties of such materials, including their colour, to be tuned. The performance of such films is dependent on the deposition method and on both the concentration and distribution of the nitrogen atoms within the matrix. Reciprocating tribocorrosion tests using an alumina pin in an artificial sweat solution of pH 4.5 show that the preferential dissolution of the M2 tool steel substrate governs the degradation, which typically occurs by delamination. This highlights the dangers of the use of thin coatings which are, or become, porous during tribo-contacts and subsequently allow substrate corrosion.

Subsequent work by Mathew [22] investigated the tribocorrosion behaviour of single-layered titanium oxycarbide, TiC_xO_y, thin films on a reciprocating sliding tribometer, and in the presence of artificial sweat solution at room temperature. Again, these films are being considered as decorative coatings on various components. The films were produced by direct current (DC) reactive magnetron sputtering. The corrosion studies demonstrated higher corrosion resistance compared to ZrO_xN_y-based films. Films with an oxygen fraction of 0.55 ($TiC_{0.36}O_{0.45}$) and 0.79 ($TiC_{0.40}O_{1.52}$) demonstrated the better tribocorrosion resistance. The mechanisms of the wear–corrosion process were found to be influenced by the wear particles and corrosion products at the contact zone. The tribocorrosion process was found to be dominated by mechanical processes.

6.8 Conclusions

As demonstrated within this chapter, material performance under tribocorrosion conditions is highly system dependent and modest changes in environmental conditions can have significant impact on wastage rates and friction levels. The discussion throughout this whole chapter also confirms that our level of understanding of the mechanism involved and controlling parameters is too low to enable informed material or coating selection for such aggressive duties. The multifunctional aspects required of surfaces subjected to tribocorrosion have been outlined and examples given where just considering wear or corrosion is now not good enough for modern designs need added functionality such as biocompatibility, decorative stability, self-healing (quickly repassivating) and self-sensing. Therefore selection must be accompanied by either relevant experience of good material performance under similar conditions or relevant experimental results from carefully constructed laboratory testing that has proven to simulate field conditions to allow material and coating screening. Some design principles are emerging to generate negative synergy surfaces but designer systems for multifunctional tribocorrosion surfaces are likely still a decade away.

Acknowledgements

Thanks goes to nCATS and in particular for Dr. Julian Wharton and Dr. Terry Harvey for their help with illustrations and discussions that informed this work.

References

1. Oxford English Dictionary Online, Oxford University Press (2006).
2. H.C. Meng, K.C. Ludema, Wear models and predictive equations: their form and content, *Wear* 181–183 (1995), 443–457.
3. F. Gao, H. Liang, Material removal mechanisms in electrochemical-mechanical polishing of tantalum, *Electrochimica Acta* 54 (2009), 6808–6815.
4. M.R. Thakare, J.A. Wharton, R.J.K. Wood, C. Menger, Exposure effects of strong alkaline conditions on the microscale abrasion–corrosion of D-gun sprayed WC–10Co–4Cr coating, *Tribol. Int.*, 41(7) (2008), 629–639.
5. X. Hu, A. Neville, The electrochemical response of stainless steels in liquid-solid impingement, *Wear* 258 (1–4) (2005), 641–648.
6. D.C. Silverman, The rotating cylinder electrode for examining velocity-sensitive corrosion—A review, *Corrosion* 60 (11) (2004), 1003–1023.
7. R.J.K. Wood, Erosion-corrosion, in Milne, I., Ritchie, R.O., and Karihaloo, B. (eds.), *Comprehensive Structural Integrity*, Vol. 6 (2007), 395–427.
8. K. Sasaki and G.T. Burstein, Observation of a threshold impact energy required to cause passive film rupture during slurry erosion of stainless steel, *Phil. Mag. Lett.*, 80(7) (2000), 489–493.
9. S. Mischler, S. Debaud, D. Landolt, Wear-accelerated corrosion of passive metals in tribocorrosion systems, *J. Electrochem. Soc.* 145 (March (3)) (1998) 750–758.

10. R.J.K. Wood, Chapter 15: Tribocorrosion, In *Developments of Electrochemistry*, D. Pletcher, Z.-Q. Tian and D Williams (eds.), Wiley (2014), 281–307.

11. B. Bozzini, M.E. Ricotti, M. Boniardi, C. Mele, Evaluation of erosion-corrosion in multiphase flow via CFD and experimental analysis, *Wear* 255 (2003), 237–245.

12. ASTM Standard G 119-93, Standard Guide for Determining Synergism between Wear and Corrosion, (1993).

13. R.J.K. Wood, S.P. Hutton, The synergistic effect of erosion and corrosion: trends in published results, *Wear* 140 (1990), 387–394.

14. R.J.K. Wood, J.C. Walker, T.J. Harvey, S. Wang, S.S. Rajahram, Influence of the microstructure on the erosion and erosion-corrosion characteristics of 316 stainless steel, *Wear* 306 (2013), 254–262.

15. R.J.K. Wood, *Tribocorrosion*, Volume 2, Chapter 15, Shreir's Corrosion, Elsevier, (2010), 1005–1050, http://dx.doi.org/10.1016/B978-044452787-5.00041.

16. Y. Yan, A. Neville, D. Dowson, Biotribocorrosion—an appraisal of the time dependence of wear and corrosion interactions: I. The role of corrosion *J. Phys. D Appl. Phys.* 39(15) (2006), 3200–3205.

17. P.E. Sinnett-Jones, J.A. Wharton, R.J.K. Wood, Micro-abrasion-corrosion of a CoCrMo alloy in simulated artificial hip joint environments, *Wear* 259 (2) (2005), 898–909.

18. D. Sun, J.A. Wharton, R.J.K. Wood, L. Ma, W.M. Raithforth, Microabrasion-corroison of cast CoCrMo alloy in simulated body fluids, *Tribol. Int.* 42(1) (2009), 99–110.

19. S.A. Alves, R. Bayón, V. Saénz de Viteri, M.P. Garcia, A. Igartua, M.H. Fernandes, L.A. Rocha, Tribocorrosion Behavior of Anodic Titanium Oxide Films and Assessment of Cell-Materials Interactions, *eCM J.*, 26(6) (2013), 104.

20. M.T. Mathew, V.A. Barão, J.C.-C. Yuan, W.G. Assunção, C. Sukotjo, M.A. Wimmer, What is the role of lipopolysaccharide on the tribocorrosive behavior of titanium?, *J. Mech. Behav.Biomed. Mater.* 8 (2012), 71–85.

21. S.C. Ferreira, E. Ariza, L.A. Rocha, J.R. Gomes, P. Carvalho, F. Vax, A.C. Fernandes, L. Rebouta, L. Cunha, E. Alves, Ph. Goudeau, J.R. Riviere, Tribocorrosion behaviour of ZrOxNy thin films for decorative applications. *Surf. Coat. Technol.* 200 (2006), 6624–6639.

22. M.T. Mathew, E. Ariza, L.A. Rocha, A.C. Fernandes, F. Vaz, TiCxOy thin films for decorative applications: Tribo-corrosion mechanisms and synergism. *Tribol. Int.* 41 (2008), 603–615.

Chapter 7

Multifunctional Nanostructured Metallic Coatings by Electrodeposition

Chee Tong John Low and Frank C. Walsh

Electrochemical Engineering Laboratory, Faculty of Engineering and the Environment, University of Southampton, Southampton, SO17 1BJ, UK

c.t.j.low@warwick.ac.uk

This chapter illustrates the use of electrodeposition to produce multifunctional nanostructured metallic coatings. Such coatings can provide several engineering functionalities, including wear and corrosion resistance to provide an effective tribological performance. This chapter is divided into three parts. The first part provides a concise overview of the principles of electroplating and anodising. The next part explains the strategies adopted to electroplate nanostructured metallic coatings, which include (a) grain refinement, (b) application of a modulated current, (c) inclusion of nanoparticles into the coating and (d) use of nanoporous templates. The final part provides examples of multifunctional tribological properties of coatings illustrated by research findings from our laboratory and the literature. Multifunctional nanostructured metallic coatings include (a) nanocrystalline, (b) functionally graded and (c) nanocomposite types. Such coatings can provide improved corrosion and wear resistance, decreased friction and longer service lifetimes.

Multifunctional Materials for Tribological Applications
Edited by Robert J. K. Wood

ISBN 978-981-4463-90-4 (Hardcover), 978-981-4463-91-1 (eBook)
www.panstanford.com

List of Symbols

d	=	grain size [nm]
f	=	frequency of pulsed current [s^{-1}]
F	=	Faraday's constant [C mol^{-1}]
I_a	=	average pulsed current [A dm^{-2}]
I_p	=	peak pulsed current density [A dm^{-2}]
H	=	hardness [Hv]
j	=	current density [A cm^{-2}]
k_σ	=	strengthening coefficient [dimensionless]
k_H	=	strengthening coefficient [dimensionless]
M_{Ni}	=	molar mass of nickel [g mol^{-1}]
n	=	constant in Hall–Petch equation [dimensionless]
t	=	time [s]
t_a	=	pulsed reverse anodic current duration [s]
t_c	=	pulsed reverse cathodic current duration [s]
t_{on}	=	pulsed current at-time [s]
t_{off}	=	pulsed current off-time [s]
x	=	deposit thickness [cm]
z	=	number of electrons [dimensionless]
ρ_{Ni}	=	density of nickel [g cm^{-3}]
ϕ	=	current efficiency for nickel deposition [dimensionless]
σ	=	yield stress [Pa]
γ	=	duty cycle for pulsed current [dimensionless]

7.1 The Principles of Electroplating and Anodising

Electrochemical methods have the scope to form a wide variety of coatings. Such methods may include (a) cathodic, (b) anodic and (c) mixed electrode deposition.

Cathodic deposition via electron gain can include

- electrodeposition of metals and alloys [1],
- electrodeposition of composites containing two different phases [2],
- electroforming of metal onto a shaped structure [3],
- high-voltage electro-spinning of fibres [4].

Anodic deposition via electron loss can involve [5]

- anodic electrodeposition of a metal oxide [6],
- high-voltage plasma electrolytic oxidation to form oxides [7],
- electrophoretic deposition of coatings [8].

Mixed electrode deposition can include

- electroless deposition of metal [9],
- deposition of conversion coatings such as chromates or phosphates [10].

This chapter focuses on electroplating; other electrochemical surface treatment methods are considered elsewhere [11, 12]. Electroplating is an electrochemical process that involves the application of an electrical current to deposit coatings via electron transfer. The coatings can be, for example, metals, alloys, composites, metal oxides or conducting polymers. Electroplating refers to the electrodeposition of smooth, adherent and compact coatings submicron to millimetres in thickness.

Electroplating is a wet-chemical processing method, in which soluble electroactive metal ions are cathodically or anodically electrodeposited onto an electrically conductive substrate. Five essential components are conventionally required for the electroplating of metals:

(1) A cathode work-piece to be electroplated, that is, a substrate
(2) An anode, which may be soluble or insoluble
(3) An external electrical connection (via a power supply)
(4) An ionically conductive liquid solution, that is, an electrolyte
(5) Soluble electroactive metal ions to be plated, the cathodic reactant

Figure 7.1(a) shows the essential components of an electroplating process while Fig. 7.1(b) shows anodising of a metal to form a metal oxide coating. An example is the anodising of titanium to form titanium dioxide nanotubular arrays, via simultaneous electrochemical formation and chemical dissolution of oxides, as described later in this chapter

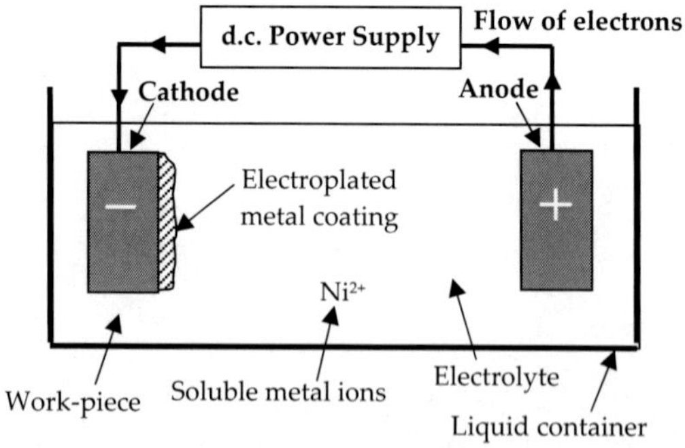

(a)

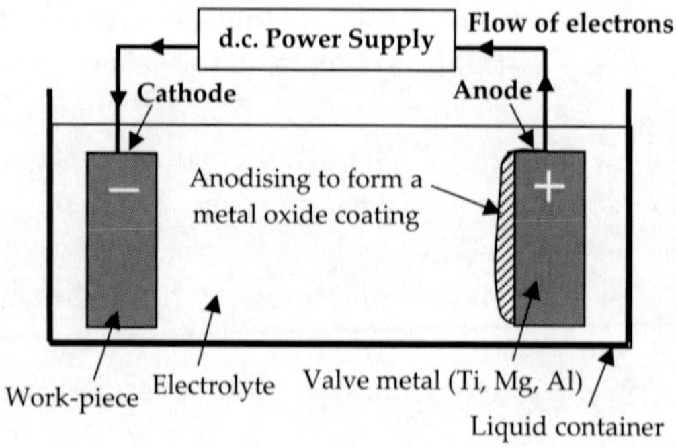

(b)

Figure 7.1 (a) The electroplating of a metal coating; (b) the anodising of metal to form a metal oxide coating.

For the electroplating of metal, in the presence of an external electrical current, soluble metal ions in an aqueous bath are electroplated as a solid metal film. The work piece to be electroplated is the cathode. The anode can be soluble, that is, made of the metal to be electroplated or insoluble, that is, an inert, electrically conductive material. Both electrodes are immersed in an ionically conducting electrolyte (the bath), which contains electroactive metal ions to be electroplated together with a mixture of additives and surfactants to provide appropriate bath chemistry.

Electrolyte additives and surfactants are multifunctional but can be grouped according to their main purpose, as

(1) brighteners to produce a matte, semi-matte or bright surface finish,
(2) surface wetters to reduce surface tension between the solid work-piece and the liquid electrolyte, to reduce coating porosity and aid the free detachment of gas bubbles,
(3) inhibitors or levellers to inhibit to growth of metal deposits in regions of a highly localised current density or promote the growth of metal deposits in regions of low localised current density,
(4) buffers to control localised changes in surface pH and help prevent deposition of surface oxides and hydroxides,
(5) conductive electrolytes to improve ionic conductivity, and
(6) chemical complexants to enable the formation of complex metal ions for the electroplating of alloys.

Surfactants can provide functionality and act as wetting agents, emulsifiers, foam suppressants or dispersants. They can be divided into (a) hydrocarbon surfactants and (b) fluorinated surfactants. Surfactants consist of a hydrophobic tail and hydrophilic head. They may be cationic, anionic, non-ionic or mixed. Fluorinated surfactants are synthetic organofluorine compounds containing a fluorinated tail and a hydrophilic head, which are more effective in lowering the surface tension of water than hydrocarbon surfactants. The stable carbon–fluorine bond in fluorinated surfactants makes it a more chemically stable surfactant since hydrocarbon surfactants can be electrochemically oxidised or reduced.

Switching on the external power supply will drive electrochemical reactions as follows: (a) at the cathode, the metal ions in solution will be transported to the surface of the cathode then reduced to form metallic nickel, and (b) at a soluble anode, which can oxidize to a higher oxidation state, either forming oxides or soluble cations or oxygen can evolve at an inert anode.

The following considers nickel electroplating from an acid sulphate (usually known as a 'Watts Nickel') bath:

At the cathode
Electroplating of nickel

$$Ni^{2+} + 2e^{-} \rightarrow Ni \tag{7.1}$$

is accompanied by hydrogen evolution as a secondary reaction

$$2H^+ + 2e^- \rightarrow H_2 \tag{7.2}$$

At the anode

Oxygen evolution at an inert, insoluble electrode

$$2H_2O - 4e^- \rightarrow O_2 + 4H^+ \tag{7.3}$$

or dissolution of nickel from a soluble electrode

$$Ni - 2e^- \rightarrow Ni^{2+} \tag{7.4}$$

At the cathode, Ni^{2+} ions are reduced to metallic nickel. At the anode, a soluble nickel electrode is oxidized to Ni^{2+}. These electrode reactions are often accompanied by secondary reactions such as hydrogen evolution (at the cathode) and oxygen evolution (at the anode). Both secondary reactions can lead to localised changes in pH near the surface of the electrodes. For example, at the surface of the cathode, the evolution of hydrogen (from water electrolysis) could cause a rise in localised surface pH, leading to the deposition of nickel oxides (hydroxides), which could cause pitting, hydrogen embrittlement and a porous nickel coating.

A common strategy is to deposit a compact and high-quality metallic coating by careful formulation of the electrolyte compositions and suitable choice of operating process parameters. Electrolyte additives are commonly deployed to control grain size, crystallographic properties, surface hydrophobic/hydrophilic properties and mechanical properties of the coating. Operational process parameters include control of the applied current, cell voltage, temperature, hydrodynamic flow environment, etc. A clean surface prior to electroplating, which can involve chemical or electrochemical methods of surface pretreatment, is important in realising high quality adherent metallic coatings.

7.2 Strategies to Achieve Nanostructured Coatings

Nanostructured metallic coatings can demonstrate coating properties superior to macrostructured or polycrystalline metallic coatings. These nanostructured metallic coatings can be tailored to

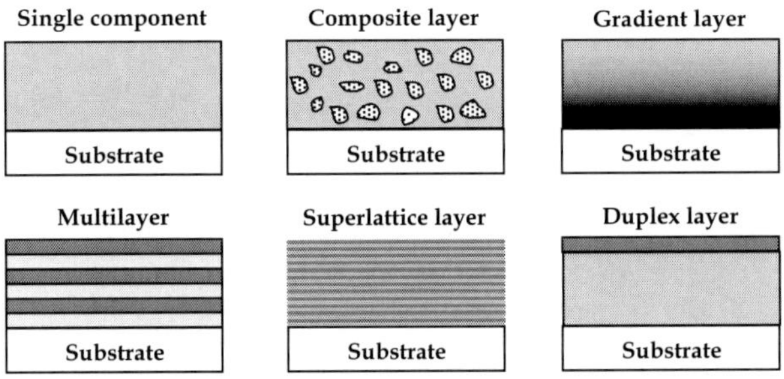

Figure 7.2 Examples of metallic surface coatings (single function or multifunctional layers) electroplated from aqueous liquid electrolyte.

simultaneously possess several engineering functionalities; they can be termed 'multifunctional' nanostructured metallic coatings.

Electroplating strategies to deposit nanostructured coatings include

(a) grain refinement via the use of additives (including surfactants),
(b) modulated (pulsed, alternating) current electroplating methods,
(c) inclusion of functional nanoparticles into a host matrix, and
(d) electroplating into soft or hard, nanosized porous templates.

Strategies (a) to (d) can be used to design multifunctional nanostructured metallic coatings. Figure 7.2 shows a number of examples of metallic surface coatings that could be electroplated from an aqueous electrolyte, which may be tailored to show a single function property of multifunctional properties. A wide range of surface protection and tribological functionalities could be designed by incorporation of various components into a coating; to produce hybrid structures, novel coating architectures and nanoscale materials such as nanocomposite, functionally graded, multilayered, superlattice, duplex and adaptive, self-responsive or even self-healing coatings.

For example, to design a multifunctional hard, self-lubricating metallic coating for wear resistance application, a metallic coating could be electroplated to contain nanosized crystallites and lubricative nanosized particles is entrapped within the coating. The first function is the nanocrystalline provides an enhanced coating hardness and improved mechanical property and the second

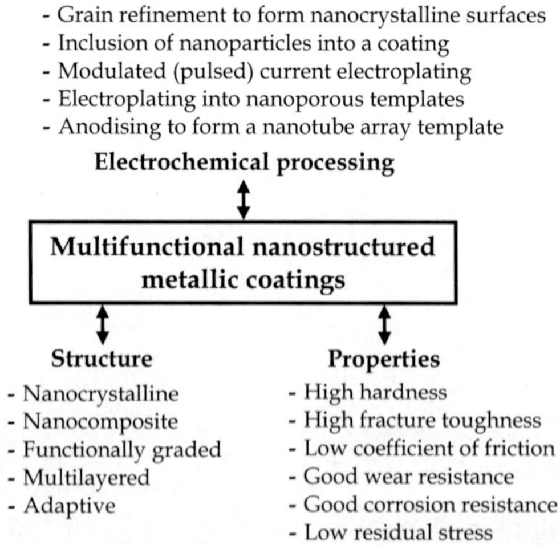

Figure 7.3 Relationships among processing, structure and properties in the design of multifunctional and nanostructured metallic coatings.

function is the lubricative nanoparticles provide a self-lubricity under wear.

Figure 7.3 shows the relationship between processing, structure and properties in order to design multifunctional nanostructured metallic coatings, via electrochemical methods. These coatings need to exhibit a range of tribological properties to meet the required application. Key advantages of nanostructured coatings over conventional coatings are enhanced mechanical hardness, resilience and toughness, improved high resistance against fatigue, oxidation, corrosion, lower friction and potential for coatings to produce lubricious tribofilms.

Hardness and toughness are important functions to determine the effectiveness of coatings for tribological applications. The Archard Wear Law [13] includes hardness (H) and Young's elastic modulus (E) [14] is often used to quantify the wear properties of materials. High H/E ratio generally implies (enhanced ability to absorb elastic strain energy) excellent fracture toughness and good wear resistance [15]. For coatings to achieve a long lifetime, they must have strong adhesion to the substrate. Adhesion can determine the durability of coatings and their load-bearing capacity.

Indeed, strong adhesion of the coating to the substrate is an essential prerequisite for most mechanical and tribological applications, even if the coating is hard and tough but has poor adhesion. The coating can fail rapidly due to delamination from the substrate. High levels of physical and chemical interaction between the coating and substrate can improve the adhesion.

Nanocrystalline coatings generally have improved yield strength, hardness and toughness compared with polycrystalline ones. The Hall–Petch relationship can be used to observe this behaviour. These coatings can be metals or alloys and tend to exhibit several functionalities for tribological applications, for example, wear, corrosion, erosion, etc. Nanoparticles can be embedded into a coating, resulting in a composite coating. The particles can provide a functional property, for example, enhanced hardness or enhanced lubrication; while the coating provides another functionality such that composite coatings often have multifunctional properties.

A gradient coating has progressively different composition or structural properties through its thickness to realise multifunctional properties. This concept usually offers to improve the load carrying capacity by introducing a smoother transition in properties from the coating (in contact with the environment) to the substrate. Thus, the contact load can be distributed functionally throughout the coating and over larger areas, which contribute to reduce contact tensile or compressive stress and the stresses induced at the coating–substrate interface.

Regular multilayered coatings may contain periodically repeated layers; typically involve two or more layers. Each individual layer can provide a characteristic functional property. Layer's thickness can vary between tens of nanometre to several or even sub-micron. Multifunctional properties are realised when these layers are sandwiched periodically. Multilayered coatings can prove to have better tribological performance than homogeneous coatings of the same materials since the layers can be tailored to have properties to combat wear by having, for example, dislocation glide and crack propagation.

Duplex coatings are multilayered (but the layers are not repetitive); two or more layers are used in conjunction to create superior properties. Multilayered coatings of materials with similar crystal structures could form columnar crystals that tend to extend throughout the coating. Such coatings are known as superlattice coatings and the individual multilayers must be sufficiently thin, typically several nanometre. By selecting a suitable combination of materials for the multilayered structure, superlattice, multifunctional properties is possible to improve the resistance against wear, corrosion, oxidation, high friction, etc. Nanocomposite and multilayered coatings can also be designed in such a way that when used in lubricated sliding applications, they can lead to the formation of self-replenishing and self-lubricating secondary films on the sliding surfaces and thus provide much lower surface friction.

Current R&D has placed emphasis on the production and uses of superhard and low-friction coatings. A superhard coating may have a hardness value exceeding 40 GPa when measured by Vickers hardness test; while a low-friction coating may offer surface friction as low as PTFE (polytetrafluoroethylene) which is notable for one of the lowest friction coefficients of material when measured sliding against stainless steel. Much research investigates the fundamental tribological mechanisms that control friction, corrosion and wear. The application conditions of future tribological environments are likely to be even more demanding than present service environments. To meet future challenges, new and more exotic coatings that can further improve durability and performance, whilst self-adaptation to an increasingly harsh and rapidly changing application conditions is needed. The ultimate goal is to design a universal coating that can work under a wide and changing range of environmental and load conditions.

7.2.1 *Grain Refinement to Achieve Nanocrystalline Deposits*

Metal coating electroplated from simple metal salt electrolytes often exhibit macroscopically rough, large polycrystalline and faceted surfaces that are associated with epitaxial or columnar growth. Additives and *surfactants can be added to the metal salt electrolytes to influence the* growth of grains, its sizes and boundary and affect the orientation of crystallite leading to a reduction in facet size. They are used to affect the formation of metal nuclei, its

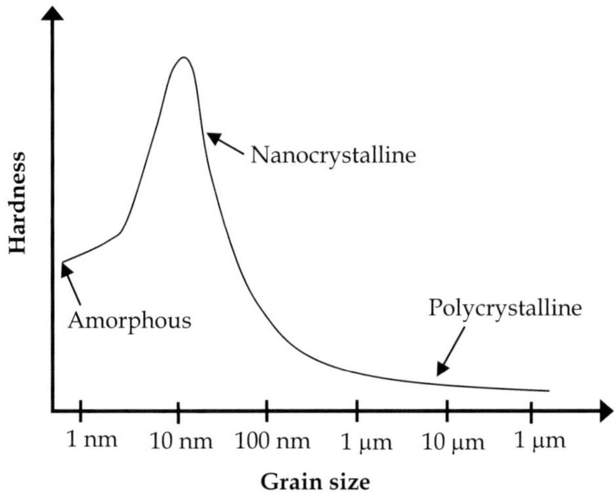

Figure 7.4 Hardness of a material as a function of grain size. Adapted from Refs. [16] and [17].

growth mechanism into a solid metal coating (either progressive or instantaneous deposit formation), to suppress undesirable secondary reactions (e.g., hydrogen or oxygen gaseous evolution), to promote co-deposition of alloys, dissolution of soluble metal (counter electrode), to influence the adsorption and desorption of chemical compounds on the electrode surface.

Enhancement in hardness and yield stress can be characterised via the Hall–Petch relationship, see Fig. 7.4. This relationship relates yield stress (σ), hardness (H) and grain size (d) [16, 17]:

$$\text{Yield stress } \sigma = \sigma_o + k_\sigma d^{-n} \tag{7.5}$$

$$\text{Hardness } H = H_o + k_H d^{-n} \tag{7.6}$$

where k_σ and k_H are strengthening coefficients and $n \approx 0.5$. Typically, a reduction in grain size, for example, from polycrystalline to nanocrystalline can hinder the multiplication and crack propagations within a coating system, leading to enhanced hardness. This effect is prominent for grain size down to tens of nanometres. However, an inverse Hall–Petch relationship exists, where further reduction in grain size less than tens of nanometres can bring about a decrease in hardness. The behaviour of material can be associated

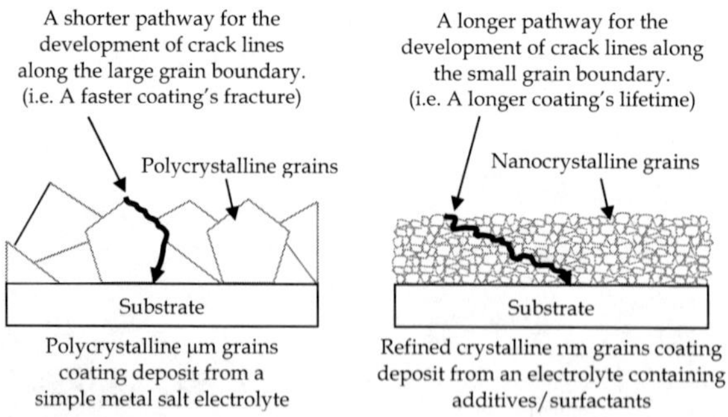

Figure 7.5 Grain refinement via electrolyte additives and surfactant use; polycrystalline vs. nanocrystalline coating.

with changes in the deformation mechanism from classic dislocation motion to mechanisms such as diffusion creep and grain boundary sliding.

Figure 7.5 shows grain refinement of nanostructured metallic coatings: polycrystalline versus nanocrystalline metal coating. Polycrystalline deposits (typically containing faceted micrometre-scale grains) from an electrolyte in the absence of additives or surfactants; while nanocrystalline deposits from an electrolyte-containing additives or surfactants. The presence of additives or surfactants could induce grain refinement mechanisms, which limit the faceted dimension of grain growth. By reducing the grain size and its boundary to nanoscale, a longer pathway could be introduced for the development of crack lines against an external load, leading to an enhanced coating performance and a longer coating lifetime. Nanosized grain offers the potential for resisting crack initiation. For example, even if a crack has initiated within a grain, strong grain boundaries within a nanocrystalline structure can potentially arrest or deflect it back to the grain, preventing crack growth.

7.2.2 *Modulated (Pulsed) Current Methods*

Steady, constant direct current is the most commonly deployed external electrical signal to deposit a metallic coating. Recent years have seen the increasing use of pulsed, alternating current to achieve nanostructured coatings. This is also known as the modulated current method, where the potential (or current) is often alternated between two values to give pulses of predetermined amplitude, duration, polarity, frequency, etc.

Pulsing the deposition current has complex effects; it can affect the diffusion layer next to the electrode surface that is in contact with the liquid solution. This will influence the deposition mechanisms of metal deposits such as altering the nucleation process and the subsequent growth of the deposit. Pulsed current can also enable the incorporation of nanoparticles to a high content in the coating as well as producing a wider range of alloys, deposit compositions and material properties.

By regulating the pulse parameters, the structural, physical and mechanical properties of coating deposit, for example, grain size, its boundary and orientation, crystallite structure, continuity, deposits porosity and uniformity of microstructure could be controlled. Modulated current has been increasingly used to produce coatings with nanoscale crystalline and/or tailor the uniformity and distribution of coating thickness on complex three-dimensional substrates.

Modulated current can be classified into (a) unipolar, where all the pulses are in one direction with no change in polarity and (b) bipolar, where a mixture of anodic and cathodic pulse is used. Many types of current waveforms can be used; the most common being square-wave pulsed direct current and pulsed reverse current, Fig. 7.6.

In direct current (dc) plating, the variable for the applied current is its magnitude. In modulated current, many parameters could be

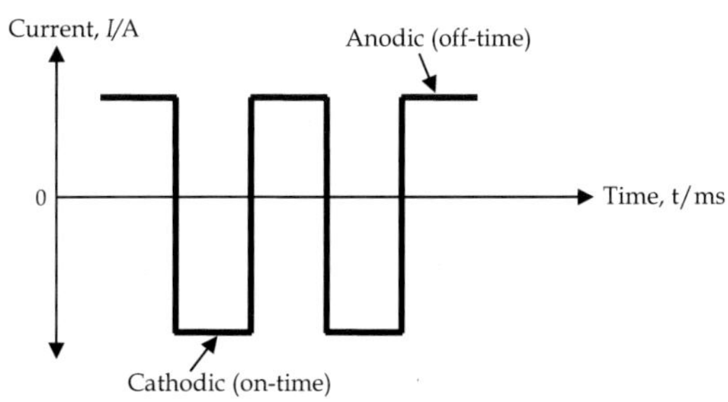

Figure 7.6 A typical square wave pulsed current waveform for electroplating, showing the anodic (off-time) and cathode (on-time).

varied, for example, pulse direct current: on-time t_{on}, off-time t_{off}, peak current I_p; pulse reverse current: anodic current duration t_a, cathodic current duration t_c, peak current I_p. A number of expressions can be defined as follows:

The duty cycle, γ is the fraction of total cycle time:

$$\text{Pulsed direct current } \gamma = \frac{t_{on}}{t_{on} + t_{off}} \tag{7.7}$$

$$\text{Pulsed reverse current } \gamma = \frac{t_c}{t_a + t_c} \tag{7.8}$$

Frequency, f is the reciprocal of cycle time:

$$\text{Pulsed direct current } f = \frac{1}{t_{on} + t_{off}} \tag{7.9}$$

$$\text{Pulsed reverse current } f = \frac{1}{t_a + t_c} \tag{7.10}$$

Pulsed current results in the metal deposition at the same rate as direct current provided its average pulse current I_a equals the magnitude of the mean direct current. The average pulsed current can be defined as:

$$\text{Pulsed direct current } I_a = I_p \gamma \tag{7.11}$$

$$\text{Pulsed reverse current } I_a = \frac{I_c t_c - I_a t_a}{t_a t_c} \tag{7.12}$$

Pulse current electrodeposition often uses a much higher instantaneous current density in comparison to direct current electrodeposition, thus pulse current electrodeposition has been found to be an effective means of perturbing the adsorption–desorption processes occurring at the metal deposit–electrolyte interface and hence offers an opportunity of controlling the microstructure of metal deposits, leading to the production of nanostructures and finer grain size coating, plus enhance the amount of nanoparticles incorporated within a metal composite coating. For example, in a reported work for the electrodeposition of Ni–Co alloys [18], the variation of peak current density has significant impact on the composition, microstructure, surface morphology, grain size, microhardness and tensile strength of the coating. For example an increase in peak current density could promote the production of deposits with less Co content, finer grain structure increased hardness and improved tensile strength.

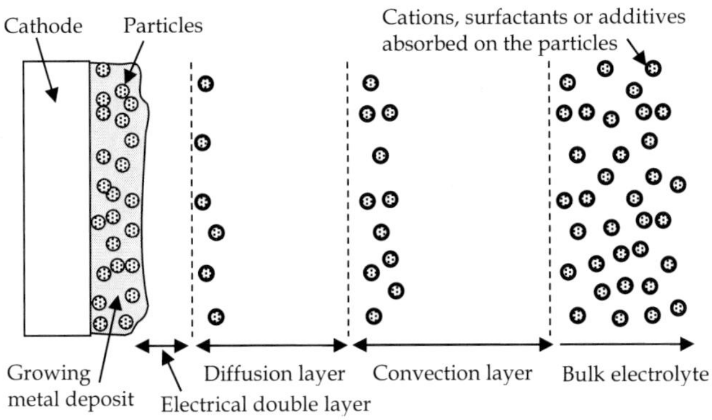

Figure 7.7 Processes involved in the co-electrodeposition of insoluble particles into a growing metal matrix to form a composite metal coating.

7.2.3 *Inclusion of Nanoparticles into a Host Matrix*

A wide range of nanosized particles, typically tens of nanometre in size and dimensions, can be incorporated into metal deposit to form a composite metallic coating [19, 20]. Figure 7.7 shows the processes involved in electroplating of particles into a growing metallic matrix.

The incorporation of nanosized particles into metallic coatings can be achieved by a number of processes including convection of the particles towards the cathode surface, mechanical entrapment of particles into the growing metal matrix and/or electrophoretic migration of particles to the growing metal deposit. The steps can be identified as

(a) charged particle formation, for example, ions and surfactants adsorbed on particle surfaces,
(b) physical transport of particles through a convection layer,
(c) mass transport of particles via a diffusion layer,
(d) migration driven by the potential gradient across an electrical double layer and
(e) physical embedding of particles into the growing metal deposit.

Co-deposition of nanosized particles into metal deposits is dependent on many process parameters, including characteristics of the particle (e.g., concentration, surface charge, type, shape, size), electrolyte composition (e.g., electrolyte concentration, additives,

Figure 7.8 Cross-sectional images of an electrodeposited nickel coating containing nanoparticle inclusions.

temperature, pH, surfactant type and concentration), applied current density (e.g., direct current, pulsed current, pulse time, duty cycle, potentiostatic control), flow environment inside the electroplating tank (e.g., laminar, mixed, turbulent regimes) and shape/size of electroplating tank and electrode geometry (rotating disk electrode, rotating cylinder electrode, plate-in-tanks, parallel plate electrodes and many variations of electroplating tanks). Examples of cross-sectional view of composite metal coatings are shown in Fig. 7.8. The electrodeposited nickel-containing nanoparticles of silicon carbide (SiC) have a diameter from 150 to 500 nm and concentration from 5 to 50 g/dm^3 and the titanates/TiO_2 nanotubes have an internal diameter of 5 nm, a tube length of 30 to 500 nm, and concentration from 10 to 20 g/dm^3. The particles can be agglomerated or, preferably, well dispersed within the coatings.

The particles in solution can be made to contain a surface charge to help dispersion and embedment into the growing metal matrix. This surface charge can arise from ionisation of surface groups, chemical binding or physical adsorption of ions. A charged particle suspended in a liquid can be surrounded by an ionic cloud, resulting from chemical absorption with electroactive species. For example, the negative surface charge of SiC particles can lead to adsorption of nickel(II) ions and alter the surface charge of particles from negative to positive in a solution containing nickel(II). This 'nickel(II)-SiC' combination can lead to more positively charged particles, which can electrophoretically promote effective transport of particles to the cathode. Coatings can consist of randomly included or well-aligned nanoparticles, especially in the case of elongated particles, such as rods, tubes or fibres. The incorporation of well-aligned nanoparticles into coating is attractive since this could potentially lead to more controlled functional properties, which could be tailored to react to a multitude of tribological environments.

The dispersion of particles in the electrolyte can affect their distribution in the metal matrix. If the nanosized particles are not well dispersed in the liquid solution, they tend to be incorporated into the metal matrix as aggregates. Particles embedded into the metal matrix can serve many functional purposes, commonly reinforcement. The reinforcement mechanisms for composite metal matrices include dispersion strengthening and particle-strengthening concepts. For the dispersion-strengthening mechanism, the matrix carries the load and the particles hinder the dislocations motion. For the particle-strengthening mechanism, the load is carried by both the matrix and the particles [21].

7.2.4 *Electroplating through Nanosized Porous Templates*

Templated electroplating is a process in which materials are deposited into the free space of a template. This technique allows dimension-controlled synthesis of nanomaterials and coatings with well-defined nano-architectures. The templates can be either soft or hard; and the free-space is typically in the nanosize dimension. Hard templates are fabricated nanoporous membranes or self-assembled colloidal crystal structure. Soft templates are self-ordered nanoporous structures of surfactants, block copolymers or liquid crystal materials. Materials deposited into these templates can produce high surface area per volume, for example, nanowires, nanotubes or nanoporous structures.

Figure 7.9 shows the typical steps for soft templated electrodeposition. Surfactant molecules can aggregate into micelles with hydrophobic and hydrophilic ends in aqueous solution. At

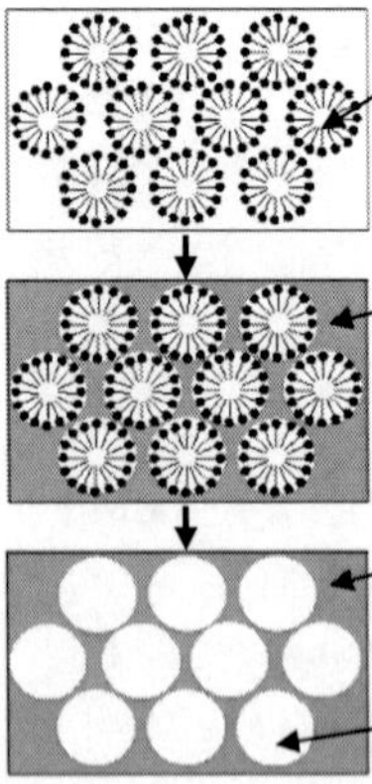

Figure 7.9 Soft template deposition, showing the steps involved.

higher concentrations of surfactant, these micelles can form an orderly packed porous structure. Materials can be deposited into the free space of this structure, which has been commonly known as templated electrochemical deposition in lyotropic crystal template. Electroplating into this structure usually resulted in mesoscale porosity and is often of 'toothpaste' like consistency. They can be categorised into three major types according to the surfactants properties, namely hexagonal phases, cubic phases and lamellar phases. Hexagonal phases consist of aligned cylindrical micelles in a hexagonal array; cylindrical micelles in cubic phases are interconnected on a gyroid lattice and lamellar phases consist of lamellar structured micelles.

This is known as a soft template because the lyotropic crystal template is soft in nature and can be washed off with a liquid (often water). The lyotropic phases are categorised into three major types: hexagonal phases, cubic phases and lamellar phases. Hexagonal phases consist of well-aligned cylindrical micelles in a hexagonal array. In cubic phases, the cylindrical micelles are interconnected on a gyroid lattice. Lamellar phases consist of lamellar structured micelles. The most commonly deployed template is the hexagonal phases of lyotropic liquid crystals for the synthesis of hexagonally ordered mesoporous materials.

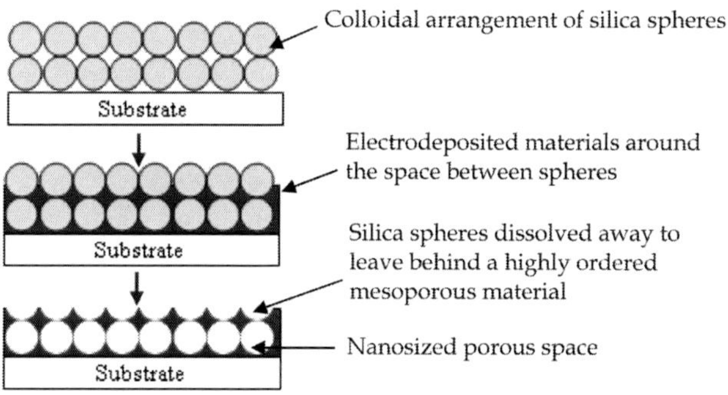

Figure 7.10 Hard template deposition, showing the steps involved.

Figure 7.10 shows the steps for hard template material deposition using a colloidal crystal array of silica spheres. This is a hardtemplated electrodeposition process since the templates are solid in nature and must chemically dissolve to reveal the nanoporous material. Methods can be used for assembly of colloidal crystals involve sedimentation and vertical deposition. Sedimentation is a simple self-assembly method based on gravitational force leading to the formation of multilayer structures. Vertical deposition is a method using attractive capability forces to arrange particles vertically along the wall of a surface leading to controlled deposition and thickness. Colloidal crystals are usually regular and periodic particle arrays that display translational symmetry. The optimal packing structure of spherical particles in colloidal crystals is the facecentred cubic lattice which is a close-packed structure [22].

Another hard template method employs nanotubular arrays as a hard template. The well-aligned nanotubes can be formed by anodising of valve metals to metal oxides. Figures 7.11 and 7.12 show the formation of TiO_2 nanotubes [5], via simultaneous electrochemical formation and chemical dissolution of oxides, reactions (7.15) and (7.16).

$$\text{Oxide formation Ti} + 2H_2O \rightarrow TiO_2 + 4H^+ + 4e^- \qquad (7.13)$$

$$\text{Oxide dissolution } TiO_2 + 6F^- + 4H^+ \rightarrow TiF_6^{2-} + 2H_2O \qquad (7.14)$$

In the initial stage of anodising, a compact layer of titanium dioxide is formed on the titanium foil. This layer could be dissolved

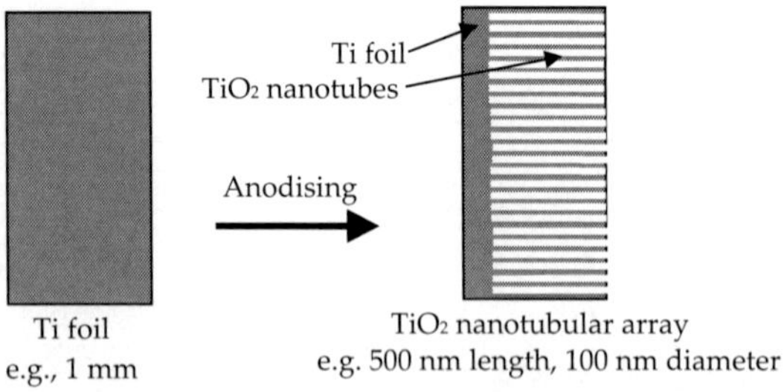

Figure 7.11 Anodising of metal to metal oxide nanotubular array.

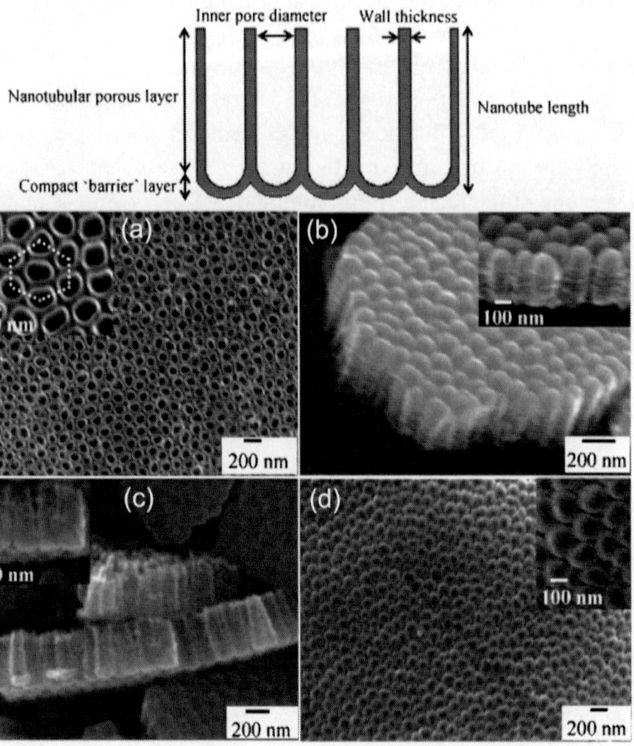

Figure 7.12 Anodising to form TiO_2 nanotube arrays, which are then used as a hard template. (a) top view: open top, (b) side view: ripples along wall, (c) cross-sectional view: individual tubes and (d) bottom view: closed cup. Wall thickness (10–50 nm), tube length (100–1000 nm), inner pore diameter (50–500 nm).

to form a soluble hexafluorotitanium complex anion in the presence of fluoride ion. This is a localised dissolution process where breakdown of the oxide layer occurred randomly on the surface and disordered pores are formed. At this stage, a compositional gradient is established between the top and bottom of the pores. This gradually results in the formation of ordered pores distributed homogeneously over the top surface. These pores continued to grow inward to the titanium foil leading to the formation of a nanotubular array. A 'barrier' layer remains between the pore bottom and the titanium substrate. At the pore mouth, dissolution of the oxide continued. In general, electroplated nanotubes and nanowires of materials can be synthesised using anodised nanotube structures while an array of silica spheres or closely packed polystyrene beads facilitates the electroplating of porous structures having spherical holes [23].

Nanotubular arrays can be capped with other materials, typically metals and alloys to form composite nanotubular structure, for example, sputter deposition of Au nanoparticles on TiO_2 nanotubular array, see Fig. 7.13. These porous composites could potentially offer a three-dimensional (3D) surface having a, large surface-to-volume ratio for sliding contact. In the past, gold nanoparticles have been added to oil used in engine lubrication or gold-coated alumina as high temperature, thermally and chemically stable lubricant coating; such surfaces can also realise high-temperature materials for electrical contacts.

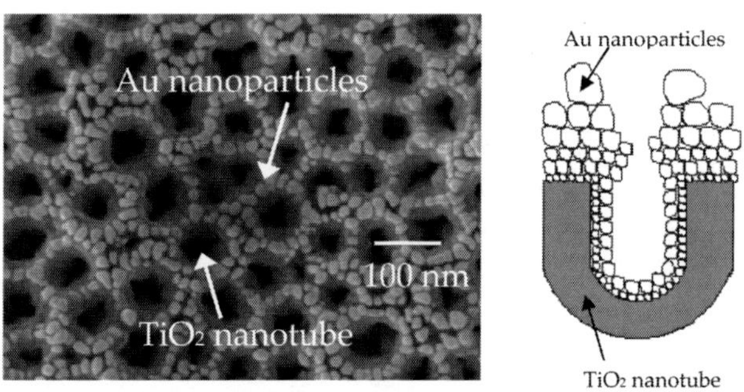

Figure 7.13 3D composite structure of an Au–TiO_2 nanotubular array.

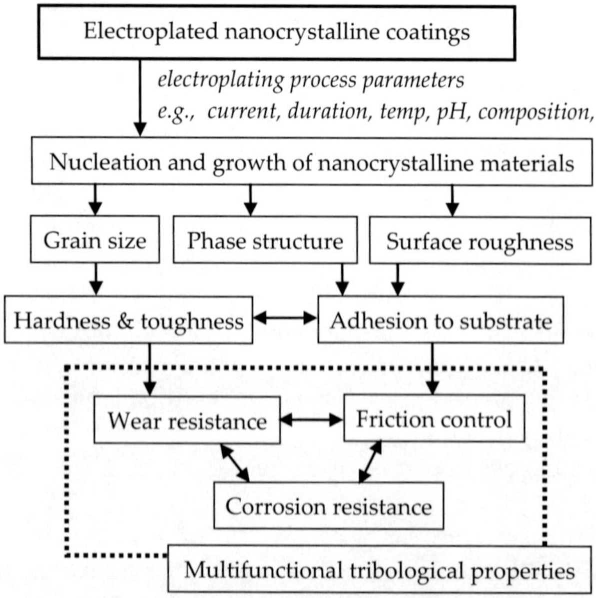

Figure 7.14 Design factors to be considered during the electroplating of multifunctional nanocrystalline coatings for tribological applications.

7.3 Examples of Multifunctional Nanostructured Coatings

7.3.1 *Nanocrystalline Coatings*

Many design factors must be considered for electroplating good-quality multifunctional nanocrystalline coatings for tribological applications. Figure 7.14 shows a comprehensive understanding of relationship between electroplating process–structure–property is critically needed to realise multifunctional coatings, for example, those which combat wear and corrosion while simultaneously lowering friction [24–26]. The nucleation and growth of nanocrystalline deposits can be tuned by controlling process parameters such as electrolyte composition, pH, temperature, current density, current waveform, duration, etc; all these parameters can manipulate the grain size, phase structure and surface roughness, which in return can design a coating to have multifunctional tribological properties, for example, wear resistance, corrosion resistance, friction control, etc.

For coatings to be effective for tribological applications, they must meet several criteria, including good adhesion, high mechanical hardness and toughness, high thermal and chemical stability, uniform and adequate thickness, smooth surface finish and low friction. The coatings must exhibit a dense morphology (i.e., no porosity, open columnar grain boundary) and a strong cohesive bonding among grains. Coatings having a porous columnar structure tend to wear rapidly as poor cohesion between grains can lead to fracture-dominated sliding wear.

Coatings that are nanocrystalline and have high fracture toughness can often resist crack initiation along a sliding contact surface since the grains are extremely small and the morphology is often not necessarily anisotropic or columnar. Highly anisotropic and columnar structure can often result in a structurally weak coating susceptible to deformation or fracture. Nanosized grains with a large-grain boundary network can provide a condition for resisting crack initiation, for example, if a crack initiates within a grain, the strong and thin grain boundaries may deflect and branch-out, reducing the stress concentration area and arresting further crack growth.

Figure 7.15 shows that nanocrystalline Co–W alloy coating performed well in a tribological environment for friction control and wear resistance. Coatings were tested via a ball-on-disk UMT-3MT tribometre using a 4.0 mm diameter GCr15 steel ball, at 22°C, 5 N load, 2.41 GPa contact force and 0.335 $m\,s^{-1}$ sliding speed. Nanocrystalline Co–W alloy coatings were electroplated by dual-pulsed current from a liquid solution containing 0.4 $mol\,dm^{-3}$ cobalt sulphate, up to 0.2 $mol\,dm^{-3}$ sodium tungstate, 0.3 $mol\,dm^{-3}$ boric acid, 0.15 $mol\,dm^{-3}$ sodium dodecylbenzenesulfonate, 0.2 $mol\,dm^{-3}$ sodium citrate, 0.7 $mol\,dm^{-3}$ sodium sulphate, 4.0 $mol\,dm^{-3}$ saccharin, pH 6 and 40°C liquid temperature [27]. The dual-pulse electroplating can continuously supply positive and reverse square wave pulses with different period, impulse-wide and pulse amplitude. Cathodic current density was varied from 2 $A\,dm^{-2}$ to 8 $A\,dm^{-2}$ and anodic current density from 0.2 $A\,dm^{-2}$ to 0.8 $A\,dm^{-2}$.

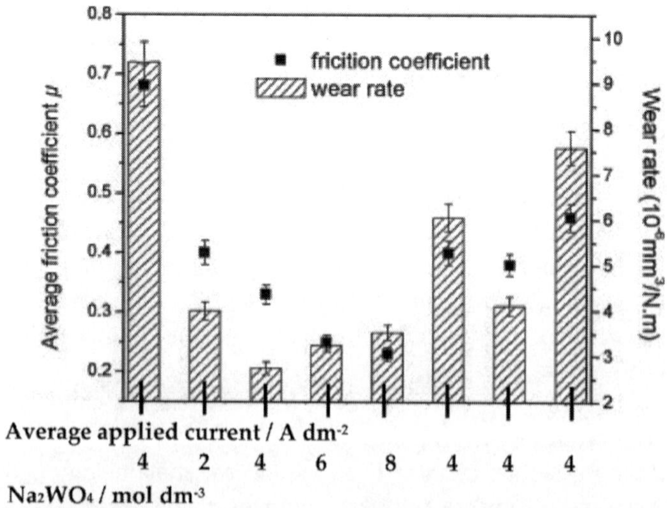

Figure 7.15 Functional properties of nanocrystalline Co-W coatings at room temperature with a relative humidity of 40–60% under dry sliding conditions. Reprinted from Ref. [27], Copyright (2012), with permission from Elsevier.

The primary function (friction control) and secondary function (wear protection) could be manipulated by changing the dual-pulse current values and concentration of sodium tungstate in the electrolyte. It was found that the coating's microhardness was affected by the coating grain size, W content and crystal orientation. Smaller grain size, higher W content and strong hcp(100) orientation improved the coating hardness, for example, microhardness over 1000 kgf mm^{-2} can be obtained, as comparable to the hardness of hard Cr coating. Co-W coating produced at an applied current density about 4.0 A dm^{-2} and with 0.10 mol L^{-1} Na_2WO_4 in the electrolyte displayed the best tribological properties.

7.3.2 *Functionally Graded Coatings*

The properties of multilayered or graded coating can be tuned by varying the properties of individual layers that make up the bulk coating. Typically, the hardness may be influenced by the levels of internal stresses within these coatings. A high level of internal stress

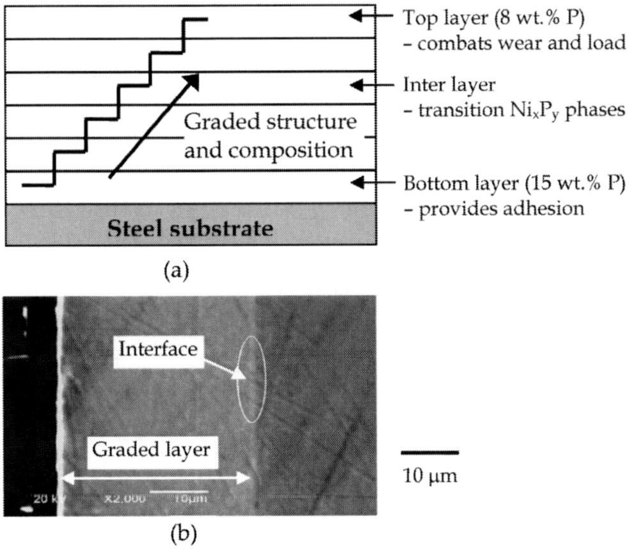

Figure 7.16 Graded Ni–P coatings: (a) The concept and (b) an SEM cross-sectional image. Reprinted from Ref. [29], Copyright (2006), with permission from Elsevier.

may be beneficial to the hardness of a coating, but it can also be detrimental to the fatigue and wear resistance of coatings.

Figure 7.16(a) suggests that the top layer provides a tribological function (e.g., wear and friction), intermediate layers provide transitional function (via changing the composition and structural properties of coatings) and bottom layer promotes adhesion between the coating and substrate. A gradient Ni–P alloy was electroplated on steel substrates with a graded change of *P* content in the direction of coating thickness. This coating was prepared from an electrolyte containing 240 g dm^{-3} nickel sulphate, 30 g dm^{-3} nickel chloride, 30 g dm^{-3} boric acid, 20 g dm^{-3} phosphorous acid, bath was kept at 70°C and a pH of 1.5 [28, 29].

Six-graded layer was electroplated by gradually changing the current density from 5 to 30 A dm^{-2}. Figure 7.16(b) shows the scanning electron microscope (SEM) cross-sectional image for this coating. Results showed that the wear resistance of a graded Ni–P alloys coating was about twice as good as an ungraded

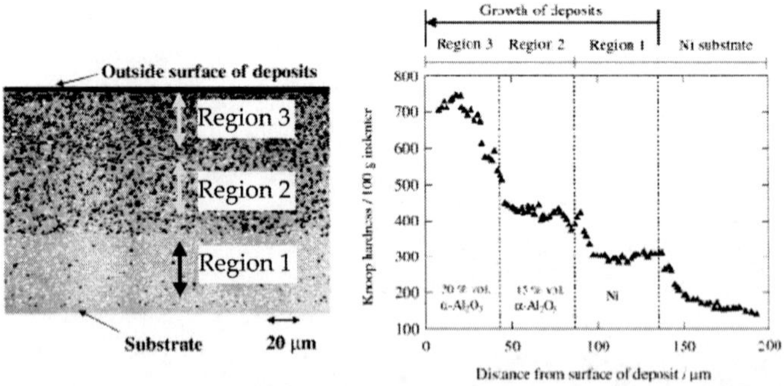

Figure 7.17 Layered composite nickel coating containing 1–4 μm Al conductive particles and 60–80 nm α-Al_2O_3 non-conductive nanoparticles. Reprinted from Ref. [19], Copyright (2006), with permission from Elsevier.

coating, for example, 4×10^{-5} $mm^3\,N^{-1}\,m^{-1}$ for graded Ni-P; 8×10^{-5} $mm^3\,N^{-1}\,m^{-1}$ for monolayer Ni-P (8 wt.% P). Graded Ni-P deposits exhibited better wear resistance and lower friction coefficient than that of hard Cr deposits up to 600°C. The formation and propagation of through-thickness cracks within heat-treated Ni–P deposits were effectively inhibited due to the introduction of graded composition and structures.

In addition to electroplating multilayered metals and alloys, the multilayers can also contain nanosized particles. For example, single and mixed particles (1–4 μm Al conductive particles and 60–80 nm α-Al_2O_3 non-conductive nanoparticles) in a nickel deposit were produced by varying the current density and concentration of particles in the electrolyte [19], see Fig. 7.17 for a cross-sectional image of three layers and changes in the Knoop hardness profile for a gradient distribution of Al and α-Al_2O_3 particles in a nickel deposit. Each layer provides a specific function, for example, Ni (Region 1) for coating's adhesion to substrate, Ni-15 vol.% Al_2O_3 (Region 2) as a transition layer and Ni-30 vol.% Al_2O_3 (Region 3) for enhanced hardness property.

Another example is graded SiC particles in a Ni–Co alloys coating. Figure 7.18 shows that an increase in the applied current density from 2 mA cm^{-2} to 15 mA cm^{-2} led to several effects including (a) an increase in the weight percent of SiC particles in the coating, (b) an increase in the composition of cobalt in the Ni–Co alloys and

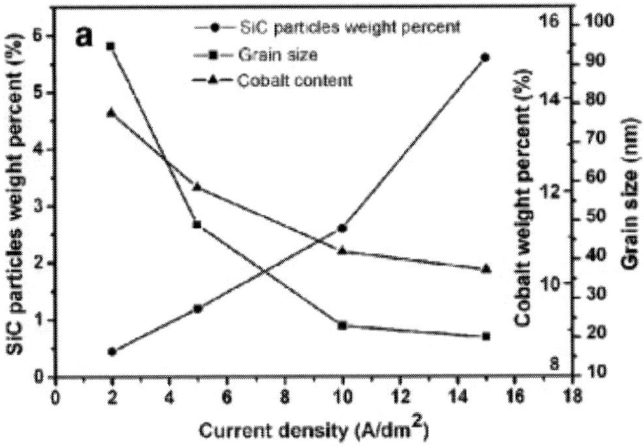

Figure 7.18 Effect of applied current density on (i) SiC particles in coating, (ii) grain size and (iii) composition of cobalt in Ni–Co alloys. Reprinted from Ref. [30], Copyright (2012), with permission from Elsevier.

(c) a decrease in the grain size of coating. Based on these results obtained for uniform electrodeposits, a multifunctional graded Ni–Co/SiC coatings were electrodeposited by incrementally increase in the current density from 2 mA cm^{-2} to 15 mA cm^{-2} [30].

7.3.3 *Nanocomposite Coatings*

Nanocomposite coatings have the potential to offer multifunctional properties, since they can draw properties found in the coating and the particles employed. It is possible to design coatings that can form self-replenishing or self-lubricating tribofilms on their sliding surfaces; these are often called adaptive (or responsive) coatings. A multifunctional nanocomposite coating may adapt and follow the changing tribological environmental. For example, one component may provide corrosion resistance while a second component may provide wear resistance; or one component is active and provides both corrosion and wear resistance at room temperature while a second component may become more active and provide corrosion and wear resistance at high temperatures.

Electroplating of nanoparticles into coatings is a popular method to extend the versatility of coatings and to confer with multi-

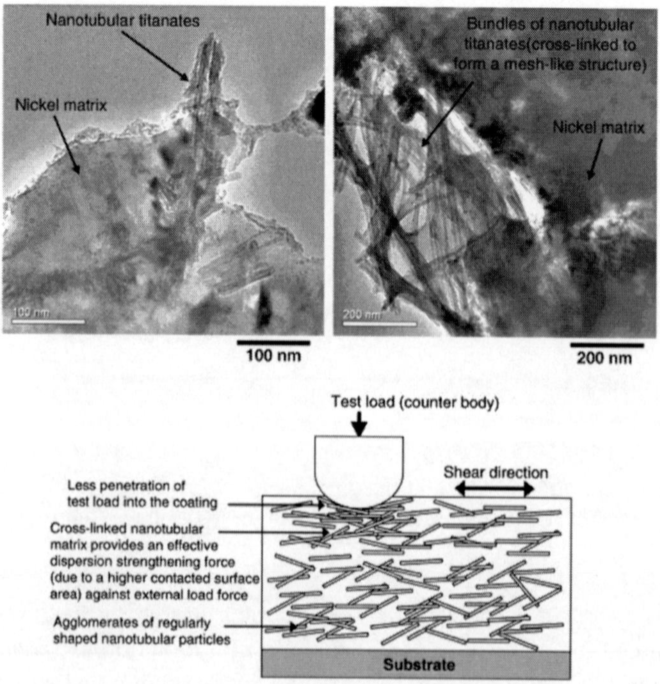

Figure 7.19 Top: TEM images of electroplated nickel containing titanates nanotubes. Bottom: strengthening and support mechanisms.

functional properties. For example, Fig. 7.19 shows transmission electron microscopy (TEM) images of the cross-section of an electrodeposited nickel coating containing nanotubular titanium oxide particles. Nanotubular titanates are clearly embedded into the nickel coating with a reasonably uniform dispersion [2].

When compared with a nickel coating containing irregularly shaped nanosized titanium dioxide particles, this composite coating has demonstrated several functionalities including 22% reduction in surface friction against steel, (b) 29% enhancement in wear resistance in a three-body water-based abrasive wear test (steel counter body and 5 μm SiC particles), (c) 102% improvement in coating hardness and (d) 26% improvement in elastic modulus. It is postulated that the nanotubes in the nickel coating agglomerated to form a cross-linked and mesh-like matrix that led to an enhancement in the dispersion-strengthening mechanism against external load force. The tribological properties of such coatings were effectively improved compared with a nickel coating reinforced with irregularly shaped titanium dioxide particles.

Composite coatings may be engineered to incorporate the properties of coating and lubricants, in such a way that they become compatible or complementary to one another's performance and characteristics. If the phases are selected strategically, they can achieve self-lubricating capacity and provide low-friction and low-wear coefficients over a broad range of test conditions, for example, one phase provides extreme resistance to wear and deformation, a second or third phase may preferentially react with the additives in oils to provide high lubricity and protection against wear. In such coating architectures, the properties of phases are strategically selected so that they may become lubricious through tribological–chemical interactions with the chemical species in operating environments, thus enabling tools to operate even in the absence of additional liquid or other lubricants.

These coatings are highly desirable for many engineering applications but not easily electroplated. A simple concept for self-lubricating composite coating was realised by electroplating of metal with oil-containing microcapsules, see Fig. 7.20 [31, 32]. The copper-microcapsule composite coating was by electroplated from an acid sulphate bath consisting of 200 g dm^{-3} copper sulphate, 60 g dm^{-3} sulphuric acid, 60 mg dm^{-3} hydrochloric acid and 20 vol.% microcapsules emulsion. They are hydrophobic capsules, 2 μm to 20 μm in size. The liquid core material was organic siloxane while the shell material was methyl cellulose. Electrochemical analysis showed that the corrosion current through the copper-microcapsule coating was lowered to 1.5 μA cm^{-2}, compared to a plain copper coating at 60 μA cm^{-2}. Ageing trials showed that the microcapsules slowly released their oil content to form a film on the coating surface for corrosion protection.

The liquid inside the capsules could be tailored for functional purposes. For example, capsules could be filled with, for example, (a) mineral oils, for self-lubricating ball bearings, (b) coloured liquids, fluorescent liquids or scented oils could facilitate a colour change or odour release when they are damaged. Several different types

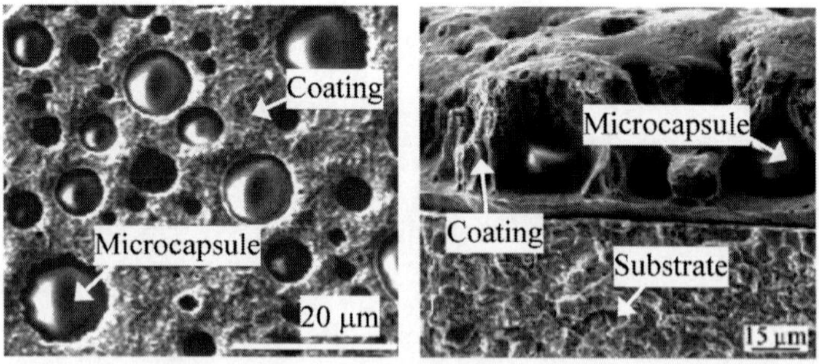

Figure 7.20 SEM images of electroplated copper containing liquid siloxane microcapsules. Left: surface view. Right: cross-sectional view. Reprinted from Ref. [32], Copyright (2011), with permission from Elsevier.

of capsules could be incorporated inside a metal layer to display multifunctional properties. Coloured or scented particles in the top coating's layer might signal wear or damage, while oil particles or chemically reactive particles in the deeper coating layer could self-repair or prevent further damage onto the substrate. Such architectures could sense the extent of wear and provide a warning prior to coating failure. For example, fluorescent particles can be placed in a particular area and known depth. If the coating is breached, illumination of this area could provide distinctive optical spectra, allowing an in situ detection of the state of wear.

7.4 Conclusions

Electroplating and anodising are simple and versatile wet-processing methods to fabricate multifunctional nanostructured metallic coatings including nanocrystalline, functionally graded and nanocomposite ones. These nanostructures can be realised by methods such as grain refinement, modulated current, inclusion of nanoparticles into coating and the use of hard or soft nanoporous templates. These approaches employ near-ambient temperature and pressure. The coatings can provide several functional properties and can be attractive as tribologically resistant surfaces. Some coatings can be designed to self-provide diagnostic information on their integrity/lifetime or even to self-adapt to increasingly harsh and rapidly changing service conditions.

References

1. C.T.J. Low, F.C. Walsh, *Surf. Coat. Technol.*, 202 (2008), 1339.
2. C.T.J. Low, J.O. Bello, J.A. Wharton, R.J.K. Wood, K.R. Stokes, F.C. Walsh, *Surf. Coat. Technol.*, 205 (2010), 1856.
3. X.Y. Wei, Z.G. Zhu, P.D. Prewett, K. Jiang, *Microelectron. Eng.*, 84(5–8) (2007), 1256.
4. C. Dechakiatkrai, C. Lynam, K. J. Gilmore, J. Chen, S. Phanichphant, D. V. Bavykin, F. C. Walsh, G. G. Wallace, *Adv. Eng. Mater.*, 11(7) (2009), B55.
5. C.T.J. Low, M. de la Toba Corral, F.C. Walsh, *Trans. Inst. Met. Finish.*, 89(1) (2011), 44.
6. C.T.J. Low, D. Pletcher, F.C. Walsh, *Electrochem. Commun.*, 11 (2009), 1301.
7. F.C. Walsh, C.T.J. Low, R.J.K. Wood, K. T. Stevens, J. Archer, A. R. Poeton, A. Ryder, *Trans. Inst. Met. Finish.* (2009), 87(3), 122.
8. K. Helle, F.C. Walsh, *Trans. Inst. Met. Finish.*, 75 (1997), 53.
9. A.H. Nahlé, C.Kerr, B.D.Barker, F.C. Walsh, *Trans. Inst. Met. Finish.*, 76 (1998), 29.
10. X. Shi, G. Jarjoura, G. J. Kipouros, *Magnesium Technol., TMS* (2006), 273.
11. F.C. Walsh, *A First Course in Electrochemical Engineering*, Electrosynthesis Company, UK, 1993.
12. D. Pletcher, *First Course in Electrode Processes*, 2nd edition, Royal Society of Chemistry, UK, 2009.
13. J.F. Archard, *J. Appl. Phys.*, 24 (1953), 981.
14. A. Leyland, A. Matthews, *Wear*, 246 (2000), 1.
15. Z. Zeng, L. Wang, A. Liang, J. Zhang, *Electrochim. Acta*, 52 (2006), 1366.
16. E.O. Hall, *Proc. Phys. Soc. London*, B64 (1951), 747.
17. N.J. Petch, *Journal of Iron Steel Institute* (1953), 174, 25.
18. Y. Li, H. Jiang, W. Huang, H. Tian, *Appl. Surf. Sci.*, 254(21) (2008), 6865.
19. C.T.J. Low, R.G.A. Wills, F.C. Walsh, *Surf. Coat. Technol.*, 201(1–2) (2006), 371.
20. J. Lapinski, D. Pletcher, F.C. Walsh, *Surf. Coat. Technol.*, 205(21–22) (2011), 5205.

21. A. Sohrabi, A. Dolati, M. Ghorbani, A. Monfared, P. Stroeve, *Mater. Chem. Phys.*, 121 (2010), 497.
22. M.A. Ghanem, P.N. Bartlett, P. de Groot b, A. Zhukov, *Electrochem. Commun.*, 6 (2004), 447.
23. M. Lai, D.J. Riley, *J. Colloid Interface Sci.*, 323(2) (2008), 203.
24. L. Wang, Y. Gao, T. Xu, Q. Xue, *Mater. Chem. Phys.*, 99 (2006), 96.
25. L. Wang, J. Zhang, Y. Gao, Q. Xue, L. Hu, T. Xu, *Scr. Mater.*, 55 (2006), 657.
26. P. Herrasti, C. Ponce de León, F.C. Walsh, *Rev. Metal.*, 48(5) (2012), 377.
27. F. Su, P. Huang, *Mater. Chem. Phys.*, 134(1) (2012), 350.
28. L. Wang, Y. Gao, Q. Xue, H. Liu, T. Xu, *Surf. Coat. Technol.*, 200 (2006), 3719.
29. L. Wang, Y. Gao, T. Xu, Q. Xue, *Appl. Surf. Sci.*, 252 (2006), 7361.
30. S.M. Lari Baghal, M. Heydarzadeh Sohi, A. Amadeh, *Surf. Coat. Technol.*, 206(19–20) (2012), 4032.
31. X. Xu, H. Liu, W. Li, L. Zhu, *Mater. Lett.*, 65 (2011), 698.
32. X. Qing, Z. Qun, L. Ping, L. Cong, *Transactions of Nonferrous Metals Society of China*, 21 (2011), 2210.

Chapter 8

Low-Friction Materials and Coatings

Ali Erdemir
Energy Systems Division, Tribology Section, Argonne National Laboratory, Argonne, IL 60439, USA
erdemir@anl.gov

8.1 Introduction

Low-friction materials and their coatings have been in use for many decades to control friction and wear in a wide range of tribological applications. Due to their multifunctional nature, they have also been very effective in combating corrosion, fatigue, scuffing and fretting-related mechanical failures under harsh operating conditions [1–9]. Latest self-lubricating coatings consist of nanocomposite and/or -layered morphologies that afford much longer wear life as well as superior resistance to oxidation and heat-related degradations. These novel coating architectures are made of two or more distinct phases with each phase providing a range of additional functionalities besides providing low friction and wear. With recent advances in deposition technologies, production of such coatings has now become rather easy and affordable for a wide range of mainstream applications.

Multifunctional Materials for Tribological Applications
Edited by Robert J. K. Wood

ISBN 978-981-4463-90-4 (Hardcover), 978-981-4463-91-1 (eBook)
www.panstanford.com

Some of the earliest references for low-friction materials and coatings go back to the 1920s [7]. Since then, these materials have become the key constituents of many types of self-lubricating coatings and metallic, ceramic, and polymeric composites for all kinds of industrial uses [1–3, 6, 8]. Low-friction materials are primarily intended for use under extreme tribological conditions that often involve sub-zero or high temperatures, compressed gasses, high vacuum or space-like environments, radiation, severe contact pressures, steam or water, etc., where liquid lubricants fail to provide good performance or long durability. They are very versatile in the sense that they can be utilised in finely ground powder forms, prepared as pastes, applied as a thin or thick coating or blended with oils and greases as particulate additives to enhance anti-friction, -wear and -seize properties of sliding surfaces. Thin solid films or thick bonded or plasma-sprayed coatings of low-friction materials have been around for many decades and used by industry extensively in aerospace, microelectronics, manufacturing and transportation fields [1–8].

Although they are effective in combatting friction and wear under severe operating conditions, most low-friction materials and their coatings suffer from various shortcomings. Except for soft metals, most of them are poor thermal and electrical conductors; hence, they cannot dissipate heat from sliding interfaces or prevent electrical charge build-up. Some are vulnerable to environmental degradations and aging, especially if there is high humidity, sour gases or radiation or if the ambient temperature is very high. For example, at high relative humidity (Rh), the friction coefficients of certain solids (i.e., MoS_2) may increase substantially, and their wear lives become very short [9]. After wearing out (especially in the case of thin solid films), low-friction materials are not easy to replenish or re-supply, in contrast to their liquid counterparts.

Low-friction materials exist in all kinds of physical shapes, chemical compositions and structural morphology. Most popular ones rely on a layered or lamellar crystal structure for high lubricity. Some of these include transition-metal dichalcogenides (such as MoS_2 WS_2, and NbS_2) [1, 2, 4–10] as well as graphite, hexagonal boron nitride (h-BN) and boric acid. Molybdenum disulphide (MoS_2), graphite and boric acid exist naturally and can be extracted

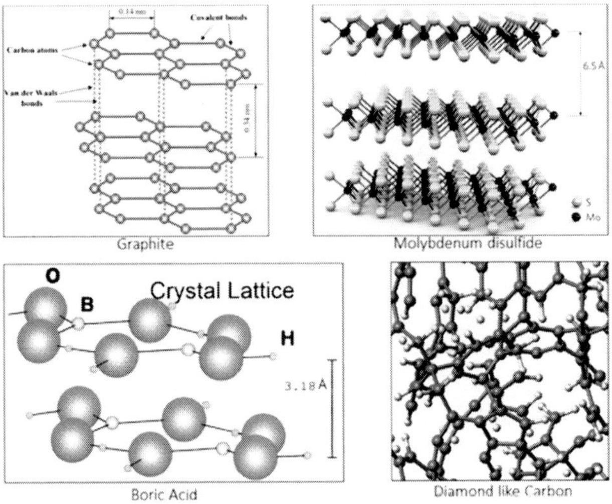

Figure 8.1 Schematic illustration of atomic structures of low-friction materials. Graphite crystal structure is courtesy of Dmitri Kopeliovich (www.substech.com); MoS_2 crystal structure is reproduced with permission from Ref. [10], Copyright (2011) Nature Publishing Group.

from mineral deposits; others are synthetically produced. Besides the layered solids, numerous other types are also self-lubricating, such as soft metals (In, Ag, Pb, Sn, etc.), polytetrafluoroethylene (PTFE), diamond and diamond-like carbon films and certain oxides (B_2O_3, V_2O_5, MoO_3, etc.), fluorides (such as CaF_2, CeF_3) and sulphates [6]. Figure 8.1 illustrates the atomic structure of some of the low-friction materials mentioned. Most often, these materials are referred to as solid lubricants in the tribology literature, mainly because of their inherent and intrinsic self-lubricating capabilities. For the remainder of this chapter, the term "solid lubricants" will be used to represent the broader family of "low-friction or self-lubricating materials."

With recent advances in nanomanufacturing using top-down or bottom-up approaches, numerous low-dimensional nanomaterials with 0 to three-dimensional (3D) architectures have been synthesised and used as novel self-lubricating materials [10–14]. Low-dimensional nature of such materials has enabled them to offer added multifunctionality and performance in demanding

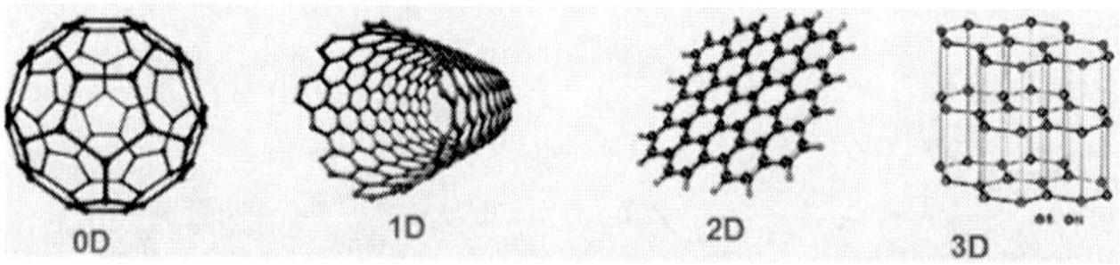

Figure 8.2 Molecular-scale representation of some of the nanomaterials (0D-buckyballs, 1D-nanotubes, 2D-grapbene, 3D-nanoclusters of h-BN, etc.) used as lubricants.

tribological applications. Figure 8.2 shows the molecular-scale structure of some of them. Nanomaterials investigated for use as possible solid lubricants include 0D nano-onions, nano-diamonds and amorphous carbon nano-spheres; 1D nano-tubes, -fibres, -rods and -whiskers; and 2D graphene, h-BN and metal dichalcogenides as well as larger nano-scale particles with various 3D architectures. Such low-dimensional materials were shown to afford much superior friction and wear properties when used as nano-colloidal lubrication additives or in the making of self-lubricating bulk nanocomposites and thin coatings [14]. In particular, several research groups have explored the tribological properties of nano-fullerenes of transition metal dichalcogenides (such as MoS_2 and WS_2) and other nanomaterials as anti-wear and lubricity additives in oils under severe boundary-lubricated sliding conditions [11–18]. At the same time, much attention has been directed to carbon

Nano-onions, nano-tubes, graphene, nano-diamond, etc. and some of these have emerged as potential novel solid lubricants due to their remarkable lubricity as a thin overcoat or in dispersion with oils [12, 18]. Despite all these positive developments, most of these nanomaterials are somewhat expensive and are, therefore, not suitable for general purpose uses in high-volume applications. The main characteristics of traditional and emerging solid lubricants are summarised in Table 8.1.

As is clear from the foregoing, the field of solid lubrication research has been very active for many years. This chapter is intended to provide an overview of these lubricants and recent advances in the field. Obviously, due to their very classic yet evolutionary nature, self-lubricating materials and their coatings have been the subject of many other reviews, book/handbook

Table 8.1 Main characteristics of solid lubricants in tribological applications

Application environment and/or condition	Solid lubricants
Inert or vacuum	Transition-metal dichalcogenides (i.e., MoS_2, WS_2) provide some of the lowest friction coefficients. Graphite, h-BN, boric acid do not work well in such environments.
Pressure	They all can endure very high pressures.
Temperature	Transition-metal dichalcogenides (i.e., MoS_2, WS_2) may oxidise at high temperatures; h-BN is stable up to 1000°C. They can also function at very low and temperatures. Lubricious oxides are also very effective at high temperatures.
Electrical conductivity	Some (graphite) are good electrical conductors, while others (h-BN, diamond-like carbon [DLC]) are insulators.
Radiation	They all are relatively insensitive to nuclear radiation or cosmic rays.
Wear	They all afford excellent wear performance or durability at low speeds and under fretting conditions. Lifetime is determined by lubricant film thickness and wear rate.
Friction	Friction can be very low depending on environment; but high friction is also feasible under certain conditions.
Thermal conductivity and heat dissipation capability	These factors are excellent for metallic lubricants, poor for most inorganic or layered solids.
Aging	They all can be stored for very long times without degradation in performance; however, transition-metal dichalcogenides are affected by moisture.
Industrial hygiene	Since they are in solid states, they pose no industrial hygiene problems due to little or no hazardous emissions.
Compatibility with non-ferrous surfaces	They are all compatible with hard-to-lubricate surfaces (i.e., Al, Ti, stainless steels and ceramics).
Resistance to aqueous and chemically aggressive environments	Most are relatively insensitivity to aqueous environments, chemical solvents, fuels and lubricants.

chapters and technical articles in the past [e.g., 1–4, 6, 9, 17, 19–32]. This chapter covers the most recent developments, with special emphasis on DLC and other multifunctional nanocomposite solid lubricant coatings as well as advanced deposition technologies that seem to afford much superior performance and durability to these coatings under severe application conditions.

8.2 General Characteristics and Classification of Solid Lubricants

The self-lubricating property of most popular solid lubricants (i.e., graphite, h-BN, boric acid, and transition-metal dichalcogenides) is primarily attributed to their lamellar or layered structures, as illustrated in Fig. 8.1. Exceptions are DLC and PTFE, because their structure is not layered; instead, their lubricity is primarily attributed to high chemical inertness, transfer-film-forming abilities and unique tribochemical interactions with test environments and sliding surfaces. Metal-based solid lubricants (In, Ag, Pb, Sn, Au, etc.) owe their lubricity to a multiple slip system and their unique tendency to not work-harden during shear deformation, especially at elevated temperatures. The lubricity of certain oxides, fluorides, and sulphates is largely due to their ability to soften at elevated temperatures and thus attain a very favourable shear property.

In most solid-lubricated surfaces (apart from DLC and other similar solids), after a brief run-in period, the solid lubricant or coating is first deformed and then smeared or transferred onto the opposing surface; hence during steady state, both surfaces are effectively covered by the solid lubricant material [29, 33]. Figure 8.3 shows the progression of transfer layer build-up on a sliding sapphire ball surface in relation to friction coefficient of a Pb-doped MoS_2 film on M50 steel in dry air. In the case of layered or lamellar solids, besides smearing or transferring to the counter face, the atomic layers are increasingly aligned along the shearing interface for much easier slip [33, 34]. There is also the possibility of some of the counter face material wearing in or blending with the transferred layers. Overall, transfer layers play an important role in the run-in and steady-state frictional behaviour of solid lubricant materials [29, 33]. Shear usually occurs within or between the original and transferred layers and determines the extent of friction and wear life.

Previous research has confirmed that most solid lubricants interact with the operating environment or the chemical species in their surroundings. In particular, the presence or absence of certain environmental species in the surrounding atmosphere can

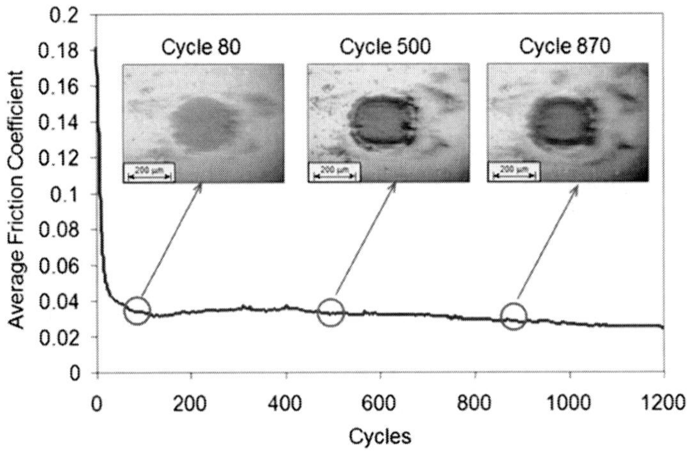

Figure 8.3 Progression of transfer film build up on sapphire ball side and its effect on friction of Pb-Mo-S film under low humidity conditions. Graph is courtesy of Kathryn J. Wahl, US Naval Research Laboratory, Washington DC, USA; inset images are adopted with permission from Ref. [36], Copyright (2007), Springer.

critically affect the friction and wear properties of solid lubricants. For example, some solids provide lower friction and wear in moist environments (such as graphite, h-BN and boric acid [4–6, 17]), while others (such as WS_2 and MoS_2) rely on lack of humidity and oxygen in the surrounding environment for ultra-low friction and wear [1, 6, 9, 30, 32–34]. As shown in Fig. 8.4, the friction coefficient of a MoS_2 coating can increase dramatically when tested in humid air. For chemically stoichiometric and structurally perfect MoS_2 coatings, Martin et al. reported friction coefficients of as low as 0.002 in vacuum [35].

DLC coatings also show strong sensitivity towards the composition of the test environment. Depending on the type of DLC (whether hydrogenated, hydrogen-free or doped), some function better in humid air, alcohol and water, while others require inert or vacuum environments to provide the lowest friction and wear coefficients [26, 27, 36]. Figure 8.5 shows the frictional behaviour of a DLC film in dry and humid environments.

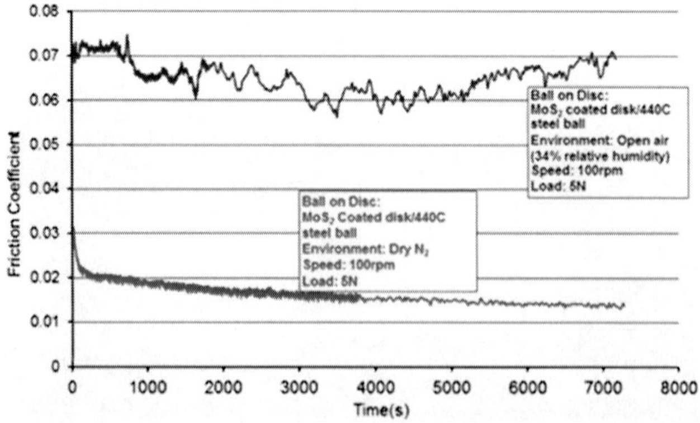

Figure 8.4 Effect of test environment on friction coefficient of MoS_2 solid lubricant coating produced on 440C steel substrate.

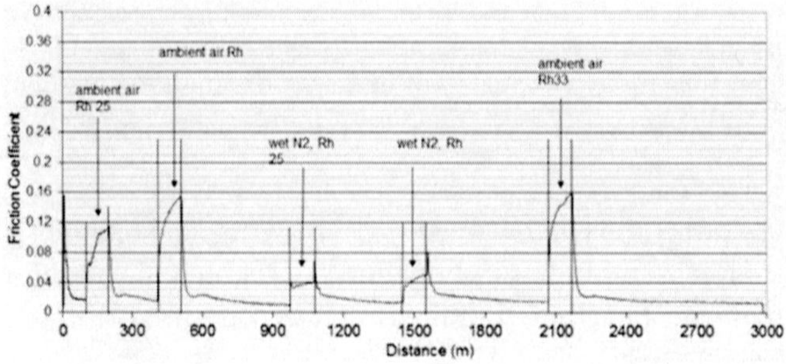

Figure 8.5 Frictional behaviour of a DLC-coated test pair in different environments including dry and wet nitrogen as well as ambient humid air with varying Rh levels. Humid air seems to be more detrimental to friction of DLC than humid nitrogen.

Systematic studies have shown that under liquid lubricated test conditions, the friction and wear behaviour of DLC films could vary a great deal. For example, under water lubricated test conditions, the frictional behaviour of DLC coatings is largely controlled by the extent of adhesive interactions that are in turn controlled by complex tribochemical reactions with water and/or counter face materials [36]. Obviously, the surface reactivity of DLC varies greatly depending on the degree of hydrogenation and counter face materials. For example, when rubbed against steel, tribocorrosion on the steel side and oxidation on the DLC side may give rise to strong tribochemical interactions and this may in turn lead to high and un-steady friction as well as rapid wear. In the case of more chemically inert ceramic counter faces, friction coefficient may be much lower mainly because of less intense tribochemical interactions [36]. With respect to the hydrogenated versus hydrogen-free DLCs, previous research has shown that hydrogen-free DLC coatings provide the lowest friction and wear coefficients under water lubrication conditions [36] presumably due to their more hydrophilic nature ensuring strong wetting and molecular-level capping of sigma bonds of surface carbon atoms as well as reduced tribochemical interactions with counter face materials.

In the case of liquid lubricants and additives, it has been shown that the presence of polar additives (glycerol and other organic friction modifiers) can help reduce boundary friction of hydrogen-free DLC films down to the 0.01 level; by contrast, hydrogenated-DLC coatings show relatively high friction (in the range of 0.05 to 0.1) under such conditions [37]. Overall, previous research has shown that the friction and wear performance of most traditional solid lubricants is very sensitive towards the chemical nature of the test environments; however, recent experimental studies with graphene have shown that its frictional properties are relatively immune or unaffected by the presence or absence of humidity or other species in test environments [38].

In addition to the operating environments, other parameters such as load, speed, type of motion and temperature can also have a strong influence on the lubricity and durability of solid lubricant coatings. Some of the solid lubricants (i.e., MoS_2, boric acid and PTFE) can provide reasonable performance at cryogenic temperatures while other solids (i.e., oxides, fluorides and sulphates) are more suitable for high-temperature applications [3, 6, 21]. Nevertheless, an ambient environment and temperature play an important role in the performance and durability of most DLC, MoS_2 and WS_2 coatings [1, 6, 27–30]. When tested in inert gases or vacuum, these solid lubricants may provide some of the lowest friction coefficients (i.e., below 0.01). Their low-friction properties may persist up to about 400°C. However, at much elevated temperatures in open air, they begin to oxidise, and hence, their easy shear character goes down and the oxide particles generated may adversely impact their lubricity and durability.

Frictional performance of other solid lubricants such as diamond and DLC coatings are also affected by elevated temperatures [39]. In particular, hydrogenated DLCs begin to lose some of their hydrogen and, hence, become hard to shear and wear out much faster due to much weaker structure. Hydrogen-free and nanoalloyed or doped DLCs show much higher resistance to heat-related degradations; hence, their wear lives might be much longer, but their frictional properties may degrade due to increased sigma (or covalent) bond interactions.

Oxide-based solid lubricants work the best at elevated temperatures. In fact, their solid lubricating properties are not that good at room temperature but improve dramatically at elevated temperatures. As explained in previous literature, their shear rheology and, hence, lubricity is strongly related to their ionic potentials, which seem to govern their ability to soften and hence shear easily at sliding contact interfaces to provide low friction [20, 40]. Table 8.2 provides a list of popular solid lubricant materials with their ranges of friction coefficients and proposed lubrication mechanisms.

8.3 Lubrication Mechanism

In general, solid lubricants and their coatings work essentially the same way as their liquid counterparts do. Specifically, when introduced into a sliding contact interface, they first form a thin tribolayer on rubbing surfaces. Under the influence of sliding motion, they then shear easily through interfilm sliding, intrafilm flow and film/substrate or interface slip to accommodate velocity [29, 41]. Figure 8.6 illustrates these velocity accommodation modes. Depending on the test conditions and lubricant type, a combination of all of these modes may be feasible. For example, for a layered solid lubricant such as MoS_2, an interlayer shear mechanism is more prevalent, while for DLC, the lubrication mechanism may be due to interface slip (especially if both surfaces are coated and the DLC type

Table 8.2 General classification of solid lubricants

Classification	Key examples	Typical range of friction coefficient*	Proposed lubrication mechanisms
Lamellar solids	MoS_2	0.002–0.25	Interlayer shear
	WS_2	0.01–0.2	
	h-BN	0.150–0.7	
	Graphite	0.07–0.5	
	H_3BO_3	0.02–0.2	
Carbon-based solids	Diamond	0.02–1	High chemical inertness and ability to form low-shear tribofilms
	DLC	0.003–0.5	
	Graphene	0.15	
	Fullerenes	0.15	
Polymers	ultrahigh molecular weight polyethylene (UHMWPE)	0.1–0.2	High chemical inertness
	PTFE	0.04–0.15	
Soft metals	Ag	0.2–0.35	Shear deformation due to multiple slip systems; lack of work-hardening
	Pb	0.15–0.2	
	Au	0.2–0.3	
	In	0.15–0.25	
	Sn	0.2	
Oxides	B_2O_3	0.15–0.6	Shear deformation; high ionic potential
	Re_2O_7	0.2	
	MoO_3	0.2	
	$CuO - Re_2O_7$	0.3–0.1	
	$CuO - MoO_3$	0.35–0.2	
	$Cs_2O\text{-}MoO_3$	0.18	
	$NiO - MoO_3$	0.3–0.2	

*Friction values given in this table represent friction measurements made on each solid lubricant over a wide range of test conditions, environments and temperatures. The objective here is to show how friction varies depending on test conditions as well as from one solid to another.

is highly hydrogenated). In the case of soft metals and lubricious oxides, the lubrication mechanism may involve intrafilm flow or shear deformation. Transition from one mode to another is also a possibility when and if the test condition or environment changes.

If a solid lubricant is present as a thin film on one of the sliding surfaces, some of it transforms to the opposing side and forms a transfer layer. As sliding goes by, the sliding usually takes place between the original and transferred solid lubricant. Back

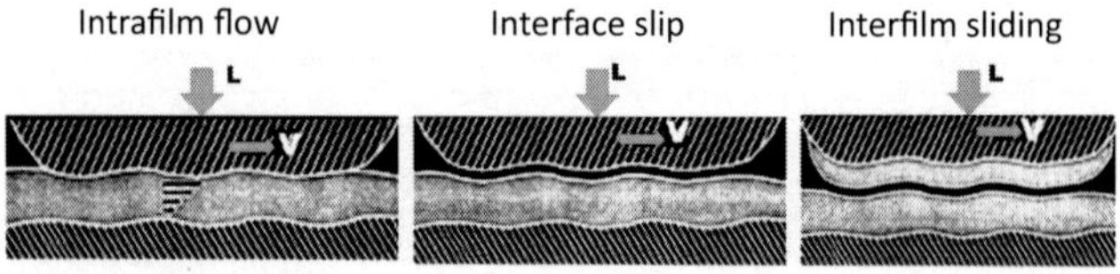

Figure 8.6 Velocity accommodation modes for solid lubricant materials at sliding interfaces.

transfer to the original surface is also possible, and such a process increases the longevity of solid lubricant at the sliding interface. This is thought to be the main reason for long wear life and good frictional behaviour of most layered solid lubricant coatings (otherwise, they would wear off quickly). Under the influence of shear deformation in the tangential direction during sliding motion, atomic layers of solid lubricant film slide over one another with relative ease and provide low friction [1, 2, 6, 21, 29]. Dedicated transmission electron microscopy (TEM) and tribological studies with layered solid lubricants have confirmed that atomic layers indeed assume a more preferred alignment essentially parallel to the sliding direction and thus achieve much easier shear along their atomic layers [33, 34].

Intrinsically, within the layered crystalline structures of transition metal dichalcogenides, graphite, h-BN and boric acid, the interatomic bonding is very strong and mainly of the covalent type. Figure 8.7 shows the planar atomic sheets of h-BN, graphite and WS_2

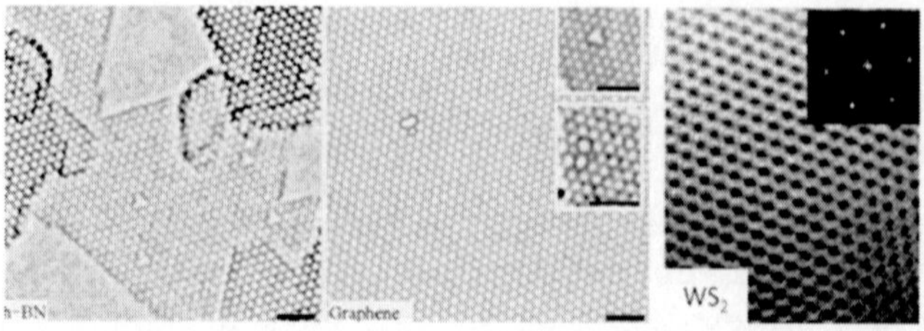

Figure 8.7 Atomic-scale TEM images of h-BN, graphite and WS_2 showing interatomic array within their layers. H-BN and graphene are reprinted with permission from Ref. [42], Copyright (2009) American Chemical Society and WS_2 reprinted with permission from Ref. [43], Copyright (2013) Nature Publishing Group.

[42, 43]. The bonding between the atomic sheets is of the weak van der Waals type and, hence, very weak. When such layered solids are present at a sliding contact interface, they can shear easily along the weak atomic layers and thus provide low friction. However, it is important to remember that a lamellar structure alone may not be sufficient for effective and long-lasting lubrication. In fact, many other lamellar or layered solids (mica, perovskite-type solid oxides, etc.) do not provide low friction when used on sliding surfaces. Furthermore, the presence or absence of other attractive forces (electrostatic, capillary, etc.) and intercalates can markedly affect the lubricity of layered or lamellar solids. In those metal dichalcogenides with easy shear character (such as MoS_2 and WS_2), electrostatic or capillary forces are either absent or very weak, especially in inert or vacuum environments.

In the case of other dichalcogenides such as NbS_2, TiS_2, VS_2 and TaS_2, shear along the layers is more difficult as these solids do not provide low friction despite their layered crystalline structures. Jamison and Cosgrove [19] proposed that the poor lubricating performance of NbS_2, TiS_2, VS_2, TaS_2, etc. is primarily due to the fact that in their layered crystalline structures, a region of negative electrical charge not only concentrates above the sulphur atoms of the layer but further extends into the regions where next or neighbouring layer's chalcogen or sulphur atoms lay. Apparently, this can create an electrostatic attraction since the bottoms of the sulphur atoms are positively charged and attracted to the negative charge extending from the other layer. In the case of MoS_2 and WS_2, the region of negative electrical charge is contained within the layers, and the surfaces of the chalcogen atoms in opposing layers are positively charged, creating an electrostatic repulsion between the layers and thus making these solids very slippery compared to other chalcogenides [19]. In general, these previous studies have shown that not all layered solids are good solid lubricants. The type and magnitude of interlayer bonding are also important.

Besides the intrinsic factors discussed above, the lubrication performance of solid lubricants and their coatings may be strongly affected by such extrinsic factors as the composition of the test environment, temperature and other test conditions [6, 27, 36]. As briefly mentioned earlier, the lubricity of MoS_2 and DLC is strongly affected by the presence or absence of water vapours, oxygen or gaseous species in the test environments. In particular, certain DLCs provide low friction only in humid air, while other DLCs and MoS_2 work the best in inert or vacuum-like environments. This unusual lubrication behaviour is one of the most studied subjects in the open literature on these materials, though not yet fully understood. It turns out that the underlying tribochemical interactions in the presence of certain environmental species are rather complex. So far, suggestions have been made that the enhanced lubricity of hydrogen-free DLCs in a humid environment may be related to the weakening effect that water molecules have on the strong covalent or σ-bonds between the carbon atoms of opposing surfaces. In the absence of humidity (such as inert or vacuum test chambers), the friction coefficient of these DLCs could be as high as 1. The poor lubricity of MoS_2 in a humid environment is believed to result from the tribochemical reaction of water molecules with MoS_2, creating a secondary film that is not as easy to shear. In depth surface studies have confirmed that MoS_2 and other dichalcogenides form MoO_3 or other types of complex oxides and oxysulphides when oxygen and water molecules are present in surrounding environments. Other factors that can play major roles in the lubricity of solid materials include ambient temperature, contact pressure, film thickness and basal plane orientation (in the case of thin solid lubricant coatings), aging and purity [6].

As mentioned above, numerous metallic, polymeric and ceramic-based solid lubricants do not have a layered crystal structure, yet they too can afford very low friction and wear to sliding surfaces when applied as thin solid coatings [6]. In particular, In, Pb, Ag and Sn can provide fairly low friction coefficients when used on sliding surfaces [44]. The self-lubricating property of these solids has been attributed to the existence of multiple shear systems in their crystals and the lack of work-hardening. Numerous polymeric materials can afford low friction and wear to sliding surfaces [25]. PTFE and UHMWPE are prime examples. Diamond-like carbon coatings are amorphous but can afford some of the lowest friction and wear coefficients [27, 45]. The lubricity of these materials is largely due to their high chemical inertness (due to hydrogen termination of carbon sigma bonds) and their ability to interact with sliding surfaces or surrounding environments, forming low shear tribolayers [26, 27].

Achieving and maintaining smooth and long-lasting operation at high-temperatures have always been very challenging in industry. Certain oxides and carbide-based ceramics can provide low friction and wear to sliding contact interfaces at high temperatures. In particular, the oxides of rhenium, molybdenum, vanadium, lead and boron are well-known for their low friction at elevated temperatures [3, 20, 21]. Attempts have been made to produce nanocomposite coating architectures that can offer stable friction and wear performance across broad temperature ranges and environmental conditions [46, 47]. One of the most natural choices for lubricant in high-temperature, open-air environments is those oxides that can provide reasonable friction and wear performance at elevated temperatures. Many lubricious oxides have in the past been developed and demonstrated to provide adequate lubricity up to 800°C [20, 21]. However, sub-stoichiometric oxides of several transition metals, known as Magneli phases, have planar lattice faults that result in favourable crystallographic shear planes with reduced shear strength at elevated temperatures and hence low friction. There are also several transition metal oxides (MO_x, WO_x, VO_x, etc.) and other molybdate and tungstate compounds for which Erdemir suggested that the ionic potentials of such metal oxides and their compounds are related to their lubricity [20, 40].

8.4 Fabrication Methods for Solid Lubricant Coatings

8.4.1 *Traditional Methods*

Traditional burnishing or spraying of powdered, aerosol or colloidal forms of solid lubricants still represent the easiest and the least-expensive fabrication methods [2, 6, 8, 21, 22]. However due to poor adhesion, non-uniformity in thickness and high porosity, coatings produced by these methods may not be suitable for more demanding applications where higher performance, durability and efficiency are of great importance. Fabrication of solid lubricants in a paste, resin-bonded or composite form is also used to achieve much longer wear life. Epoxy resins, plastics and polymers can be blended with powders of MoS_2, graphite and PTFE at appropriate concentrations to fabricate bulk self-lubricating composites [22]. In the case of resin-bonded thick coatings, solid lubricant powders (called "pigments") can be mixed with a binder material (in liquid or powder form) and then applied on surfaces by spraying, dipping, brushing, sol–gel, etc. as thick coatings. Most often a special surface treatment may be needed to achieve strong bonding. Binder materials are typically made of thermoplastic and/or thermosetting resins while self-lubricating ingredients could be made of any type of known solid lubricants in powder and fibre forms. Binders prevent wear or easy removal of the lubricating solids and protect them against environmental degradation or corrosion. Due to their thickness, bonded solid lubricant coatings can last for a very long time.

8.4.2 *Physical and Chemical Vapour Deposition Methods*

During the last three decades, with the emergence of very robust physical and chemical vapour deposition (PVD and CVD) technologies, most solid lubricants are now applied on a tribological surface as a strongly bonded, chemically unique and structurally dense tribo-coating. In these cases, for increased multifunctionality and long durability, film thickness, surface roughness, structural morphology and chemical stoichiometry have to be controlled very closely as they play critical roles in overall tribological performance [6, 24, 29]. Super-low frictional values (i.e., 0.002) are often provided by films with very dense morphology, strong bonding, preferred basal orientation, low amount of impurity and residual stress and high chemical stoichiometry [35]. Again, such super-low friction values are achieved in an ultra-high vacuum where there is little or no oxygen, water molecules or other contaminants on the sliding surfaces. In moist air, the friction coefficients are substantially higher (i.e., 0.05–0.25) [9, 35].

In the past, conventional sputtering or ion-plating methods were generally used to deposit solid lubricant films on tribological surfaces; however, these methods often led to an anisotropic coating structure with open columnar morphology and hence lacked durability and multifunctionality in actual applications. When used under severe loading conditions, the resultant coatings could easily deform or fracture through the weak column boundaries [24]. In practice, films with open columnar structures also tended to wear out rather quickly, and they were more vulnerable to oxidation and other types of environmental degradation as well. To circumvent these problems, researchers have developed novel approaches such as intense ionic bombardment to achieve much denser film morphology during deposition or built a nanostructured and composite coating architecture in which the grain size is very small and the morphology is not necessarily open columnar [47]. As a result, these coatings are not easily deformed or fractured along their grain boundaries under loading (the average grain size of such nanostructured and nanocomposite coatings was in a range that is not well suited for crack initiation and propagation); as a result, these coatings could last for a very long time in tribological applications.

8.4.3 *Advanced PVD Methods*

With recent advances in deposition technologies such as pulse DC, arc-PVD, pulsed laser deposition (PLD) and high-power impulse magnetron sputtering (HIPIMS) [48–50], the latest films exhibit some of the highest chemical and structural qualities; hence, much improved multifunctionality and durability in industrial applications. Because of the highly energetic nature of their gas discharge plasmas, these techniques are able to establish much stronger bonding between solid lubricant films and underlying substrate materials and hence much longer durability. The film microstructure is also highly refined and very dense, which are essential for superior resistance to corrosion and oxidation. The quality of such films was further improved with the use of sophisticated computer simulation and modelling methods, which help more closely match the coating properties with those of the substrate materials through graded interface engineering and better internal stress control. Accordingly, in today's more demanding applications, modern PVD and CVD methods are used extensively to achieve a range of functionalities besides very long wear lives and much lower friction coefficients in applications [1, 6, 23, 51]. These methods enable the deposition of high-density nanostructured and nanocomposite coatings with excellent adhesion and surface finish, which are very important for film durability and frictional performance [47–51].

Among the many methods mentioned above, magnetron sputtering is perhaps the most widely used commercially for the deposition of many types of solid lubricant films. With high-rate sputtering capability through a closed-field unbalanced magnetic field, very thin or thick solid lubricant coatings are feasible in monolithic or composite forms consisting of nano-to-micro scale hard and/or soft layers and phases [51, 52]. These layers or phases can be sputtered from one of the several targets in sequence or all together to build the nanolayered or nanocomposite films, affording a wide range of multifunctionality in actual applications. Figure 8.8 shows a schematic illustration of multitarget PVD systems. One major drawback of conventional magnetron sputtering is that the available ionisation efficiencies and ion energies are relatively low; hence, atomic-scale mixing at the film-substrate interface is not at a desirable level. Grain morphology is most often columnar and the density is somewhat poor [24]. In tribological applications, inadequate adhesion, low density and columnar morphology may adversely impact coating durability, especially under severe application conditions such as bearing or gear applications where strong film-to-substrate adhesion and high structural density are of great importance.

In most magnetron sputtering protocols, argon gas is used for sputtering purposes. In the plasma, it is ionised by constant collisions with electrons trapped by the magnetic field in the vicinity of targets. These ions, due to their positive charge, are then accelerated to the surface of negatively biased target material. In the case of RF and pulsed DC sputtering, reverse polarity is generated to discharge any electrical build-up and hence enable sputtering of insulating materials. In pulsed DC, the sputtering takes place during negative voltage pulses. During low-voltage positive pulses, electrons are attracted to the target surface and, thus, help discharge any accumulated positive charge from the target surface. The duration of negative and positive pulses on the target surface delineates the duty cycle, which is an important parameter to adjust for control of sputtering efficiency.

Besides magnetron sputtering, the cathodic arc PVD method is also used to deposit certain solid lubricant films, such as DLC. Unlike magnetron sputtering in which high voltage and low current are used, arc PVD involves the use of high-current/low-voltage power sources, with typical range being 50–1000 A and 20–200 V. Under such conditions, high electrical currents create arc spots, which swiftly move on the surface of the targets due to overlapping electrical and magnetic fields. As a result, the rapid heating vaporises atoms and/or clusters from these arc spots in the form of highly ionised species, which are then accelerated towards the negatively biased surfaces of substrates [47, 48]. One major drawback of this method is that along the ionised species, it can also dislodge larger nano-to-micro scale particles or droplets from the surface of the target material and these particles can also be incorporated into the growing films, with the result that the film has a very inhomogeneous morphology and rough surface finish. For most tribological applications, these properties are not acceptable; hence, most often a post-deposition polishing is required. Droplets can be eliminated by filtering techniques that use special magnets to deflect the stream of ionised species away from the path of the droplets at an angle so that the substrates are saved from line-of-sight trajectory of massive droplet impingements.

To circumvent the deficiencies of conventional magnetron sputtering, a HIPIMS method was developed and like arc-PVD, has been demonstrated to afford very strong film-to-substrate bonding. Because the fraction of ionised species (i.e., 30 to 90%) is much higher than that of conventional sputtering techniques, the structural density of the resultant coatings is very high. Most often, dense film morphology results in much better mechanical and, hence, tribological properties. This technique also overcomes the droplet formation issue, which is a handicap for arc-PVD. Accordingly, for more demanding tribological applications, HIPIMS holds great potential in terms of achieving superior film durability and multifunctionality under severe operating conditions. HIPIMS uses high power in repeated short pulses for the cooling channels to maintain the target temperature below the melting point. HIPIMS relies on special power supplies that use sophisticated pulse-forming networks and circuitry based on recent breakthrough developments in high-power electronics, which can deliver short but high-current and high-voltage pulses [50]. These power sources are portable and can be adopted to an existing magnetron sputtering system to achieve HIPIMS capability.

Besides the methods mentioned above, PLD is used for the deposition of solid lubricant films. Due to problems in scale-up, surface irregularities and substrate selection, PLD is often used as a research tool and in a few specialty applications [27, 53].

8.4.4 *Advanced CVD Methods*

In addition to the traditional and emerging PVD techniques mentioned above, CVD techniques are frequently used for the deposition of certain solid lubricant coatings. In particular, plasma-activated or -enhanced CVD has been routinely used for many years for the deposition of DLC films [26]. Other CVD methods (metal–organic CVD, atomic layer deposition [54, 55] and hot-filament or microwave CVD [56]) are also available for the deposition of very thin or thick self-lubricating coatings such as MoS_2 and micro- and nanocrystalline diamond films. Compared to DLC, diamond coatings require very high deposition temperatures (as high as 1000°C) in both hot-filament and microwave CVD methods. The same methods may also be used to deposit nanocrystalline diamond films with grain sizes in the range of 2–10 nm, affording very low friction and wear in tribological applications [57].

8.4.5 *Deposition of Nanolayered, Doped or Nanoalloyed Solid Lubricants*

With the use of advanced PVD and CVD deposition technologies mentioned above, it has also become easy to dope or alloy such lubricants at nanoscales to further increase their multifunctionality and thus improve their tribological performance in practical applications [33, 34, 46, 47, 51, 58–60]. The same technologies are also used in the deposition of hard diamond and DLC coatings with

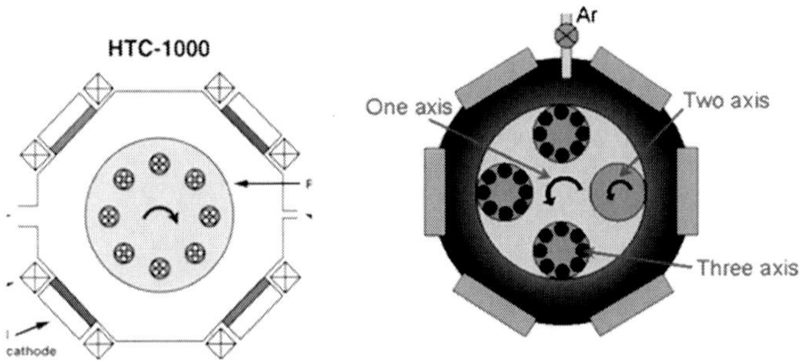

Figure 8.8 Schematic illustration of modern PVD methods with four and six target configurations. Targets can be operated in hybrid sputtering and arc-PVD modes as well as with HIPMS. Rotational capability for each sample holder with three independent axis can help achieve very uniform coverage and good stoichiometry in coatings. Reprinted with permission from Ref. [51], Copyright (2009), Taylor & Francis Group, LLC.

various alloying or doping elements. Such DLC coatings have become very popular in recent years due to their much superior functionality in industrial applications [52]. Systematic tribological studies have confirmed that these nanoalloyed coatings are truly multifunctional as they were shown to be far more durable than their monolithic analogs in comperative testing and relatively insensitive towards the presence of moisture and other environmental species in surrounding environments. Furthermore, doped coatings can have a highly disordered and amorphous film morphology, which, in turn, greatly improves their wear lives and resistance to environmental degradation [33, 34]. It is believed that doped or nanoalloyed solid lubricant coatings have favourable basal plane alignments, increased hardness and stiffness, much denser structure and higher resistance to oxidation, all of which can positively influence wear life and lubricity of such coatings. In addition to doping, solid lubricant films can be produced in nanolaminate or layered fashions using hybrid systems with multiple targets (see Fig. 8.8). For example, while a thicker layer is made of a solid lubricating phase, thinner phases are made of metallic or ceramic layers in an alternating fashion [58]. Films produced according to this method may possess much superior endurance and friction coefficients compared to

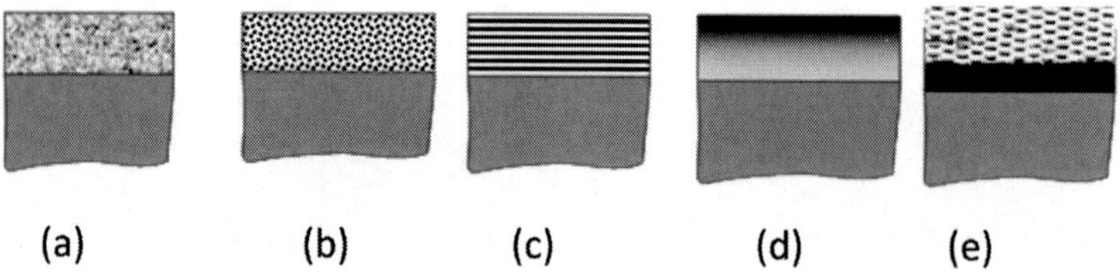

Figure 8.9 Schematic representation of various design architectures for nanostructured and composite coatings that are essential for greater multifunctionality: (a) doped, (b) nanocomposite, (c) nano-layered, (d) gradient and (e) nano-textured. Reprinted with permission from Ref. [51], Copyright (2009), Taylor & Francis Group, LLC.

single-phase or monolithic coatings that are sputtered from a target. In these hybrid systems, cathodic-arc and HIPIMS power sources may be used for cleaning the substrate surfaces prior to deposition and forming a fairly thick graded interface, which is key to achieving strong adhesion [47–50]. Figure 8.9 illustrates some of the design considerations for nanocomposite and nanostructured solid lubricant films that are in use today.

Most of the deposition methods reviewed above are available in industrial-scale production capacities. To increase the deposition rate and overcome some of the other deficiencies of single-source deposition methods, systems incorporating more than one source are now available. These hybrid deposition processes may encompass conventional magnetron sputtering along with arc-PVD or HIPIMS in one system (see Fig. 8.8). These systems have multiple targets, some of which are attached to the pulse-DC and/or arc-PVD power sources to provide flexibility in producing the exotic composite coating architectures that are more desirable for some special application. These targets can be employed selectively or in combination to alloy solid lubricant coatings with a variety of metals (Au, Ni, Ti, Pb, C, etc.) and compounds. Overall, recent advances in PVD and CVD deposition technologies have paved the way for the development of more robust and longer lasting solid lubricant coatings. As a result, chemical stoichiometry, structural morphology, surface topography and the density of these films improved tremendously, all of which resulted in increased durability and better tribological performance.

Figure 8.10 Examples of industrial applications for solid lubricants in various piston assemblies, fuel injectors, bushings, polymeric seals, bearings, gears, sleeves and liners.

8.5 Industrial Practice

Since the beginning of the last century, solid lubricants have successfully met many critical application needs in various industrial sectors, including aerospace, transportation, manufacturing, microelectronics, power generation and high-temperature forming or forging operations. Figure 8.10 shows some of the mainstream applications. At present, several kinds of metal-, ceramic- and polymer-matrix composites are available with graphite, MoS_2 disulphide, PTFE and h-BN particles. Electroless nickel, chromium and nickel–phosphorus coatings that contain small amounts of nanoscale graphite, MoS_2, PTFE and diamond particles have been in use for a long time because of their self-lubricating properties. For example, common self-lubricating polymer composites are blended with graphite and MoS_2 particles. Many sealing materials (especially the dynamic kinds) function flawlessly; thanks to self-lubricating PTFE, carbon or graphite fillers.

Cold and hot spraying can produce fairly thick solid lubricant films on tribological surfaces of moulds, dies and rollers of many metal-forming or stamping operations. Besides these, simple powder lubrication, burnishing, aerosol spraying and sprinkling methods are also common practices for the application of solid lubricants. In oil-lubricated mechanical components, thin film or bonded forms of solid lubricants are often used as a run-in or backup lubricant; specifically, when and if the oil film fails or loses its effectiveness, the self-lubricating coating can still provide low friction and wear. In the case of nano- or micro-composites, the major phases or matrices are selected from ceramic, metallic or polymeric materials; the minor phases or constituents are made up of the self-lubricating solids.

Regardless of the solid lubricant films, strong cohesion between the grains and strong adhesion or bonding to the substrate surface are critically important for achieving long endurance life, high load-bearing capacity and desirable performance in actual applications. Poor cohesion between the grains or layers or different phases, high levels of internal stresses and weak adhesion between the coating and substrate may cause adhesive delamination and/or cohesive fracture or fragmentation of coatings. In particular, if the level of internal stress is too high, the film fails or delaminates at the interface. Coating adhesion can be improved by high energy bombardment of the substrate surface at the onset of the deposition process to remove much of the contaminant or oxide layer, followed by physical intermixing and chemical reaction between the coating and substrate materials. With the latest techniques such as arc-PVD and HIPIMS, which promote high levels of interdiffusion or mixing between coating and substrate atoms at the interface, the adhesion and cohesion of the coatings are much better and less prone to failure.

In recent years, much emphasis has been placed on the design and deposition of solid lubricant coatings with gradient, duplex, multilayered and nanocomposite architectures for greater multifunctionality and hence superior performance as illustrated in Fig. 8.9. For example, self-lubricating MoS_2 films are used routinely by the aerospace industry; however, in single-phase or monolithic form, MoS_2 may not last very long. Accordingly, some metallic ingredients such as Ni, Pb, Ag, Au, C, Ti, etc. have been introduced into the MoS_2 structure to achieve longer wear life and reduced environmental sensitivity. Most of these coatings are produced by modern sputtering systems having multiple targets and sample holders with high-precision rotational capabilities. Nano-layered coatings prepared by such systems consist of discrete or graded layers of two or more self-lubricating solids in an alternating fashion and the thickness of each layer may vary from less than 1 nm to several nanometres. Tribological studies on such coatings have confirmed much superior friction and wear properties compared to those that are provided by single-phase films. Such coatings are utilised in various dry-sliding and manufacturing applications (e.g., cold forming, stamping, self-lubricating bearings and gears for aerospace and vacuum applications). Superior performance of these coatings is largely attributed to the much higher mechanical (i.e., better hardness, toughness and stiffness), structural (i.e., more equiaxed, more uniform and denser and smaller grain morphology) and chemical (i.e., superior chemical stoichiometry, inter-mixing, etc.) characteristics. Due to the dense structure, these coatings also provide superior protection against corrosion and oxidation. They are often used in high-precision positioning devices for aerospace and nanomanufacturing applications, where surface finish and dimensional accuracy are important.

In the case of DLC coatings, many metallic dopants or alloying elements (such as W-, Ti- and Cr-doped films) have been used to increase their functionality and durability [50]. With the introduction of industrial-scale, more-robust coating systems, such multifunctional DLC films have become easy and inexpensive to produce. Overall, these and other coatings have made a significant positive impact on many industrial sectors. They are used in all kinds of manufacturing industries to prevent wear and material transfer during drawing, stamping, moulding and rolling operations. In the transportation sectors, fuel injectors, piston pins and rings, tappets, gears, etc. in high-performance engines are now coated with low-friction coatings such as DLC. Long-lasting magnetic storage media and razor blades owe their extraordinary performance to very thin DLC coatings on their surfaces [26, 27]. In coming years, there is no doubt that such coatings will be used in several other critical applications since the operating conditions are becoming far more stringent and the liquid lubricant alone is not able to meet the increasingly more demanding operational conditions.

8.6 Summary and Future Directions

This chapter attempted to provide an overview of solid lubricants and their coatings that offer low friction and high wear resistance for many demanding tribological applications. Solid lubricants fill an important industrial need and are the only choices for those applications where the use of liquid lubricants is not feasible or practical. The operating conditions of future tribosystems will undoubtedly become far more stringent, and hence, the uses of solid lubricants with greater multifunctionality (either alone or in combination with liquid lubricants) are expected to grow. Some of the major shortcomings of solid lubricants are that they cannot provide consistent and low friction over broad test conditions, temperatures and environments. In particular, the performance of most layered or lamellar solid lubricants is strongly dependent on environmental conditions. For example, graphite needs moisture for good performance, while MoS_2 provides good lubricity in inert or vacuum environments. Nano-scale doping or alloying can improve their functionality and reduce sensitivity to environmental effects. In most modern solid lubricant coatings, nanoalloying has become a very routine practice in recent years.

Solid lubricants can be used in their most primitive forms as pellets, flakes or powders; more preferably, they can be incorporated into bulk materials or coatings as self-lubricating ingredients in a composite form. For solid lubricant coatings, strong adhesion to substrate material is a must for long service life and greater multifunctionality. Modern PVD methods are quite capable of providing strong bonding between solid lubricant films and their substrates. Without strong bonding, coatings delaminate or wear out very quickly. For applications involving high temperatures, most transition metal dichalcogenides become unsuitable. There exist some lubricious oxides, fluorides and other compounds that can provide reasonable lubricity and wear performance. A series of adaptive or chameleon coatings with strategically selected ingredients may afford greater multifunctionality and lubricity over broader temperature ranges.

By and large, traditional lamellar solid lubricants such as graphite, h-BN and transition metal dichalcogenides constitute the great majority of the solid lubricants in use today. Diamond, DLC and other coatings are also making a strong entry into industrial fields as self-lubricating coatings. In particular, DLC has been widely adopted as a good alternative to traditional solid lubricants.

Polymeric materials such as PTFE and UHMWPE are also used as solid lubricants for applications where cost, weight, corrosion and biocompatibility are the major considerations. In short, solid lubricants have been around for a very long time, and they have fulfilled some important and critical tribological needs in the past. There is no doubt that with the use of advanced PVD and CVD methods, they will attain even much higher multifunctionality and hence superior performance and durability in more demanding tribological applications.

Acknowledgement

This work is supported by the U.S. Department of Energy, Basic Energy Sciences, Office of Energy Efficiency and Renewable Energy, under contract # DE-AC02-06CH11357.

References

1. T.W. Scharf, S.V. Prasad, Solid lubricants: A review, *J. Mater. Sci.* 48 (2013), 511–531.
2. J.K. Lancaster, *CRC Handbook of Lubrication: Theory and Practice of Tribology*, Vol. II, (ed. E. R. Booser), *Solid Lubricants*, Boca Raton, FL: CRC Press, Inc., (1984), pp. 269–290.
3. H.E. Sliney, Solid lubricant materials for high temperatures: a review, *Tribol. Int.,* 15 (1982), 293–302.
4. R.H. Savage, Graphite lubrication, *J. Appl. Phys.*, 19 (1948), 1–10.
5. J.M. Martin, T. LeMogne, C. Chassagnette, M.N. Gardos, Friction of hexagonal boron nitride in various environments, *Tribol. Trans.*, 35 (1992), 463–472.
6. A. Erdemir, *Handbook of Modern Tribology*, (ed. B. Bhushan), *Solid Lubricants and Self-lubricating Films*, Boca Raton, FL: CRC Press, (2001), pp. 787–818.
7. *Memorandum on Solid Lubricants,* Great Britain Department of Scientific and Industrial Research, Bulletin No. 4, 1920.
8. M.E. Campbell, J.B. Loser, E. Sneegas, *Solid Lubricants: A Technology Survey*, NASA SP-5059, (1966).

9. W.O. Winer, Molybdenum disulfide as a lubricant: A review of the fundamental knowledge, *Wear*, 10 (1967), 422–452.
10. B. Radisavljevic, A. Radenovic, J. Brivio, V. Giacometti, A. Kis, Single-layer MoS_2 transistors, *Nat. Nanotechnol.*, 6 (2011), 147–150.
11. R. Tenne, Inorganic nanotubes and fullerene-like nanoparticles, *Nat. Nanotechnol.*, 1 (2006), 103–111.
12. N. Matsumoto, K. Mistry, J.-H. Kim, O.L. Eryilmaz, A. Erdemir, H. Kinoshita, N. Ohmae, Friction-reducing properties of onion-like carbon under high contact pressure in liquid lubricant, *Tribology*, 6 (2012), 116–120.
13. L. Rapoport, N. Fleischer, R. Tenne, Applications of WS_2 (MoS_2) Inorganic nanotubes and fullerene-like nanoparticles for solid lubrication and for structural nanocomposites, *J. Mater. Chem.*, 15 (2005), 1782–1788.
14. J.M. Martin, N. Ohmae, *Nanolubricants*, West Sussex, UK: John Wiley & Sons, Ltd., (2008).
15. M.K. Dubey, J. Bijwe, S.S.V. Ramakumar, PTFE-based nanolubricants, *Wear*, 306 (2013), 80–88.
16. Y. Kimura, T. Wakabayashi, K. Okada, T. Wada, H. Nishikawa, Boron nitride as a lubricant additive, *Wear*, 232 (1999), 199–206.
17. A. Erdemir, *Nanolubricants*, (eds., J.-M. Martin, N. Ohmae), *Boron-Based Solid Nanolubricants and Lubrication Additives*, West Sussex, UK: John Wiley & Sons, Ltd. (2008), pp. 203–224.
18. L.J. Pottuz, N. Ohmae, *Nanolubricants*, eds., J.-M. Martin and N. Ohmae, *Carbon-based Nanolubricants, in Nanolubricants* West Sussex, UK: John Wiley & Sons, Ltd. (2008), pp. 93–142.
19. W.E. Jamison, S.L. Cosgrove, Friction characteristics of transition-metal disulfides and diselenides, *Tribol. Trans.*, 14 (1971), 62–72.
20. A. Erdemir, A crystal-chemical approach to lubrication by solid oxides, *Tribol. Lett.*, 8 (2000), 97–102.
21. M.D. Kanakia, M.B. Peterson, *Literature Review of the Solid Lubrication Mechanisms*, Southwest Research Institute, Interim Report, BFLRF #213, San Antonio, TX, 6–186, (1987).
22. R.M. Gresham, *CRC Handbook of Lubrication: Theory and Practice of Tribology*, Vol. III, ed., E. R. Booser, *Bonded Solid Lubricants*, Boca Raton, FL: CRC Press (1997), pp. 167–181.
23. A. Erdemir, Engineered tribological interfaces for improved boundary lubrication, *Tribol. Int.*, 38 (2005), 249–256.

24. T. Spalvins, A review of recent advances in solid film lubrication, *J. Vac. Sci. Technol. A.*, 5 (1987), 212–219.
25. K. Friedrich, Z. Lu, A.M. Hager, Recent advances in polymer composites tribology, *Wear*, 190 (1995), 139–144.
26. C. Donnet, A. Erdemir, *Tribology of Diamondlike Carbon Films: Fundamentals and Applications,* eds. C. Donnet, A. Erdemir, *Diamondlike Carbon Films: A Historical Overview*, (Springer), (2008), pp. 1–10.
27. A. Erdemir, C. Donnet, Tribology of diamondlike carbon films: current status and future prospects, *J. Phys. D: Applied Physics*, 39 (2006), R311–R327.
28. P. Sutor, Solid lubricants: overview and recent developments, *MRS Bull.*, 16(5) (1991), 24–30.
29. I.L. Singer, *Fundamentals of Friction: Macroscopic and Microscopic Processes*, eds. I.L. Singer, H.M. Pollock, *Solid Lubrication Processes,* NATO-ASI Series, Vol. 220, London: Kluwer Academic Publ. (1992), pp. 237–261.
30. C. Donnet, A. Erdemir, Historical developments and new trends in tribological and solid lubricant *Coatings, Surf. Coat. Technol.,* 180–181 (2004), 76–84.
31. E.L. McMurtrey, *Lubrication Handbook for the Space Industry. Part A: Solid Lubricants*, NASA TM-86556 (1985).
32. A.R. Lansdown, *Molybdenum Disulphide Lubrication,* Tribology Series, 35, (ed. D. Dowson), Amsterdam: Elsevier (1999).
33. K.J. Wahl, D.N. Dunn, I.L. Singer, Wear behavior of Pb-Mo-S solid lubricating coatings, *Wear*, 230 (1999), 175–1832.
34. T. Polcar, F. Gustavsson, T. Thersleff, S. Jacobson, A. Cavaleiro, Complex frictional analysis of self-lubricant W-S-C/Cr coating, *Faraday Discussions*, 156 (2012), 383–401.
35. J.M. Martin, C. Donnet, T. LeMogne, T. Epicier, Superlubricity of molybenum disulfphide, *Phys. Rev.* B48 (1993), 10583–10586.
36. D.C. Sutton, G. Limbert, D. Stewart, R.J.K. Wood, The friction of diamond-like carbon in a water environment, *Friction*, 1 (2013), 210–221.
37. M. Kano, DLC coating technology applied to sliding parts of automotive engine, *New Diamond Front. Carbon Technol.*, 16 (2006), 201–210.
38. D. Berman, A. Erdemir, A.V. Sumant, Grapnen: a new emerging lubricant, *Materials Today* (2013), in press.
39. A. Erdemir, G.R. Fenske, Tribological performance of diamond and diamondlike carbon films at elevated temperatures, *Tribol. Trans.,* 39 (1996), 787–794.

40. A. Erdemir, A crystal chemical approach to the formulation of self-lubricating nanocomposite coatings, *Surf. Coat. Technol.*, 200 (2005), 1792–1796.
41. S.D. Dvorak, K.J. Wahl, I.L. Singer, Role of third bodies in friction and wear of protective coatings, *J. Vac. Sci. Technol. A.*, 21 (2003), 232–240.
42. C. Jannik, J.C. Meyer, A. Chuvilin, G. Algara-Siller, J. Biskupek, U. Kaiser, Selective sputtering and atomic resolution imaging of atomically thin boron nitride membranes, *Nano Lett.*, 9 (2009), 2683–2689.
43. A. Berkdemir, H.R. Gutiérrez, A.R. Botello-Méndez, N. Perea-López, A.L. Elías, C.-I. Chia, B. Wang, V.H. Crespi, F. López-Urías, J.-C. Charlier, H. Terrones, M. Terrones, Identification of individual and few layers of WS2 using Raman Spectroscopy, *Nature Scientific Reports,* 3, Article number 1775 (2013).
44. M. Maillat, A.K. Chattopadhya, H.E. Hintermann, Preparation of silver coatings to obtain low friction in alternating sliding at 570°C, *Surf. Coat. Technol.*, 61 (1993), 25–29.
45. A. Erdemir, The role of hydrogen in tribological properties of diamond-like carbon films, *Surf. Coat. Technol.*, 146–147 (2001), 292–297.
46. A.A. Voevodin, J.S. Zabinski, Supertough wear-resistant coatings with *Chameleon* surface adaptation, *Thin Solid Films*, 370, 223–231 (2000).
47. A. Erdemir, A.A. Voevodin, *Handbook of Deposition Technologies for Films and Coatings Science, Applications and Technology,* (ed. P. Martin), *Nanocomposite Coatings for Severe Applications*, Amsterdam, Netherland: Elsevier, (2010), pp. 679–715.
48. A. Anders, *Cathodic Arcs: From Fractal Spots to Energetic Condensation*, New York, USA: Springer (2008).
49. D.M. Sanders, A. Anders, Review of cathodic arc deposition technology at the start of the new millennium, *Surf. Coat. Technol.*, 133–134 (2000), 78–90.
50. J.T. Gudmundsson, N. Brenning, D. Lundin, U. Helmersson, High power impulse magnetron sputtering discharge, *J. Vac. Sci. Technol. A*, 30 (2012), 030801–35.
51. A. Erdemir, O.L. Eryilmaz, M. Urgen, K. Kazmanli, N. Mehta, B. Prorok, *Nanomaterials Handbook* (ed. Y. Gogotsi), *Tribology of Nanostructured Materials and Coatings*, Taylor & Francis Group, LLC, (2006), pp. 685–711.
52. J.C. Sánchez-López, A. Fernández., *Tribology of Diamondlike Carbon Films: Fundamentals and Applications*, eds. C. Donnet, A. Erdemir, *Doping*

and Alloying Effects on DLC Coatings, New York: Springer (2008), pp. 311–338.

53. J.S. Zabinski, M.S. Donley, S.V. Prasad, N.T. McDevitt, Synthesis and characterization of tungsten disulphide films grown by pulsed laser deposition, *J. Mater. Sci.*, 29 (1994), 4834–4839.
54. T.W. Scharf, R. Diercks, P. Gorman, S.V. Prasad, M.T. Dugger, Atomic layer deposition of tungsten disulphide solid lubricant nanocomposite coatings on rolling, element bearings, *Tribol. Trans.*, 52 (2009), 284–292.
55. G.L. Doll, B.A. Mensah, H. Mohseni, T.W. Scharf, Chemical vapor deposition and atomic layer deposition of coatings for mechanical applications, *J. Thermal Spray Technol.*, 19 (2010), 510–516.
56. P.K. Bachmann, W. Enckevort, Diamond deposition technologies, *Diam. Rel. Mater.*, 1 (1992), 1021–1034.
57. A. Erdemir, G.R. Fenske, A.R. Krauss, D.M. Gruen, T. McCauley, R.T. Csencsits, Tribological properties of nanocrystalline diamond films, *Surf. Coat. Technol.*, 121 (1999), 565–572.
58. J.S. Zabinski, M.S. Donley, S.D. Walck, T.R. Schneider, N.T. McDevitt, Effects of dopants on the chemistry and tribology of sputter-deposited MoS_2 films, *Tribol. Trans.*, 38 (1995), 894–904.
59. M.R. Hilton, G. Jayaram, L.D. Marks, Microstructure of Co-sputter-deposited metal and oxide MoS_2 solid lubricant thin films, *J. Mater. Res.* 13 (1998), 1022–1032.
60. M.R. Hilton, R. Bauer, S.V. Didziulis, M.T. Dugger, J. Keem, J. Scholhamer, Structural and tribological studies of MoS_2 solid lubricant films having tailored metal-multilayer nanostructures, *Surf. Coat. Technol.*, 53 (1992), 13–23.

Chapter 9

Thin Multifunctional Coating Materials for Tribological Contacts

Peter A. Dearnley
The national Centre for Advanced Tribology at Southampton (nCATS), University of Southampton, Highfield Campus, Southampton SO17 1BJ, UK
p.dearnley@soton.ac.uk

9.1 Introduction

The aim of this chapter is to review the multi-functional (MF) requirements of some specific exemplar tribological devices and show how these can be better realised by applying appropriate thin coating to their material surfaces. After a brief description of the nature of relevant basic tribological environments and contact mechanics, three application sectors are reviewed in depth: bio-medical device bearing surfaces, automotive piston-ring components and cutting tools for machining applications.

The idea of exploiting MF designer coatings is very attractive. In principle, a coating material can be selected that preserves the bulk property requirements of a device or component whilst enhancing its surface properties to meet the physical, chemical and geometrical challenges of any given operating environment, as schematically

Multifunctional Materials for Tribological Applications
Edited by Robert J. K. Wood

ISBN 978-981-4463-90-4 (Hardcover), 978-981-4463-91-1 (eBook)
www.panstanford.com

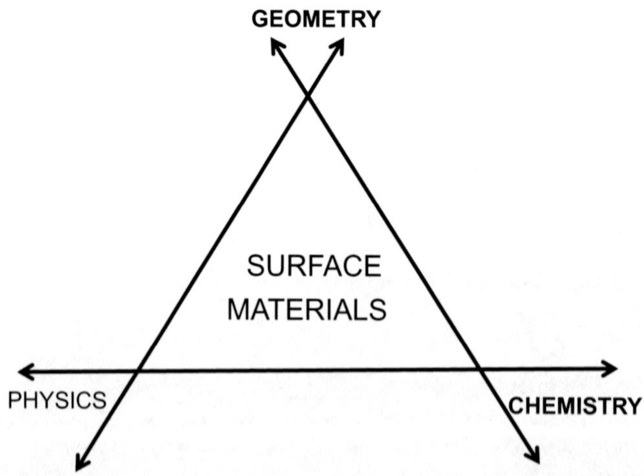

Figure 9.1 The materials surface triangle. Reprinted with permission from *An Introduction to Surface Engineering*, Copyright (2015), Cambridge University Press.

summarised by the 'surface triangle' shown in Fig. 9.1. We are still at a time when this goal is in the process of being realised and where the peculiarities of specific operating environments are being learnt from a combination of field test results and dedicated laboratory experiments. Coating material formulation and design is therefore a complex process of synthesising data and testing hypotheses in real situations. A systematic and cautious approach is required since applications can be in high risk areas as in human joint replacements, high-end aircraft and automotive engines, nuclear plant and advanced manufacturing situations. Some of the diversity for tribological functions can be gleaned from Figs. 9.2 and 9.3. Given the broadness of these activities, only three specific application sectors will be covered in detail in this chapter: piston ring materials for internal combustion (IC) engines, bearing surfaces for total hip and knee joint replacement devices and cutting tools for metal machining operations (used in high volume manufacturing). Many such applications require common immunities to be displayed by protective coatings. This ideal is depicted in Fig. 9.4, where it is envisaged that one coating material design or architecture, would be able to solve all surface requirements by providing an effective

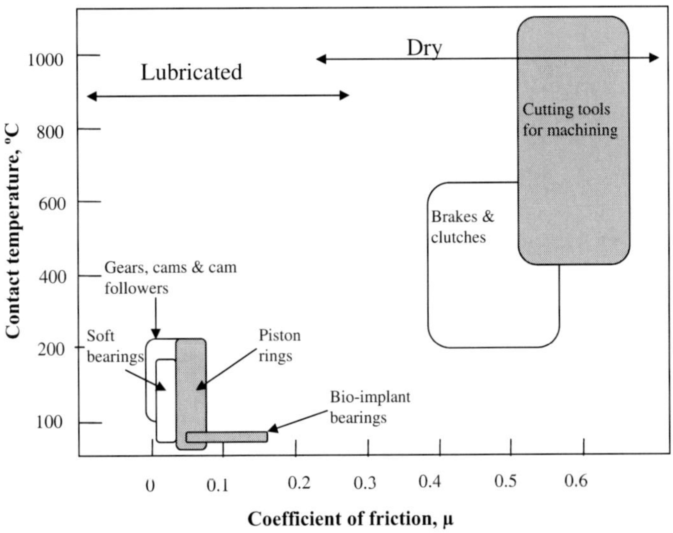

Figure 9.2 Friction-temperature regimes. Shaded fields are discussed in this review. Reprinted with permission from *An Introduction to Surface Engineering*, Copyright (2015), Cambridge University Press.

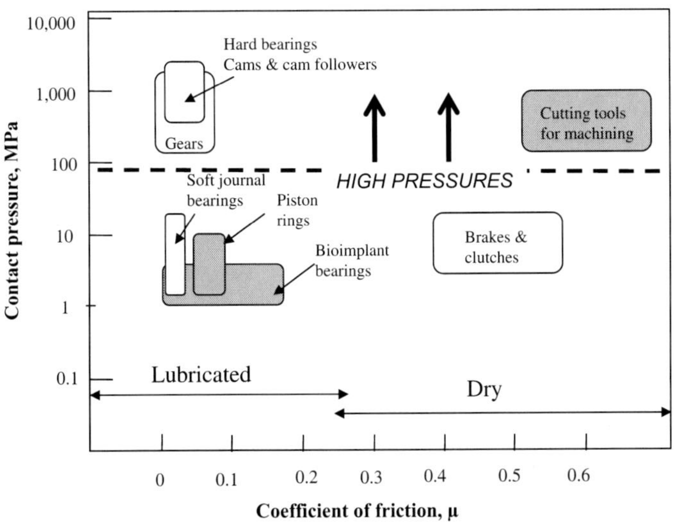

Figure 9.3 Friction-pressure regimes. Shaded fields are discussed in this review. High contact pressure devices are indicated above the dashed line. Reprinted with permission from *An Introduction to Surface Engineering*, Copyright (2015), Cambridge University Press.

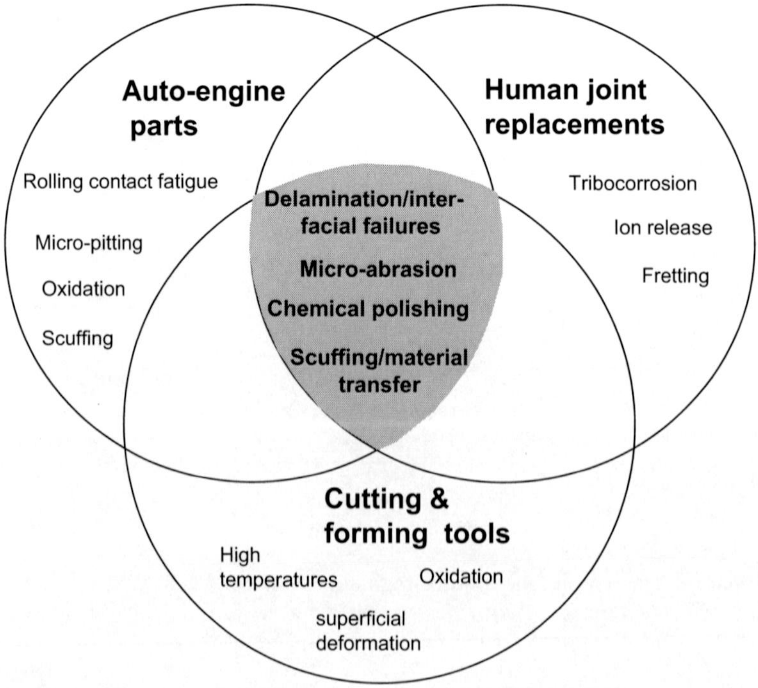

Figure 9.4 Immunities required of a multifunctional tribological coating shield.

multifunctionality 'immunity shield,' lending its use to several substrate devices used in several engineering tribology areas.

9.2 Basic Tribological Contacts

It is convenient to divide tribological behaviour into the various modes of contact and wear. Contacts between material surfaces maybe dry or in the presence of a fluid (liquid or gas). For low friction ($\mu < 0.1$) devices, oil lubrication is widely used. In specialty appliances like human joint replacements, lubrication is provided by the operating environment, which in that instance is a synovial fluid enriched in long chain protein molecules. In high friction devices ($\mu > 0.3$), as in the ubiquitous brake disc rotor/friction pad, contacts are intentionally dry. Here, the possibility of oxidation of the contacting surfaces is common, the severity of which increases with increasing sliding/rolling speed and contact pressure, due to the predominance of frictional heating. A range of friction, contact pressure and temperature is therefore possible, which varies for the tribological device concerned. Specific examples are given in Figs. 9.2 and 9.3. While sales people of various proprietary coating processes may claim that their particular coating material intrinsically lowers friction of specific devices, it is fair to say that the maintenance of a reproducible and predictable friction coefficient is a better ideal. In the exemplars chosen in this chapter, high friction ($\mu > 0.5$) is the status quo in material machining operations, whilst lower friction ($\mu < 0.3$) is the norm for joint replacement bearing surfaces and automotive piston-rings (Figs. 9.2 and 9.3). It can also be noted that very low friction can be displayed during the destruction of a material surface, as is the situation for MoS_2 and PTFE; accordingly these materials cannot be sensibly used as monolithic coatings in sliding or rolling contact situations as they have no durability.

9.2.1 *Sliding Contacts*

This is the most simple of contact situations found in many mechanical engineering devices. There are two broad types: unidirectional and bidirectional sliding, exemplified by the brake disc rotor and the IC engine piston-ring/cylinder bore contact, respectively. Joint replacements bearings are essentially bidirectional devices but in the specific example of hip joint replacements, there are non-linear variations in the direction of motion, due to abduction–adduction of the joint.

9.2.2 *Rolling Contacts*

In free rolling contacts, very low friction levels (<0.05) are possible, especially when additionally aided by a lubricant fluid film. This is the case for many roller and ball bearing/raceway contacts. It is only when stopping and starting such devices that micro-asperity contact can take place due to momentary loss of lubrication. Accordingly, rates of wear are typically very low, enabling use for several years without any degradation in performance.

9.2.3 *Rolling with Sliding*

In devices like IC engine, cam/cam followers and power transmission gears, variable contact situations exist, changing from pure rolling through to a combination of sliding and rolling contact. In the latter situation, where rolling and sliding contact co-exist, surfaces are subjected to a wear phenomenon called *rolling contact fatigue pitting* [1].

9.2.4 *Dynamic Seizure*

This is a very special situation mainly encountered in metal machining operations and is not one usually appreciated by many engineers. When cutting metals and metal matrix composites (MMCs), the tool acts as an asymmetrical wedge that removes a ribbon of material (called the chip) from the work piece. This is achieved by intense plastic shearing along the primary shear plane (OD in Fig. 9.5) and within a secondary shear zone (OL in Fig. 9.5). In the latter zone, shear strains can be greater than 20 and shear strain rates commonly exceed 1000 per second [2]. Here, the mechanical work done is sufficient to raise temperatures in the secondary shear zone and the adjacent tool to very high values, $>900^\circ$C, for example, when cutting medium carbon steels at surface speeds >200 m/min. This in combination with high contact pressures (>500 MPa) provide a very challenging environment for tool materials and their protective coatings. It should be appreciated that the chip underside is essentially bonded or seized to the cutting tool rake face and there is no prospect of lubricating the tool/chip interface at normal metal removal rates. This is a unique tribological situation and differs profoundly from machine element contacts found in normally lubricated devices, such as those found in an automotive IC engine.

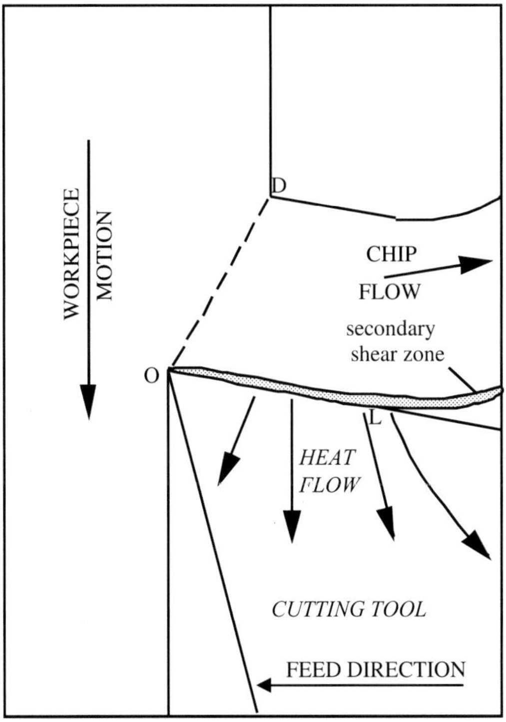

Figure 9.5 Essentials of the cutting tool workpiece interaction.

9.2.5 *General Consideration of Liquid Lubrication Regimes*

Engineering devices are often subjected to more than one lubrication regime, determined by contact geometry and relative motion (entraining velocity of the lubricant) and the lubricant dynamic viscosity. A useful summary of this behaviour is given by the Stribeck curve, shown in Fig. 9.6. Ideally, apposing sliding surfaces are separated by a fluid film and no inter-material contact takes place between micro-asperities. This is the situation under elastohydrodynamic lubrication (EHL) conditions, where the friction coefficient reaches a minimum value for the tribological contacts concerned and wear is essentially absent. On the other hand, appreciable apposing material contact takes place at the micro- and nanometre level under so called boundary lubrication

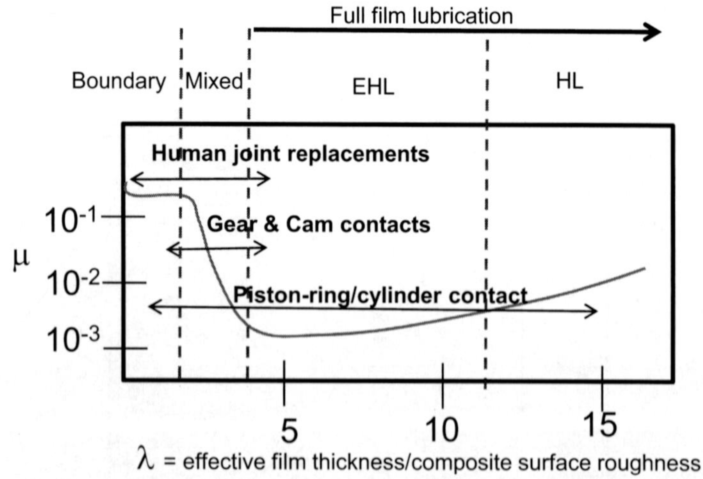

Figure 9.6 Schematic Stribeck curve illustrating the lubrication regimes for low friction auto-engine, power transmission and human joint replacement components.

regimes, where the vertical separation may only be an oil molecule length thick. More commonly mixed lubrication (EHL + boundary lubrication) takes place. When the oil entraining velocity reaches zero, as at the top dead centre of the motion of a piston ring/cylinder wall contact, starvation occurs and separation between the surfaces is zero. It is seen from Fig. 9.6 that cams/cam followers, transmission gear, piston-rings and human joint replacement devices typically experience a wide range of lubrication regime during use [3]. It is during mixed, boundary and starvation conditions that wear takes place. To obviate this effect, additives are made to lubricating oils that result in the deposition of 'soft' (low shear strength) metallo-organic and inorganic compounds on the contacting surfaces. There are two important types: (a) anti-wear additives (of the zinc dialkyl dithiophosphate [ZDDP] variant [4]), which yield Zn/P compound residue layers on iron-based surfaces and (b) friction modifiers (reducers), based on molybdenum trialkyl-dithiocarbomate (MoTDC), which form low friction MoS_2 deposition. Both Zn/P and MoS_2 compound layers are created by thermo-mechanical decomposition of ZDDP and MoTDC, respectively, a mechanism that requires a minimum operational temperature ($\sim$150°C) for initiation. Such reactions cease when interfacial temperature exceed $\sim$250°C when oil decomposition begins. The function of such deposits is complex and their role in mediating or even accelerating wear and corrosion-wear mechanisms remains a front line research challenge. Undoubtedly, they are important but their role in preventing or possibly augmenting the wear of coated materials remains poorly understood, although it is clear that functionality has already been optimised for steels surface hardened by carburising/nitriding and nitrocarburising, such materials having been used successfully in IC power train devices for many decades.

9.3 Basic Contact Mechanic Considerations

Given that many surface coatings have very high yield strengths ($>$ 2000 MPa), it is natural to consider them for use in high contact pressure devices ($>$ 200 MPa) as indicated in Fig. 9.3. However, it must always be borne in mind that the functionality of such coatings or layers is reliant on the yield strength of the substrate being high enough to support them. It should be appreciated that when subject to a frictionless *static* normal point or line contact loading situations, the maximum resolved shear stress ($\tau_{\max}$) takes place below the surface and *not* at the surface. However, where friction develops ($\mu > 0.3$) between the contacting surfaces, $\tau_{\max}$ can reach the surface. This effect also takes place at lower friction levels, during sliding contact, due to rotation of the elastic contact stress field. Hence, in some devices, like automotive cams and gear, the position of $\tau_{\max}$ maybe continuously varying. Given this scenario and that many coating materials are only a few micrometres thick, coating subsidence can take place due to yielding and plastic flow of the substrate, as depicted in Fig. 9.7, a phenomenon that is akin to skating on thin ice. A practical example of this type of failure is shown in Fig. 9.8, where a TiN coating has collapsed into a high-speed steel (HSS) substrate whilst being used to machine steel [2]. Here, the HSS below the coating became over heated and over tempered (thermally softened), during the cutting process, causing it to plastically flow and allowing the coating to break-up via a type

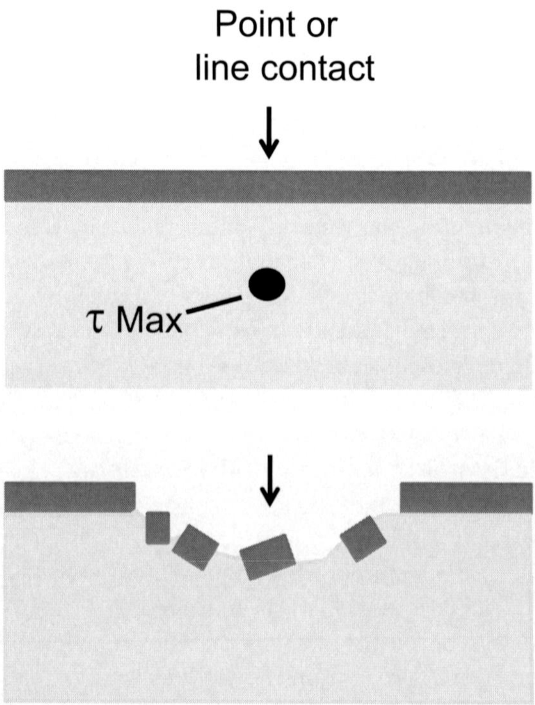

Figure 9.7 The "thin ice effect"; overloading of the sub-surface, via excessive point or line contact loading. Reprinted with permission from *An Introduction to Surface Engineering*, Copyright (2015), Cambridge University Press.

of subsidence (Fig. 9.8). It can also be seen here that the coating remained adherent to the substrate; but the coated 'system' failed due to insufficient substrate strength *at the application temperature* (in this case $> 800°C$).

The micro-hardness versus depth 'signature' for many materials with thin hard coatings, such as CrN, TiN, diamond-like carbon (DLC) or TiC applied to engineering steels, Ti, Co or Al alloys is shown in Fig. 9.9a. Here, an abrupt change in hardness is seen across the coating substrate interface. Such coatings may be applied using deposition technologies, such as electroplating, chemical vapour deposition (CVD) or physical vapour deposition (PVD). Depending upon the substrate elements, it may be possible to form a hard

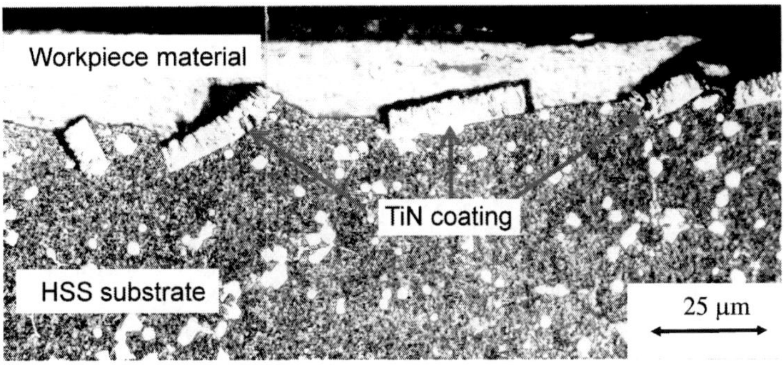

Figure 9.8 Collapse of a TiN coating on a HSS tool rake face, caused by overheating of the substrate whilst machining steel at high speed [2].

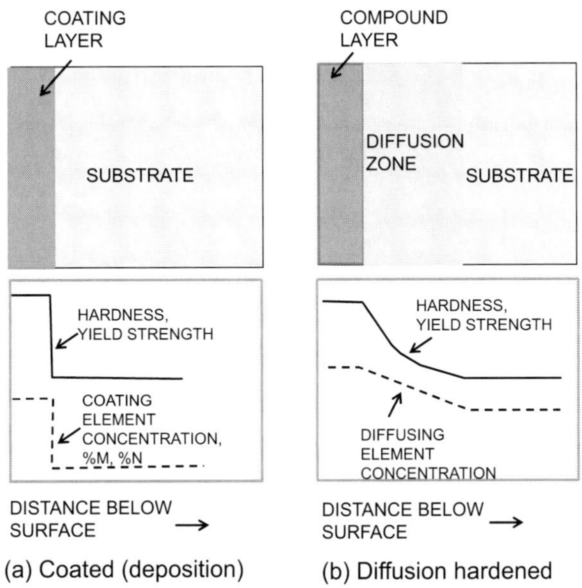

Figure 9.9 Surface hardening via (a) deposition; (b) diffusion. Reprinted with permission from *An Introduction to Surface Engineering*, Copyright (2015), Cambridge University Press.

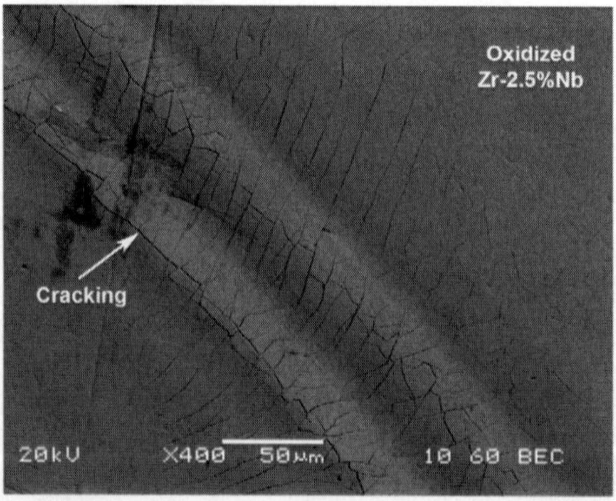

Figure 9.10 View of a damaged oxygen diffusion hardened Zr-2.5 Nb alloy (Oxinium) knee replacement component (posterior condyle) surface, harvested from an active patient, as reported by Burnell et al. [5]. There are surface cracks and folds in the ZrO_2 protective layer due to subsurface plastic deformation of the substrate caused during indentation from the adjacent Zr alloy tibial tray when the joint was fully open. No layer loss was reported.

surface layer via elevated diffusion hardening processes such as nitriding or thermal oxidation. This approach often brings about the added advantage of creating diffusion strengthened zone beneath the external compound layer (or diffusion coating). This scenario is depicted in Fig. 9.9b. The latter designs of surface engineered metals are less prone (although not immune) to the thin ice effect shown in Fig. 9.7. An application example is shown in Fig. 9.10 [5], where damage to the posterior condyle region of an oxygen diffusion hardened zirconium alloy (Oxinium) knee joint replacement bearing surface, has arisen, following adverse line contact loading, either by the adjacent titanium alloy tibial tray (in vivo) or during surgical removal. Although it is evident that the exterior m-ZrO_2 compound layer has been severely cracked, it was reported as remaining adherent to its substrate—no delamination of the layer being observed [5], indicating the diffusion hardened sub-surface had restricted collapse of its exterior oxide layer.

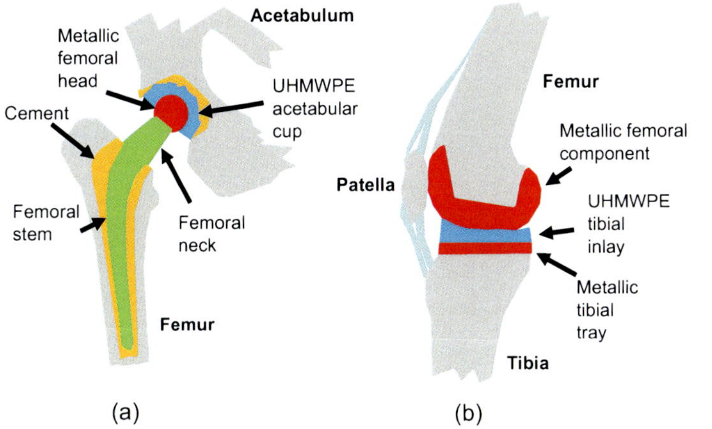

Figure 9.11 Schematic layouts showing principal components in (a) total hip replacement (THR) and (b) total knee replacement (TKR). Reprinted with permission from *An Introduction to Surface Engineering*, Copyright (2015), Cambridge University Press.

9.4 Surface Engineered Biomedical Implant Bearings

More than 50,000 patients per year receive total joint replacements in the UK [6, 7]. Such surgical procedures provide alleviation from the debilitating pain caused by the advanced stages of osteoarthritis. These remarkable clinical procedures were originally made possible through the pioneering work on total hip replacements (THRs) carried out by Sir John Charnley during the 1960s. This success led to developments in total knee replacements (TKRs) and other joint replacement devices. Basic designs of THR and TKR mobile joints are shown in Fig. 9.11. The femoral components for both devices are frequently made from special metallic alloys, typically cast or wrought Co–Cr–Mo alloys [8] or more recently in the USA from oxygen diffusion hardened Zr-2.5Nb alloy (with the proprietary name 'Oxinium'). The counter face bearing for both is typically manufactured from ultra-high-molecular-weight-polyethylene (UHMWPE), which is frequently nowadays given extra wear durability by stimulating cross-linking via irradiation. Additionally, femoral components for THRs can be made from Alumina or Zirconium toughened Alumina, but these are not practical for

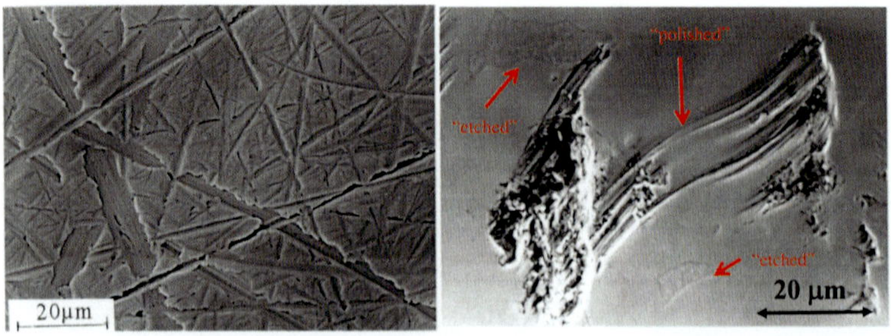

(a) (b)

Figure 9.12 View THR Co–Cr femoral head bearing surfaces: (a) After 15 years implantation (b) After 20 years implantation (note the arrowed etched and polished features discussed in text). After Dahm et al. [9].

TKRs since they are prohibitively costly to make in such relatively large sizes. Due to cost constraints in the United Kingdom, femoral THR components were until circa 2000 made from the austenitic stainless steel Ortron 90 (an Fe-20Cr-10Ni-2.5Mo-0.4N alloy) [8].

Although very successful, there are various factors that limit the lifetime use of total joint replacements. Of most concern to the implant manufacturer is the susceptibility or not of the bearing surfaces to the production of particulate debris and/or the release (solution) of metal ions into surrounding tissues. Both effects have been linked to the formation of pseudo-tumours and the loss of bone stock (due to bone cell death) that leads to aseptic implant loosening (via a process known as osteolysis) that require costly and uncomfortable surgical revision.

The examination of worn joint replacements (explants) reveals evidence of much scratching of the metallic bearing surfaces [9] in vivo (Fig. 9.12a). Associated with the scratch grooves are protruding scratch 'lips' that cause considerable plastic deformation of the apposing UHMWPE counter surfaces via micro-asperity cutting and shearing, which creates a significant source of particulate debris. Many of the metallic scratch lips become sheared off [10] and contribute to the overall volume of wear debris that collects in tissue surrounding the implant. Metal ion solution also takes place evidenced by raised ion concentrations in the human body following surgical implantation of THR and TKR devices.

A further view of a Co–Cr–Mo femoral head surface after 20 years implantation is shown in Fig. 9.12b. The surface has been severely damaged at some point—possibly by a surgical instrument or by bone cement. It is evident that the mechanical damage has become subsequently polished. Further, the metal carbides contained in the Co–Cr–Mo alloy matrix have become lightly etched (Fig. 9.12b). Both features are indicative of material being removed via anodic dissolution/corrosion.

Given the highlighted problems of mechanical and chemical action on joint replacement bearing surfaces, the prospect for improving their performance though surface engineering is appealing. In principle, surfaces can be made more resistant to scratch damage by making them harder and more resistant to dissolution by making them less chemically active *provided such coating materials are bio-compatible* within the human body. TiN coated Ti-6Al-4V femoral heads for THRs was the subject of an early (1992) patent application by Pappas [11] and there are two reports on the surface condition of related components retrieved after several years human implantation (Harmen et al. [12], Raimondi and Pietrabiss [13]). Here, TiN coating delamination via localised fracture along the coating–substrate interface (Fig. 9.13) was a concern. In another work, Taeger et al. [14] reported an scanning electron microscope (SEM) examination of DLC coated Ti-6Al-4V

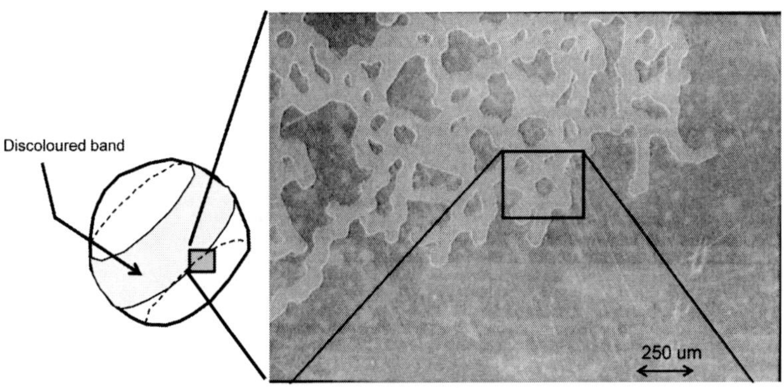

Figure 9.13 Femoral head explant showing adhesive failure of the TiN coating along the coating/substrate interface, Raimondi & Pietrabissa [13].

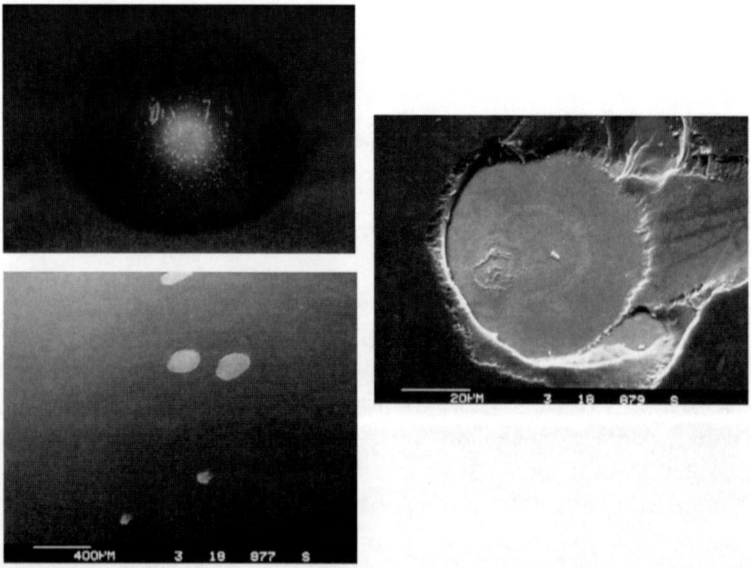

Figure 9.14 SEM images of a retrieved DLC coated Ti-6Al-4V femoral head, showing localised circular coating loss due to in vivo Type II corrosion wear, after Taeger et al. [14].

femoral heads recovered from active patients. Here, numerous circular patches of exposed substrate were observed to cover the bearing surface in one particular case (Fig. 9.14), which in the opinion of the present author indicate the decapitation of coating blisters, via Type II corrosion wear (see next paragraph).

Corrosion wear is defined as the surface degradation of a material due to the sequential or simultaneous actions of corrosion and wear [15]. This phenomenon is also termed tribocorrosion. Laboratory experiments show while such effects can be mitigated by the application of surface coatings, these might become undermined via chemical attack of the substrate. There are three basic types [15] of corrosion-wear mechanism relevant to total joint replacements (TJRs):

- Type I: The removal of the coating passive film during sliding contact and its subsequent regeneration (Fig. 9.15).

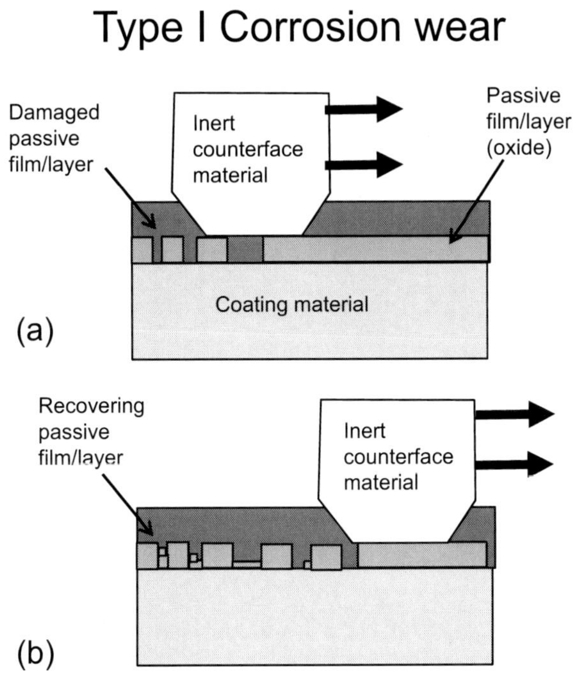

Figure 9.15 Schematic depiction of Type I antagonistic corrosion wear loss. After Dearnley & Aldrich-Smith [15].

- Type II: Galvanic attack of the substrate leading to blistering and then removal of the coating or diffusion compound layer during sliding contact (Fig. 9.16).
- Type III: Galvanic attack of the bearing surface, causing it to roughen and subsequently abrade any apposing surface with which it makes contact.

While Type I corrosion wear is pervasive in most joint replacement bearing surfaces and is probably responsible for the polishing effect observed in the uncoated Co–Cr–Mo explant shown in Fig. 9.12b, Type II corrosion wear also appears possible and is the likely explanation for the circular artefacts shown in the DLC coated Ti-6Al-4V femoral head shown in Fig. 9.14.

Similar effects, blister formation and decapitation, were also observed following laboratory tests with alumina pins sliding

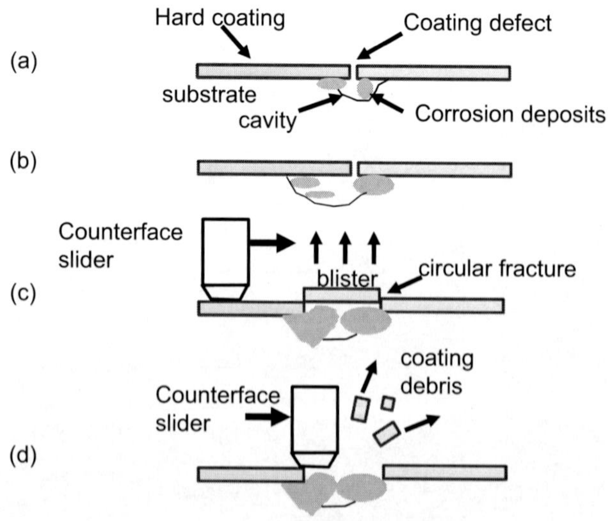

Figure 9.16 Type II corrosive wear. After Dearnley & Aldrich-Smith [15].

against DLC coated Ti-6Al-4V plates immersed in 0.9% NaCl solution at 37°C by Kim et al. [16]. These workers also demonstrated that doping DLC with up to 3 at.% Si retarded this effect. The latter types of DLC appear not to have received any medical application. Moreover, at the time of writing, very few non-doped DLC-coated implant materials remain in clinical use. It is a matter of surmise that such devices have not been successful enough to warrant further use. Whilst both TiN- and DLC-coated Ti-6Al-4V represent the design approach given in Fig. 9.9a, a more successful materials design, equivalent to that shown in Fig. 9.9b, that of oxygen diffusion hardened Zr-2.5 Nb [17, 18]—'Oxinium'—has been in widespread use since circa 1995, Fig. 9.17. It has only been in very recent years that retrieval reports, on such devices, have appeared in the open literature. Apart from a reduced incidence of scratching, the work of Heyes et al. [19] is significant since it reports pitting of the condylar surfaces of six retrieved ODH Zr-2.5 Nb TKR femoral components, Fig. 9.18a, whilst no pitting was observed in another nine retrieved parts. Most were smoothly worn, Fig. 9.18b, which is indicative of Type I corrosion wear. The work of Heyes et al. was based on a small cohort of components retrieved after a maximum implantation time

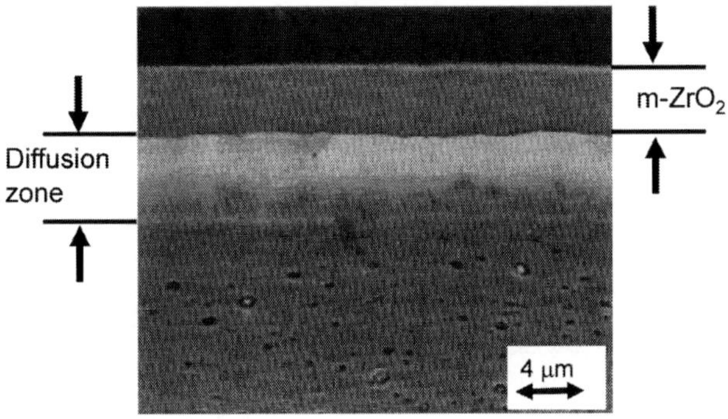

Figure 9.17 Oxygen diffusion hardened Zr-2.5Nb alloy "Oxinium", after Davidson et al. [18].

Figure 9.18 Oxinium treated TKR Zr–Nb alloy femoral component (condyle) surfaces retrieved from a patient after an unspecified time of implantation: (a) Stereo-view showing both condyles with some scratching (Scr & Str), pitting (P) and delamination (Del); (b) Microscopic view showing polishing wear, possibly due to Type I corrosion wear. Dark patches are an unidentified second phase. After Heyse et al. 2011 [19].

of 46 months. The pitting (most probably due to Type II corrosion-wear) would appear to have been mainly identified using stereo light microscopy, so it is possible that very small pits may have been present on the other examined components but defied easy identification. This [19] is important since it shows that for one small grouping of patients with Zr-2.5 Nb TKR femoral head components protected by a m-ZrO_2 layer, pitting via Type II corrosion-wear was *not* destructive to implant life in the short term (up to 46 months).

Until very recently, there was resurgence in the use of metal-on-metal THRs. At the lead in this area was the development of ~8–12 μm thick CrN coated Co–Cr–Mo femoral head components. Laboratory hip simulator tests indicate that metal ion release during articulation is at least seven times lower when using PVD-CrN coated Co–Cr–Mo femoral heads compared to standard uncoated components [6]. This work also indicates that CrN coated Co–Cr–Mo gives better overall performance in terms of wear resistance and coating adhesion compared to PVD-Cr(C,N), PACVD-DLC (hydrogenated) and PVD-TiN coated Co–Cr–Mo. To date, *no comparisons* have been made between CrN coated Co–Cr–Mo and Oxinium treated Zr-2.5Nb femoral components, which should nowadays be regarded as the industry standard for surface engineered join replacement bearing surfaces given that these are the *only* surface engineered bearing surfaces in widespread global use. Instead, almost all laboratory-based implant-bearing life/wear studies use uncoated Co–Cr alloy bearing surfaces as the datum material.

9.5 Coated Automotive Engine Components

The exploitation of thin vapour deposited coatings in auto-engine components is relatively immature. Several examples of the state-of-the-art are given in Table 9.1. All coatings to date are applied to various steel components used under differing oil lubricated conditions. Although hard coatings are the main focus of current development, 'soft' Al–Sn alloys for protecting steel journal bearings (so called bearing shells—used in apposing contact with steel or cast iron crank and cam shafts) are amongst the highest volume production items by current day original equipment manufacturers. Their sub-micrometre grain structures convey fatigue endurance well beyond that which is possible by centrifugally cast lower melting temperature Pb–Sn alloy bearing layers. Of the hard coating materials, CrN and Cr_2N deposited by PVD have seen significant commercial uptake for piston ring applications. These devices were originally brought to the market by the Japanese firm Kobelco (~1995) for use in truck diesel engines. *The success of CrN and Cr_2N coatings is they fulfil the main functional requirements of many*

Table 9.1 Coating materials applied to steel components used in engine tribology applications. Lubrication regimes are indicated in italics

Coating material	Coating technology	Journal (soft) bearings (steel) *Fluid film lubrication*	Wrist pins/gudgeon pins (steel) *Near starvation to EHL*	Piston rings (Martensitic stainless steel) *Near starvation to EHL*	Engine cams and hard bearings (Through hardened 1% carbon bearing steel.) *EHL*
CrN and/or Cr_2N	PVD	None known	None known	Good for diesel engine applications, especially when matched with specific oil additives	None known
W doped DLC or Si doped DLC	PECVD, hybrid PVD-PECVD	None known	Good—auto racing applications only	Under development	Good when matched with specific oil additives
Amorphous tetrahedral carbon	PVD	None known	Under development	Under development	Under development
Al-Sn	PVD	In large scale production	None known	None known	None known
Cr plate with Alumina particles	CKS—electroplating/electrophoresis	None known	None known	In large scale production	None known

automotive devices to resist surface degradation via a combination of chemical and mechanical processes. Since boundary lubrication occurs at the extreme displacement of the piston-ring cylinder liner contact, the durability of the *top* piston ring requires resistance to wear caused by [20]:

- Scuffing, galling (metal transfer) and seizure
- Polishing and oxidation wear
- Micro-abrasion
- Coating delamination

To assure an adequate engine service life of ~150,000 km, the CrN and Cr_2N coatings must be ~30–50 μm thick. Within Europe, such piston rings are being widely used in small passenger car diesel engines. Here, the PVD-CrN coatings are applied to nitrided 0.85%C, 18% Cr stainless substrates since it remains impractical to apply them to cast-iron rings.

The possibility of using carbon coatings, such as DLC, for use in piston rings has received limited investigation to date. A prima facia objection is that they may thermally degrade. However, it has been shown that tetrahedral carbon (ta-C) rich DLC can survive without degradation up to 1000°C provided it is *not* exposed to oxygen [21]. Once oxidised, however, degradation ensues at 450°C to 500°C. Given this scenario, there is a strong case for exploring the development of oxidation resistant DLCs made possible, for example, through doping with Si [22, 23]. Such coating materials promote the formation of a protective Si rich surface passive film [16].

In an ideal world, it would be best to test all prototype tribological coatings in an IC engine. Such tests while just possible are extremely time consuming and very expensive to carry out—requiring many thousands of hours and associated labour costs required in instrumentation monitoring as well as lengthy disassembly and reassembly of an engine. In practice, simple configuration tests are used as a first step, which may be supplemented by more specialised testing to mimic the application environment more closely. For example, in the case of piston rings a high-speed reciprocation sliding contact rig has been designed for this purpose. In the laboratory of one major piston ring manufacturer good correlations between these tests and more advanced 'fired engine' tests have been made. Here, the test contact pressures are a minimum of an order of magnitude higher than in the real application, so as to accelerate the chemical and mechanical degradation interactions found in IC engines. Typically, the following conditions are used for first level assessment of prototype coated steel piston rings:

Test duration: 23 hours

Frequency: 10 Hz

Total test cycles: 828,000

Lubricant and flow: Uncombined ester at 0.12 mL/min (with no ZDDP or MoTDC additives)

Normal force: 450 N

Contact pressure: 225 MPa

Lubricant temperature: 150°C

Ring material: High carbon martensitic stainless steel (0.9%C, 18%Cr, 1%Mo, 0.1%V German spec: ST 1.4112) with prior nitriding.

Ring dimension: 78.3 mm OD × 1.75 mm

Cylinder dimension: 108 mm ID

Cylinder liner: cross-honed flake graphite pearlitic grey cast iron

It is worth emphasising that *the 1.4112 (0.9%C, 18%Cr, 1%Mo, 0.1%V) martensitic stainless steel* piston rings are preferably nitrided prior to PVD coating. Although this provides excellent mechanical support for thin hard coatings, its prime purpose is to prevent bending fatigue fracture of the ring. Figure 9.19 shows this architecture. The simple configuration test condition cited above enable the rapid detection of the attributes and/or failings of potential new coated piston ring materials. These are illustrated as follows.

9.5.1 *Scuffing and Galling (Metal Transfer)*

An example of this behaviour is shown in Fig. 9.20. Here, transfer of cast-iron cylinder liner material from the apposing liner has resulted

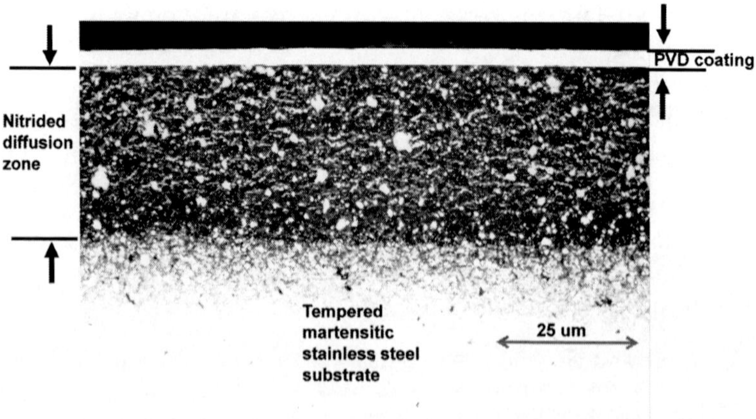

Figure 9.19 Architecture of a state-of-the-art surface-engineered IC engine piston ring, showing the nitrided diffusion zone beneath a hard PVD coating. Reprinted with permission from *An Introduction to Surface Engineering*, Copyright (2015), Cambridge University Press.

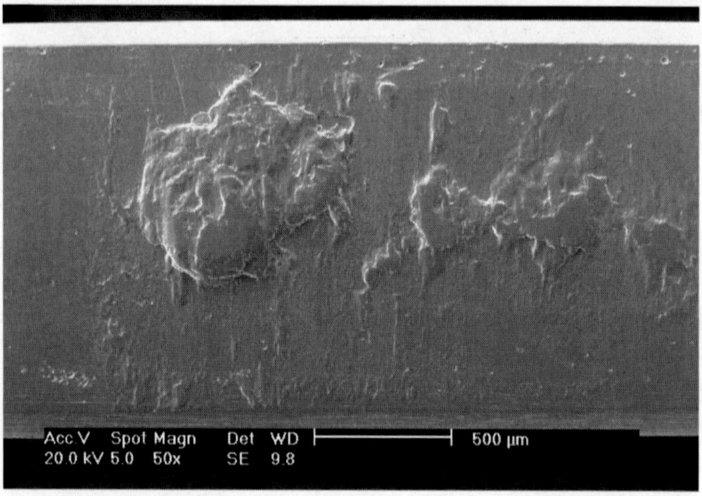

Figure 9.20 The running test surface of a PVD S-phase (18at.%N) coated stainless steel piston ring, showing a build up of cylinder liner transfer material. This phenomenon must be avoided in this application. Accordingly this coating is unsuitable. Reprinted with permission from *An Introduction to Surface Engineering*, Copyright (2015), Cambridge University Press.

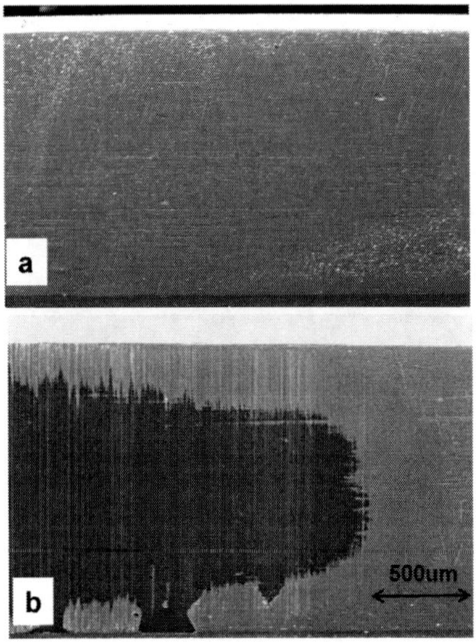

Figure 9.21 The running test surface of a PVD W-B-N coated stainless steel piston ring, showing: (a) original topography and; (b) micro-abrasion resulting form lubricated sliding contact (direction was up and down in this view) with a cast-iron cylinder liner material. Reprinted with permission from *An Introduction to Surface Engineering*, Copyright (2015), Cambridge University Press.

in a build-up of this material on a nitrogen S-phase coated piston ring.

9.5.2 *Micro-Abrasion (Third Body Abrasion)*

This phenomenon is due to cutting and/or ploughing action of hard particles against the coated piston ring surface. An example is shown in Fig. 9.21, where the external PVD coating has been completely removed over most of the contact area of the piston ring. Shallow scoring (in the direction of reciprocation sliding) can be seen to have traversed across the PVD-W-B-N coating and its parent nitrided stainless steel substrate. The hard particles maybe micron-sized soot particles created by a partial thermal decomposition of the lubricant and incomplete combustion, which require the inclusion of detergent and dispersant additives in commercial engine oils [24].

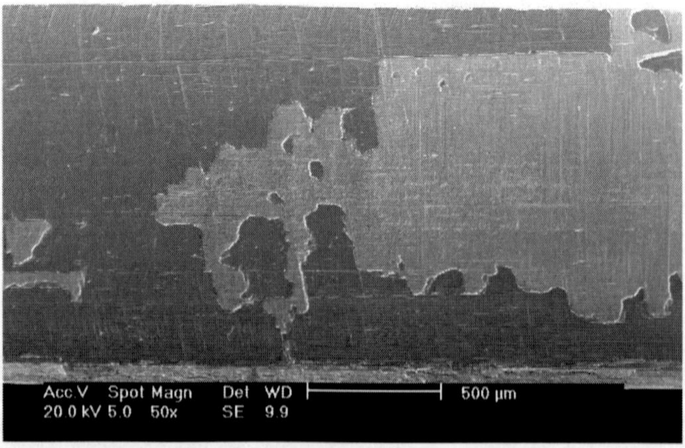

Figure 9.22 The running test surface of a PVD Cr-Ti-Al-N coated stainless steel piston ring, showing exfoliation of the coating due to insufficient bonding across the coating/substrate interface. This phenomenon must be avoided. Reprinted with permission from *An Introduction to Surface Engineering*, Copyright (2015), Cambridge University Press.

9.5.3 *Coating Delamination*

An example of this effect is shown in Fig. 9.22. Here, localised peeling away of small areas of coating material is evident. It is interesting to note that this appears to have a similar topography to that observed on the retrieved TiN-coated femoral head component shown in Fig. 9.13, which also displayed delamination of the coating.

9.5.4 *Polishing Wear*

This effect is evident in Figs. 9.23 and 9.24, which show various coating surface topographies before and after testing. Although small variances are noted in the as deposited states, all topographies have become similarly flattened and 'polished,' as the result of some kind of chemical wear process during the high-frequency reciprocation sliding tests. It is possible that the process responsible for this effect may be akin to that of Type I corrosion wear shown in Fig. 9.15. Certainly, contact temperatures would be sufficiently high to enable the onset of the early stages of hot corrosion (oxidation),

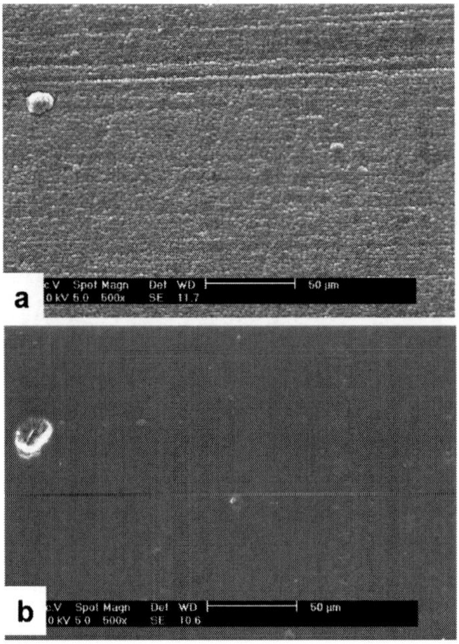

Figure 9.23 The running test surface of a PVD Ti, Al (C,N) coated stainless steel piston ring, showing (a) original topography and (b) "polishing wear" resulting form lubricated sliding contact with a cast-iron cylinder liner material. Reprinted with permission from *An Introduction to Surface Engineering*, Copyright (2015), Cambridge University Press.

which (in the tests reported here) may have been assisted by the partial decomposition (reduction) of the ester via a de-esterification, that is, via the reversible reaction: ester + water = alcohol + organic acid. It can be argued that the piston ring surfaces are quickly placed into a chemically activated (nascent) state, since any pre-existent passive film will have been mechanically disrupted during sliding contact against the cast-iron liner. Such surfaces would easily chemically adsorb oxygen containing species such as alcohol and organic acids, an effect that has been observed on metallic coating surfaces subjected to sliding contact testing in alcohol [25]. Such dynamically formed passive films, can be short lived, becoming subsequently disrupted by mechanical action, leading to material loss, via Type I corrosion wear (Fig. 9.15). Accordingly, the ability of a piston ring surface to resist polishing wear depends upon its ability

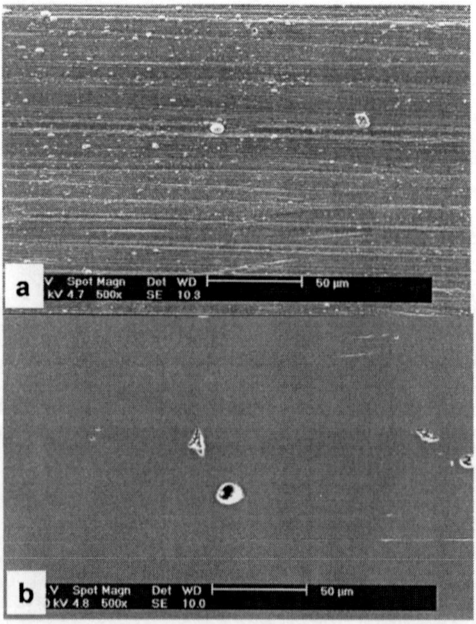

Figure 9.24 The running test surface of a PVD CrN coated stainless steel piston ring, showing (a) original topography and (b) "polishing wear" resulting form lubricated sliding contact with a cast-iron cylinder liner material. Reprinted with permission from *An Introduction to Surface Engineering*, Copyright (2015), Cambridge University Press.

to create a tenacious passive film once it is exposed to organic-based oxidising species. Electrochemically deposited Cr platings of various types have been used as protective piston ring coatings for many years [26]. This metal forms a very stable and adherent passive film. The results of one recent investigation [20] are collated in Fig. 9.25, from which it is clear that polishing wear is *not* mitigated by a high coating hardness, this can be seen more clearly for a wide range of coating materials in Fig. 9.26. The results in Fig. 9.25b show that doping Cr matrix coatings with interstitial **N** and **B** (during reactive PVD sputter deposition) has a complex effect on the rates of polishing wear during lubricated sliding contact with cast-iron cylinder liner material. In solute concentrations up to 26 atomic% boron polishing wear is retarded, beyond this amount, for example, at 33 at.% B, polishing wear is slightly increased, despite hardness being increased, Fig. 9.25a. CrN by itself is slightly more resistant to

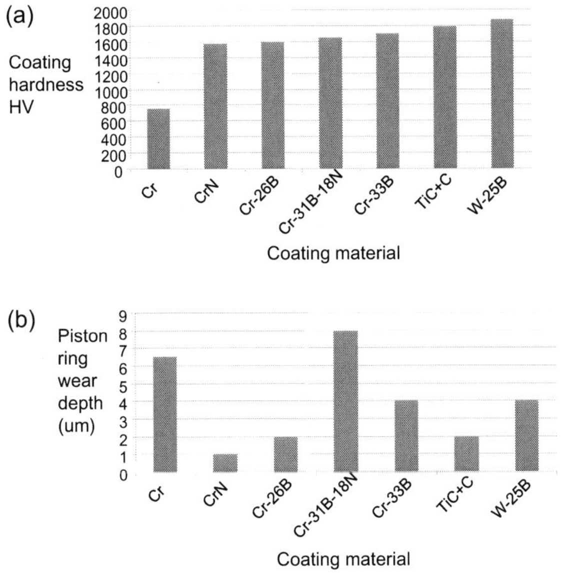

Figure 9.25 Comparison of PVD coated stainless steel piston rings [20] showing (a) hardness variation and (b) "polishing wear" depth variation (when used in lubricated sliding contact against a cast-iron cylinder liner material).

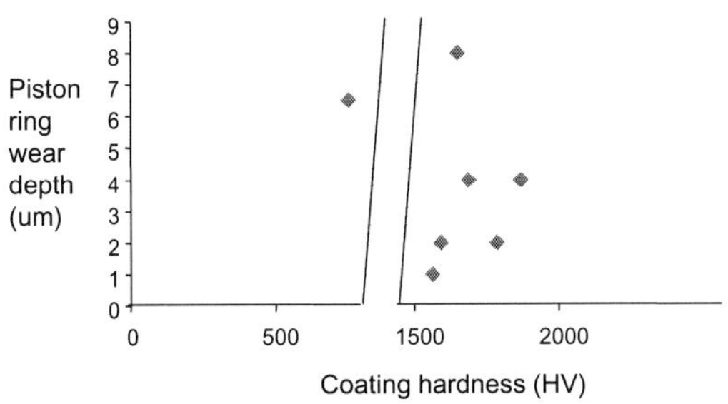

Figure 9.26 Variation of "polishing wear" depths of PVD coated stainless steel piston rings as a function of coating hardness. There is no correlation.

polishing wear than Cr-26B, which supports its present day use in mass production IC engines, whereas the introduction of B to CrN or alternately N to Cr-B coatings, considerably worsens polishing wear, for example, as seen for a Cr-31B-18N coating (Fig. 9.25b).

Tungsten-based compositional analogues of the Cr-based coating systems mentioned so far perform somewhat worse. Whilst W-25B is harder than Cr-26B (Fig. 9.25a) its resistance to polishing wear is significantly worse (Fig. 9.25b). Both effects reflect the differing rates of ion solution of these materials in the ester lubricant and/or the ability to form a stable protective passive oxide film. A *related* effect has been observed for the same materials used in aqueous saline solutions. Here, tungsten-based coating materials are unable to form a stable passive or transpassive film compared to Cr-based coatings [27]. While the exact composition of the passive films formed by the local oxidation of lubricated piston rings will differ to that formed on the surfaces of coated materials sliding against metals in aqueous environments, there would appear to be similarities. In another example, of CrN coatings doped with up to up to 3 at.% Si, the rates of piston ring polishing wear are depressed by about 25% compared to that of CrN [20]. This is similar to the role of Si in the doping of DLC's, *but*, when used in aqueous corrosion-wear environments, where it has been claimed that Si promotes the formation of a more stable passive film [16]. From this work, it can be concluded that there is significant potential for increasing the durability of the passive film of coating materials (against Type I corrosion wear) by doping them with other oxide forming elements such as Si and B.

Other approaches in the development of piston rings include the creation of nanocomposites. One promising architecture is shown in Fig. 9.27, which comprises overlapping sub-micrometre layers of TiC and C (TiC + C). Although results have been very good to date (Fig. 9.25b), they have so far failed to equal the performance of non-doped CrN and there is some evidence that greater wear of the apposing cast-iron cylinder liner takes place [20].

9.6 Coated Cutting Tools for Machining Metals and Composites

Hard-coated cemented carbide cutting tools first became available in about 1969. They now represent $>90\%$ of the all cutting tools that are manufactured globally. Since the 1980s, the coating

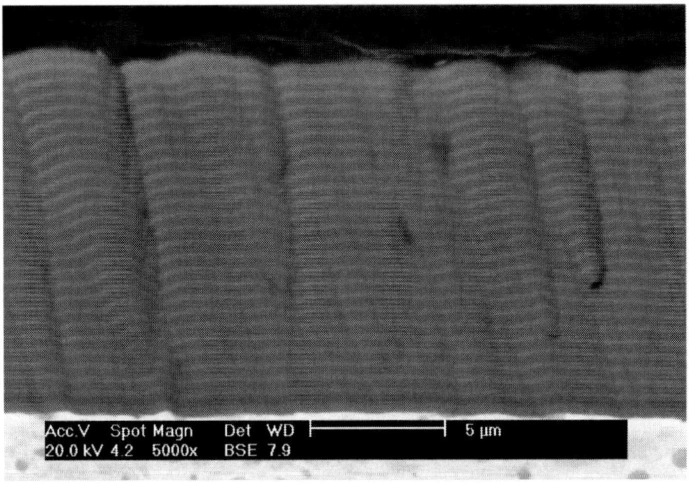

Figure 9.27 Proto-type multilayered architecture of a (TiC+C) nano-composite PVD coated stainless steel piston ring, developed at the Karlsruhe Institute of Technology, Germany, by Dr Michael Stueber.

architectures for metal cutting tools have been largely based on single or multiple layers of TiC, TiN, (Ti,Al)N, Ti(C,N) and Al_2O_3 as well as the more recently developed (circa 2000 to 2005) multicomponent nitride coatings such as (Ti, Al, Cr, Y)N and those based on the Si–Ti–N system. The latter group, in the as deposited state, yield a nanocomposites comprising nanometre-sized crystalline TiN precipitates embedded in an amorphous silicon nitride matrix. Although originally applied by CVD, PVD technologies, initially developed for application of hard coatings to HSSs to avoid necessitating post coating heat treatment, have seen niche application for drills, milling inserts and end mills. By the 1990s, PACVD also began to be exploited for this application, due to its greater throwing power compared to PVD. Yet further technological development has led to the creation of 'hybrid' coating systems that incorporate both PVD and PACVD capability enabling the creation of metal and non-metal doped DLCs.

The *MF requirements* of cutting tools is complex in that they must maintain a high coefficient of friction (> 0.5) whilst resisting wear under high temperatures ($> 800°C$) and pressures (> 500 MPa).

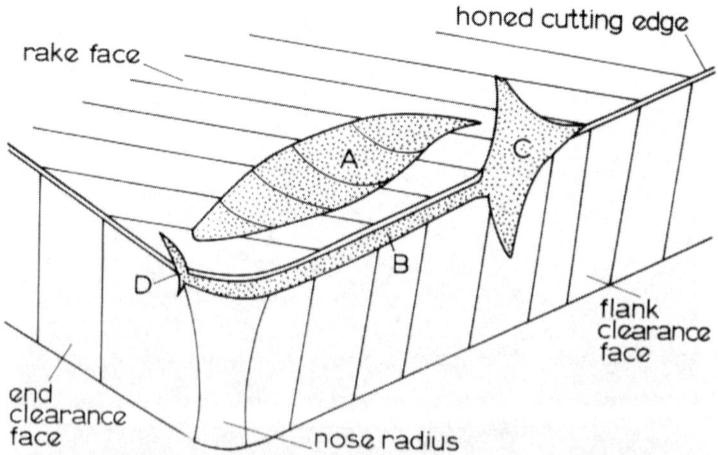

Figure 9.28 Principal wear zones on a coated metal cutting tool [27].

Coatings for improving the wear endurance of cemented carbides were developed specifically for machining steels and cast irons. Latterly, efforts were made to develop diamond coatings for machining aluminium alloys, whilst little progress has been made in efforts to achieve compatible coatings for machining Ti- and Ni-based alloys. Coatings for metal cutting tool applications must be able to combat both the individual and combined actions of mechanical and chemical wear processes. It should be appreciated that degradation takes place in several locations on a cutting tool. These are represented as Zones A to E in Fig. 9.28 [28]. The wear mechanisms that operate in these zones are not often the same, so any would be coating for cutting tools has to be versatile enough to accomplish this demand. In practice, a combination of coating materials is required since different coating materials fair better than others for each of the principal wear zones depicted in Fig. 9.28.

The following wear mechanisms have a significant impact on coated cutting tool endurance:

- Micro-abrasion
- Dissolution or polishing wear (also termed diffusion wear)
- Coating delamination

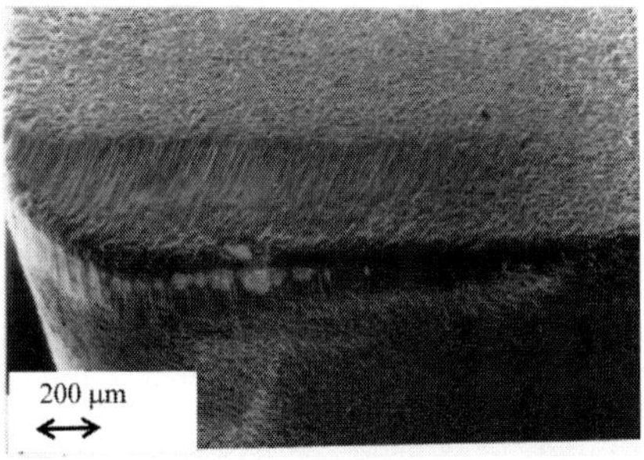

Figure 9.29 Wear of an Al_2O_3 coated cemented carbide tool after cutting 0.4%C steel at 240 m/min, showing higher wear on the flank face (zone B in Fig. 9.28) compared to the wear on the rake face (zone A in Fig. 9.28). The exposed substrate is visible on the flank face due to the higher wear rate.

- Oxidation
- Superficial or discrete plastic deformation

The effectiveness of different coating materials to resist the various forms of wear strongly depends on the specific work piece being machined (e.g., the exact steel composition, heat treatment condition, microstructural morphology and so on) and the cutting speeds and feed-rates selected. High cutting speeds and feeds assure a near laminar flow of the chip material over the tool, but such flow becomes turbulent at lower speeds when built-up-edge formation takes place. The latter should be avoided as much as possible as it leads to delamination of the coating during the periodic detachment of the BUE. Unfortunately, this is almost unavoidable when machining cast irons, which places great demands on coated cutting tools.

When cutting medium carbon steels, the wear rates for Al_2O_3 and TiN coatings are greater on the flank face (Zone B) than on the rake face (Zone A), for example, see Fig. 9.29. Conversely, TiC coatings are usually worn less severely on the flank face than the rake face, Fig. 9.30. A *summary of wear rates* due to this contrasting behaviour (determined for cutting three types of steel) is collated in Fig. 9.31 [29]. From this data, it is clear that all coatings radically reduce rake and flank face wear compared to the rates of wear on uncoated cemented carbides, while confirming that no single

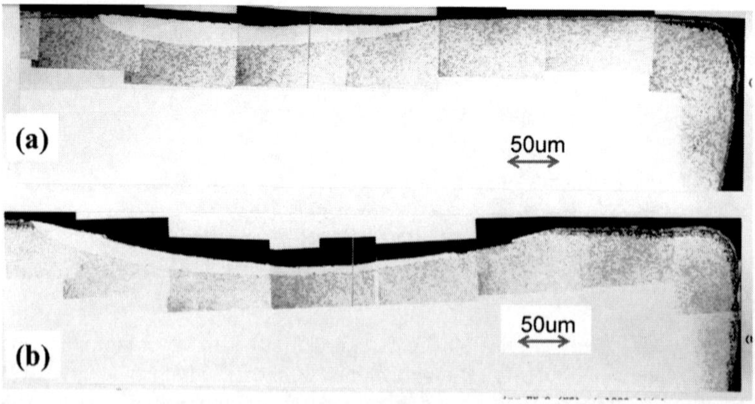

Figure 9.30 Metallographic cross-sections revealing the wear of two TiC coated cemented carbide tools after cutting annealed 0.4% C steel at 300 m/min for 5 mins, showing higher wear on the rake face (zone A in Fig. 9.28) compared to the wear on the flank face (zone B in Fig. 9.28): (a) TiC coating thickness 6 μm; (b) TiC coating thickness 4 μm. Note the longer crater in (b) due the thinner coating being worn through more quickly.

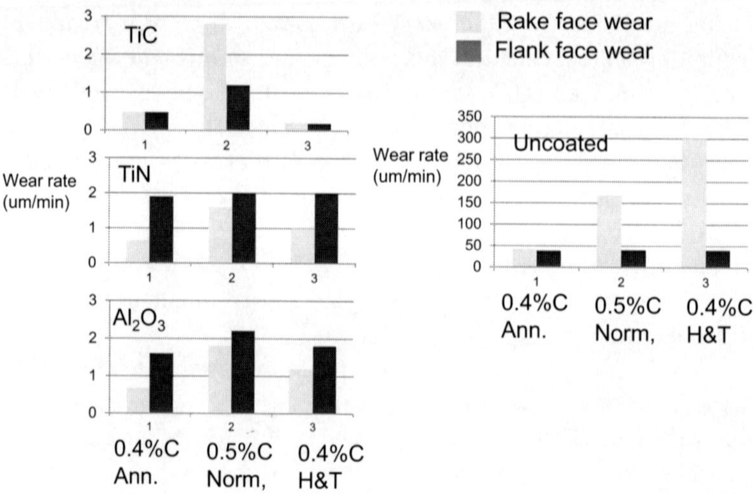

Figure 9.31 Summary of rake face (zone A in Fig 28) and flank face wear (zone B in Fig 28) depths determined after 30 minutes cutting of 0.4% annealed, 0.5% normalised and 0.4%C hardened & tempered steels at ~250 m/min, 0.25 mm/rev [28]. Rake face wear due to dissolution (polishing) and/or superficial plastic deformation, flank wear due to dissolution (polishing) and/or micro-abrasion.

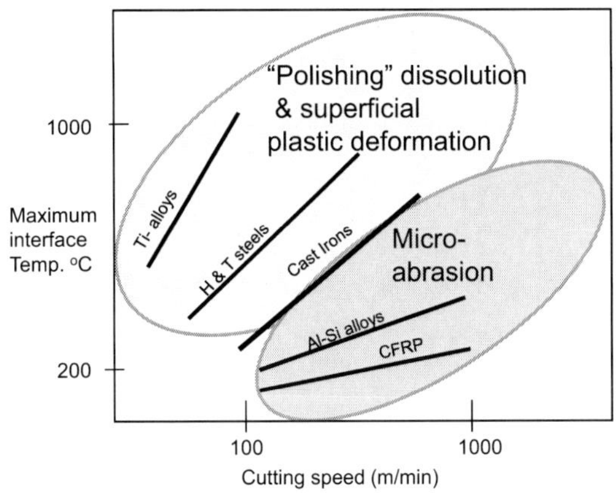

Figure 9.32 Dominant wear mechanisms as a function of cutting speed and maximum tool-workpiece interface temperature. Ranges for Al –Si and CFRP (carbon fibre reinforced polymer) are estimated.

coating material, offers the best wear resistance on all tool surfaces. Such phenomena were recognised quite early in the development of coated metal cutting tools that led to the introduction of multi-layered (multi-phased) composite coating architectures. Dissimilar rates of degradation with respect to notch wear resistance (Zones C and D in Fig. 9.28) are also observed. This phenomenon is often due to localised oxidational wear. Here, notch wear resistance for coatings made from $Al_2O_3 > TiN > TiC$ [28], which agrees with their relative oxidation resistance.

A global overview of the various responses of thin hard coatings materials to the principal wear zones and their associated wear mechanisms is cited in Table 9.2, whereas Fig. 9.32 shows how the predominance of the various rake and flank face wear mechanisms depends upon the materials being machined and the rate at which they are cut. Cutting speed is very important as it raises the operating temperature of the tool, which is highest on the rake face. Most wear processes increase with increasing temperature. Of particular note in this regard are polishing (dissolution wear) and superficial/discrete plastic deformation (Fig. 9.33b) of the coating surface asperities. It should be noted that the former process does *not* involve oxidation in any regard and should not be confused

Table 9.2 Vapour-deposited coating materials applied to cemented carbide cutting tool for wear protection when turning, drilling and milling various workpiece materials

Coating material	Coating technology	Combating rake face crater wear (zone A, Fig. 9.28).	Combating flank wear (zone B, Fig. 9.28).	Combating groove/notch formation (zones C & D, Fig. 9.28).
$\alpha + \kappa$ Al_2O_3 $\alpha-$ Al_2O_3 $\gamma-$ Al_2O_3	CVD CVD/MTCVD/PECVD PVD (at ~550°C)	When machining cast-irons. Poor when machining hardened and tempered and stainless steels.	Excellent when machining cast-irons. Poor when machining heat treated and stainless steels	Excellent when machining plain carbon steels & pearlitic grey cast-irons
TiN	CVD/MTCVD/PECVD/PVD	Good when machining steels & cast-irons above 100 m/min	Poor when machining steels & cast-irons	Good when machining steels & pearlitic grey cast-irons
Ti(Al,N)	PVD	Good when machining steels & pearlitic grey cast-irons	Good when machining steels & pearlitic grey cast-irons	Good when machining steels & pearlitic grey cast-irons
Ti(C,N)	CVD/MTCVD/PECVD/PVD	Excellent when machining stainless steels	Good when machining steels & cast-irons	Moderate when machining steels & pearlitic grey cast-irons
TiC	CVD	Good when machining steels & cast-irons	Excellent when machining steels	Poor when machining steels & pearlitic grey cast-irons
Diamond	Microwave -CVD	Variable when machining Al, Al-Si alloys and MMCs	Variable when machining Al, Al-Si alloys and MMCs	Unknown
DLC	PECVD	Variable when machining Al and Al-Si alloys	Variable when machining Al and Al-Si alloys	Unknown

Note: CVD coating technology is widely used to apply multilayered, multiphased coatings, for example, TiC/Ti(C,N)/TiN to maximise rake face and flank wear resistance.

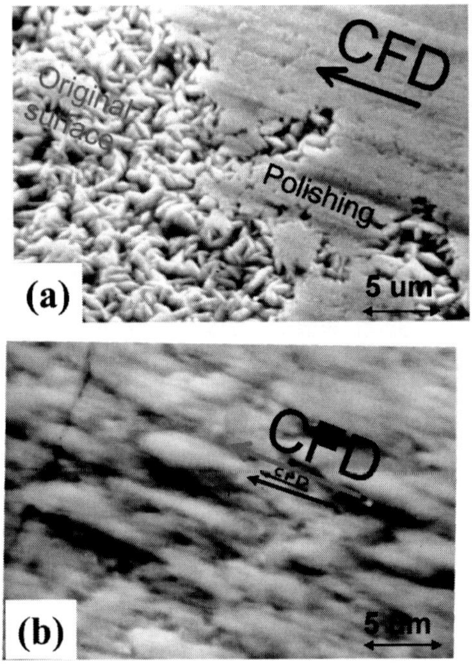

Figure 9.33 "Polishing" smooth wear (dissolution) and superficial plastic deformation on the rake faces of cemented carbide cutting tools coated with (a) TiN and (b) TiC [27]. Both morphologies produced by cutting medium carbon steel at ~200 m/min, 0.25 mm/rev. CFD = chip flow direction.

with oxidational wear processes that predominate in the notch wear zones C and D shown in Fig. 9.28.

Diamond-coated cemented carbides (produced via microwave CVD) are designed and recommended for machining hypo- and hyper-eutectic Al–Si alloys but tool life remains problematic [30]. Here, flaking or delamination of the diamond coating can take place (like that shown in Fig. 9.34). A similar problem occurs when using these types of tool to cut abrasive carbon fibre reinforced polymers (CFRP), carbon-carbon composites and MMCs [31]. In the latter situation, coating degradation is aggrevated by coarse SiC particles that are not only very hard (> 20 GPa) but are larger (> 15 μm diameter) than the thickness of the diamond coatings (~5 μm). This often necessitates the use of monolithic polycrystalline diamond

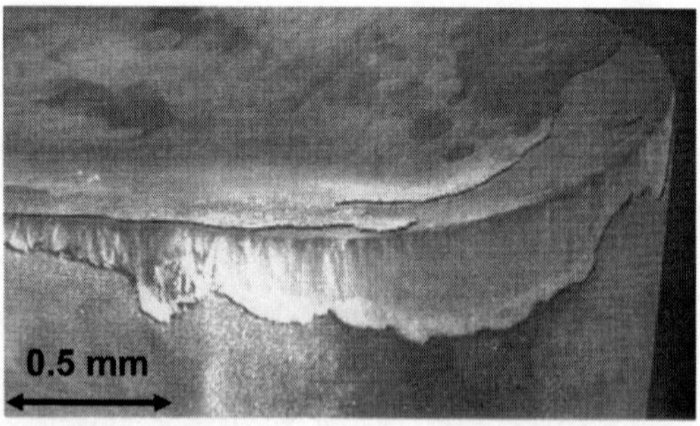

Figure 9.34 Delamination failure of 6 mm thick diamond coating from its WC-Con substrate after machining Al-8% Si truck wheels in production, at 1850 to 2500 m/min & 0.25–0.8 mm/rev. (after Reineck et al. [29]. Used with permission of the communicating author (Dr Mats Sjostrand).

or cubic boron nitride cutting tools that can be a factor of 30 to 40 times more costly per insert than coated cemented carbides. Here is an opportunity for DLC (synthesised via PACVD) coated tools to realise their full potential. Presently, their main limitation is variable DLC coating adherence to cemented carbide or HSS substrates which provides variable wear protection of the tool rake and flank faces. Nonetheless, they have sufficient adherence to prolong the life of solid cemented carbide drills when swarf blockage is the main cause of catastrophic tool failure—the low-friction quality of the DLC coatings conveyed to the drill flutes facilitating rapid swarf clearage/evacuation. This is important in enforced dry cutting operations where pressurised down hole coolants (used to aid chip clearage) cannot be used, for example, when using small diameter drills (<8 mm). Dasch et al. [32] showed that one type of DLC coating (comprising a blended mix of hydrogenated and non-hydrogenated DLC) applied to cemented carbide drills provided good performance when cutting a hypo-eutectic (6.5% Si) Al–Si alloy compared to several other types of DLC coated tools (Fig. 9.35). Of interest was that one type of diamond-coated cemented carbide drill performed very badly and only achieved four drilled holes before catastrophic fracture of the tool occurred due to swarf blockage of the drill shank flutes. Significantly, in *all* cases, coatings were rapidly removed from the contact areas and no wear protection was afforded to the tool rake or flank faces. Since poor coating

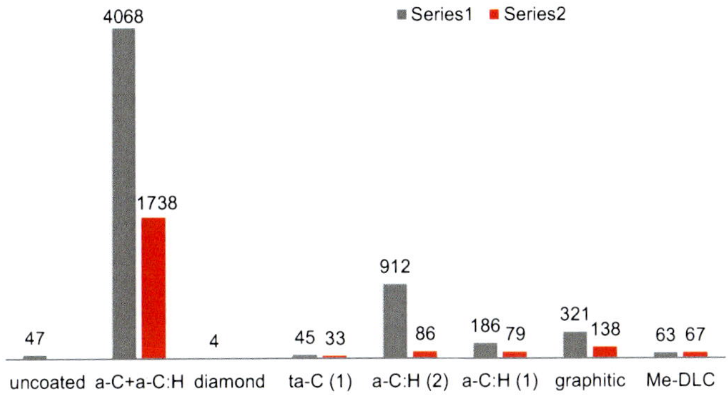

Figure 9.35 Drill lives expressed as total number of holes drilled to fracture for uncoated cemented carbides versus various carbon coated carbides. All tests made on B319 plate (Al-6.5%, Si-3.5%Cu, 0.3% Mg; BHN = 80) at a radial speed of 213 m/min and a down feed of 0.18 mm/rev. Based on the original tabular data of Dasch et al. [31] for two series of tests.

adhesion can be attributed to high internal compressive stresses (> 1 GPa) that may also lead to blistering [33], it is essential to implement new measures such as selective doping with Si (in the range of 1 to 10 atomic percentage) or vacuum annealing in order to significantly reduce internal stresses. This could easily be put into practice for cemented carbide substrates where there is no substrate degradation properties associated with such a heat treatment.

9.7 Towards MF Coating Materials for Tribology

It is worth exploring how close the state-of-the-art coated devices are to realising the target of true multifunctionality. From this review it is clear that for each of the three application sectors of cutting tools for machining materials in manufacturing and devices for automotive engines and bio-medical joint replacements there is some commonality in the mechanisms of degradation.

Tables 9.3 and 9.4 show a classification system created to encompass the *dynamic* surface degradation mechanisms encountered in the three main application sectors considered. Broadly these are divided into *mechanical* and *chemical* (major) groups, respectively;

Table 9.3 Dynamic *mechanical* material loss degradation mechanisms and modes

Applications dynamic mechanisms	Tools for metal & composite machining *dry contacts*	Human joint-replacements THRs & TKRs *natural or saline lubricant*	Piston rings for IC engines *man-made oil lubricants*
Mechanical mode 1. metal/ material transfer	For example, Material transfer build up on tool edges when machining cast irons across a wide range of cutting speeds	For example, Polymer transfer to metallic surfaces during in vitro wear testing	For example, Cylinder liner transfer to S-phase coated stainless steel (Fig. 9.20)
Most resistant MF-coating(s) identified to date	Least build up on Al_2O_3+ TiN coated cemented carbides.	ZrO_2 diffusion coated Zr-2.5Nb alloy	PVD-CrN coated martensitic stainless steels and others
Mechanical mode 2. Coating delamination	For example, CVD-TiC coated cemented carbides tools (when machining cast irons). Diamond coated cemented carbides (when machining Al-Si alloys – Fig. 9.34)	For example, THR Ti-6Al-4V femoral head bearing surfaces coated with TiN or DLC.	For example, Ti-Al(N), (Cr,Ti, Al) N coated stainless steel rings (Fig. 9.22)
Most resistant coating/substrate combination identified to date	CVD-Al_2O_3+ TiN coated cemented carbides. PVD-TiN or Ti(C,N) coated cemented carbides	ZrO_2 diffusion coated Zr-2.5Nb alloy in vivo CrN coated Co–Cr; in vitro only.	PVD-CrN coated stainless steels rings.
Mechanical mode 3. Micro-abrasion	For example, 1. Flank wear when machining steels, cast-irons and metal-ceramic composites. 2. Rake face wear when machining cast-irons, metal-ceramic composites, CFRP, C-C and polymer-ceramic composites.	For example, uncoated Co–Cr TKR Condyles. Uncoated stainless steel and Co–Cr femoral heads (Fig. 9.12a).	For example, W-B(N) coated stainless steel rings (Fig. 9.21)

Most resistant MF-coating(s) identified to date	TiC coated cemented carbides for steel cutting. Diamond coated carbides for machining Al–Si alloys and various ceramic and carbon reinforced ceramics.	ZrO_2 diffusion coated Zr-2.5Nb alloy; in vivo. CrN coated Co–Cr; in vitro only.	Si-doped and non-doped CrN coated stainless steel rings (Figs. 9.24 and 9.25b).
Mechanical mode 4. Discrete/ superficial plastic flow	For example, Rake face wear of TiC or Al_2O_3 coated carbides used to cut steels (Fig. 9.33b)	For example, in vitro tested CrN coated stainless steels but not yet identified in explants.	None known
Most resistant MF-coating(s) identified to date	Ti(C,N) and TiN coated carbides used to machine stainless steels. CrB_2 coated carbides used to machine Ti alloys.	None known	None known

Table 9.4 Dynamic *chemical* material loss degradation mechanisms and modes

Applications *Dynamic mechanisms*	Tools for metal & composite machining *Dry contacts*	Human joint-replacements THRs & TKRs *Natural or saline lubricant*	Piston rings for IC engines *Man-made oil lubricants*
Chemical mode 1. Solid (dry) state (polishing) wear	For example, (a) Oxidative notch wear when cutting steels; (b) Non-oxidizing environments: rake face dissolution wear when machining steels and cast irons—Fig. 9.33a.	None known	None known
Most resistant MF-coating(s) identified to date	(a) Al_2O_3 or TiN or Ti(Al,N) coated carbides to combat notch wear. (b) TiC to combat rake face dissolution when machining steels. CrN to be developed for notch wear?	Not applicable.	Not known.
Chemical mode 2. mechanically antagonized liquid (wet) state dissolution (polishing) wear (Type I corrosion wear) + fretting (wet)	None known	For example, Type I CW on uncoated Co–Cr femoral head (Fig. 9.12b). Modular femoral head interface fretting, for example, Ti alloy trunnion/Co–Cr Morse taper interface.	For example, Type I CW on coated and uncoated stainless steels. Also, piston-ring/piston groove fretting.
Most resistant MF-coating(s) identified to date	Not applicable	Type I: ZrO_2 diffusion coated Zr-2.5Nb alloy in vivo. CrN coated Co–Cr; in vitro only. Fretting not assessed.	Si doped and non-doped CrN coated stainless steels

Chemical mode 3. Liquid (wet) state pitting & decapitation—Type II corrosion wear	None known	For example, DLC-coated Ti-6Al-4V THR-femoral heads—severe attack in vivo. Si doped DLC-coated Ti-6Al-4V attacked in vitro (Kim et al., 2008).	None known
Most resistant MF-coating(s) identified to date	Not applicable	ZrO_2 diffusion coated Zr-2.5Nb alloy—mild attack in vivo. CrN coated Co–Cr; in vitro only.	Not applicable

the former comprising four important modes (sub-types) and the latter three modes. While most of these were described in detail in the previous sections of this chapter, it is worthwhile to bring attention to those coating materials that can mitigate more than one sub-type, that is, coating materials that approach an MF capability. Amongst the very few, are CrN coatings (applied via PVD technology). This material demonstrates excellent protection to metallic substrates when exposed to apposing sliding contacts in mildly oxidising environments, where the oxidising medium is a liquid, whether it be organic (ester based—as in the case of piston ring testing) or inorganic (water based—as in the case of saline or proteinacious fluids in human body environments). It would appear that CrN is able to form a tenaciously adherent passive oxide film (during oxidation) that does not readily break down as depicted in the Type I corrosion-wear model shown in Fig. 9.15. It also strongly bonds to its parent metal substrates (such as stainless steel), showing very good resistance to metal transfer and micro-abrasion, in lubricated piston ring cylinder liner contact tests so conveying an effective protective shield against degradation from mechanical modes 1, 2 and 3 (Table 9.3). Further, it would also appear to offer resistance to chemical mode 3 degradation when applied to Co–Cr femoral heads and subjected to hip joint simulator tests, the results of more advanced clinical trials having not been released to the public domain. Based on these observations, it would be worthwhile to evaluate CrN for: (1) Protection of the fretting wear resistance (chemical mode 2 in Table 9.4) of modular hip replacement devices (to preserve their Ti-alloy trunnion/and uncoated Co–Cr femoral head parts); and (2) the notch wear protection of cutting tools (in dry environments—chemical mode 1 in Table 9.4) of cemented carbide cutting tools, which hitherto has not been tried. It would seem that oxide layers themselves, being already oxides, offer similar capability in some regard, provided they remain adherent to their substrate and do not fracture internally. This is the case for thermally oxidised Zr-2.5Nb (Oxinium) femoral component bearing surfaces presently in use for some designs of THR and TKR, which are protected by an exterior layer of m-ZrO_2 (Fig. 9.17). These show effective protection against mechanical modes 1, 2 and 3, as well as chemical modes 2 and 3. So far m-ZrO_2 coatings (which could be created and applied using PVD technology) do not appear to have been evaluated as protective coatings for cemented carbide tools and stainless steel piston rings (to mitigate chemical modes 1 and 2, respectively) and possibly even mechanical modes 1, 2 and 3 (Table 9.3). At this time, some forms of ZrO_2 are applied to Ni base alloys (> 50 μm thick) as thermal barrier coatings [34], using elevated temperature electron beam PVD methods, but such coatings are micro-porous and not suited to tribological use. Thinner (~ 5 μm) and very dense coatings of various crystalline forms of ZrO_2 can however be synthesised and applied to metallic substrates using reactive magnetron sputtering, Wong et al. [35]. So far, these have *not* been evaluated for corrosive-wear applications where chemical modes 2 and 3 (Table 9.4) predominate. However, pre-oxidised ZrN coatings have demonstrated resistance to these phenomena in salinic solution environments, Ferreira et al. [36]. Accordingly, it is worth exploring the potential of magnetron sputtered ZrO_2 coatings for use in the corrosive-wear environments found in piston rings and orthopaedic implant bearing surface applications, where this coating material can be applied to any substrate.

9.8 Conclusions

This chapter has reviewed the functional performance of surface-engineered materials used in three important engineering sectors: (1) Bearing surfaces for bio-medical joint replacements; (2) Automotive IC engines components; and (3) Cutting tools for used for machining materials in large scale manufacturing. The importance of a given surface modified material's ability to resist surface degradation processes has been emphasised in this chapter. Within the application sectors considered here, four kinds of *mechanical* and three types of *chemical* degradation mechanisms were found to predominate. Coating materials that approach MF capability are those that provide a shield for the substrate and mitigate against several of these mechanisms. It was shown that these are presently confined to coating materials that either form a stable and a mechanically tenacious oxide passive film or are already in an oxide state; these being CrN and m-ZrO_2, respectively. The latter could be better exploited as PVD coating materials—but these require deeper tribological investigation before this is possible. There is also significant scope for the augmentation of coating behaviour to combat dynamic surface degradation mechanisms via the doping of certain coating materials, such as CrN and DLCs, with strong oxide forming elements such as Si.

References

1. D. Yonekura, R.J. Chittenden, P.A. Dearnley, Wear mechanisms of steel roller bearings protected by thin, hard and low friction coatings, *Wear*, 259 (2005), 779–788.
2. E.M. Trent, *Metal Cutting*, 2nd ed., London: Butterworths (1984), pp. 59–64, 116.
3. C.M. Taylor, Fluid film lubrication in automobile valve trains, proceedings of the institution of mechanical engineers, Part J, *J. Eng. Tribol.*, December 208(4) (1994), 221–234.
4. H. Spikes, The history and mechanisms of ZDDP, *Tribol. Lett.*, 17(3) (2004), 469–489.
5. C.D.C. Burnell, J.-M. Brandt. M.J. Petrak, R.B. Bourne, Posterior condyle surface damage on retrieved femoral knee components, *J. Arthroplasty.* 26(8) (2011), 1460–1467.
6. J. Fisher, X.Q. Hu, J.L. Tipper, T.D. Stewart, S. Williams, M.H. Stone, C. Davies, P. Hatto, J. Bolton,M. Riley, C. Hardaker, G.H. Isaac, G. Berry, E. Ingham, An in vitro study of the reduction in wear of metal-on-metal hip prostheses using surface-engineered femoral heads, *Proc. Inst. Mech. Eng., Part H: J. Med. Eng.*, 2002, 216, 219–230.
7. S. Williams, J.L. Tipper, E. Ingham, M.H. Stone, J. Fisher, In vitro analysis of the wear, wear debris and biological activity of surface-engineered coatings for use in metal-on-metal total hip replacements, *Inst. Mech. Eng., Part H*: 217 (2003), 155–163.
8. P.A. Dearnley, A review of metallic, ceramic and surface-treated metals used for bearing surfaces in human joint replacements, *Proc. Inst. Mech. Eng. Part H: J. Med. Eng.*, 213(H) (1999), 107–135.
9. K.L. Dahm, I. Anderson, P.A. Dearnley, Hard coatings for orthopedic implants, *Surf. Eng.*, 11(2) (1995), 138–144.
10. H. Minakawa, M. Stone, B. Wroblewski, J. Fisher, *J. Bone Joint Surg. Br.* 80(5) (1998), 894–899.

11. M.J. Pappas, F.F. Buechel, Prosthesis with biologically inert wear-resistant surface, *Ceramics in Substitutive and Reconstructive Surgery* (ed. P. Vincenzini), Elsevier, 1992.
12. M.K. Harman, S.A. Banks, W.A. Hodge, Wear analysis of a retrieved hip implant with titanium nitride coating, *J. Arthoplasty*, 12(8) (1997), 938–945.
13. M.T. Raimondi, R. Pietrabissa, *Biomaterials*, 21 (2000), 907–913.
14. G. Taeger et al., Comparison of diamond like carbon and alumina articulating with polyethylene in total hip arthroplasty, *Matt-wiss. U. Werkstofftech.*, 34(12) 2003, 1094–1110.
15. P.A. Dearnley, G. Aldrich-Smith, Corrosion-wear mechanisms of hard coated austenitic 316L stainless steels, *Wear*, 256/5 (2004), 491–499.
16. J.-G. Kim, K.-R. Lee, S.-J. Yang, Wear corrosion performance of Si-DLC coatings on Ti-6Al-4V, *J. Biomed. Mater. Res.* 86A (2008), 41–47.
17. J.A. Davidson, Zirconium oxide coated prostheses for wear and corrosion resistance, US Patent 5037438, 1991.
18. J.A. Davidson, Ceramic coatings for orthopaedic bearing surfaces, In *Ceramics in Substitutive and Reconstructive Surgery* (ed. P. Vincenzini), Elsevier, Amsterdam, 1991, pp. 157–166.
19. T.J. Heyse, D.X. Chen, N. Kelly, Matched-pair total knee arthroplasty retrieval analysis: Oxidized Zirconium vs. CoCrMo, *The Knee* 18(6) (2011), 448–452.
20. E.C. Contract, G5RD-CT-2000-00430, Final Technical Report - HIDUR, Improving the competitiveness and conserving the environment through high durability nanocomposite coatings, 6th September 2005.
21. Robertson, J., Diamond-like amorphous carbon, Mater Sci. Eng. R Rep, R37 (2002), 129–281.
22. Bauer, C., Leiste, H., Stüber, M., Ulrich, S., Holleck, H., Constitution, microstructure, mechanical properties and performance of magnetron sputtered carbon films with additions of silicon, *Mat.-wiss. U. Werkstoff.*, 34 No 10/11 (2003), 912–918.
23. Bendavid, A., Martin, P. J., Comte, C. E., Preston, W. Haq, A. J. Magdon F. S. Ismail, R. K. Singh, The mechanical and biocompatibility of DLC-Si films prepared by pulsed DC plasma activated chemical vapour deposition, *Diamond Relat. Mater.*, 16 (2007), 1616–1622.
24. Z. Pawlak, Tribochemistry of lubricating oils, Elsevier, Amsterdam, 2003, p. 25.

25. K.L. Dahm, P.A. Dearnley, On the nature, properties and wear response of S-phase coatings on AISI 316L, *Proc. Inst. Mech. Eng.* 214(L) (2000), 181–198.
26. *Piston Ring Handbook,* Burscheid: A.E. Goetze (1995).
27. B. Mallia, *Novel Nanostructured Coatings for Extreme Tribological Environments*, PhD Thesis, University of Leeds, 2008.
28. P. A. Dearnley, E. M. Trent, Wear mechanisms of coated carbide tools, *Met. Technol.*, 9 (1982), 60–75.
29. P. A. Dearnley, Rake and flank wear mechanisms of coated cemented carbides, *Surf. Eng.*, 1(1) (1985), 43–58.
30. I. Reineck, M. E. Sjostrand, J. Karner, M. Pedrazzini, HCDCA diamond coated cutting tools, *Diamond Relat. Mater.*, 5 (1996), 819–824.
31. C. Lane, Proceedings of the Machining of Composite Materials Symposium, *ASM Materials Week*, Chicago, Illinois, 1–5 November, 1992, (ed. T.S. Srivatsan, D.M. Bowden); The Effect of Different Reinforcements on PCD Tool Life for Aluminium Composites, *ASM International* (1992), pp. 17–28.
32. J.M. Dasch, C.C. Ang, C.A. Wong, Y.T. Cheng, A.M. Winer, L.C. Lev, E. Konca, A comparison of five catagories of carbon based tool coatings for dry drilling of aluminium, *Surf. Coat. Technol.* 200 (2006), 2970–2977.
33. X.L. Peng, T.W. Clyne, Residual stress and debonding of DLC films on metallic substrates, *Diamond Relat. Mater.* (Switzerland), 7(7) (1998), 944–950.
34. J. Nicholls, D. Rickerby (Eds.), *High Temperature Surface Engineering*, The Institute of Materials, Book Number 693, London, 2000.
35. M.S. Wong, W.J. Chia, P. Yashar, J.M. Schneider, W.N. Sproul, S.A. Barnett, High-rate reactive d.c. magnetron sputtering of ZrOx, coatings, *Surf. Coat. Technol.* 86–87 (1996), 381–387.
36. S.C. Ferreira et al., Tribocorrosion behaviour of ZrOxNy thin films for decorative applications, *Surf. Coat. Technol.* 200 (2006), 6634–6639.

Chapter 10

The Role of Viscoelastic and Plastic Deformations in the Creep Behaviour of UHMWPE: An Experiment-Based Computational Study Using an AMTI Knee Wear Simulator

Federico Quinci,[a,b] Matthew Dressler,[c] Anthony M. Strickland,[b] Amber Metcalfe,[c] Mark Taylor,[b,d] and Georges Limbert[a,b]

[a] *The national Centre for Advanced Tribology at Southampton (nCATS), Faculty of Engineering and the Environment, University of Southampton, Highfield, Southampton SO17 1BJ, UK*
[b] *Bioengineering Science Research Group, Engineering Sciences, Faculty of Engineering and the Environment, University of Southampton, Highfield, Southampton SO17 1BJ, UK*
[c] *DePuy Orthopaedics Inc., 700 Orthopaedic Drive, Warsaw, IN 46582, USA*
[d] *Medical Device Research Institute, School of Computer Science, Engineering and Mathematics, Flinders University, Adelaide 5001, Australia*
f.quinci@soton.ac.uk

10.1 Introduction

Ultra high molecular weight polyethylene (UHMWPE) has been used as bearing material in total knee replacements (TKR) since

Multifunctional Materials for Tribological Applications
Edited by Robert J. K. Wood

ISBN 978-981-4463-90-4 (Hardcover), 978-981-4463-91-1 (eBook)
www.panstanford.com

the late 1960s [1, 2]. Besides biocompatibility, UHMWPE offers excellent wear and fatigue resistance properties. However, the coupled effect of kinematics and cyclic contact stresses generated at the articulation result in plastic deformation and wear debris formation, thereby limiting the lifespan of the implant [3, 4]. The long-term complications of TKR are correlated to loosening and osteolysis [5], which are consequences of the body reaction to the polyethylene wear debris present in the periprosthetic soft tissues. A way to improve the lifetime of these implants has been to modify the microstructure of UHMWPE through irradiation-induced crosslinking combined with thermal treatments [6]. However, it has been shown that by crosslinking the material, the wear resistance improves at the expenses of other mechanical properties such as toughness, ductility, fatigue resistance and oxidation [4].

UHMWPE, like other thermoplastic polymers, has a complex time-dependent non-linear mechanical response that can be challenging to characterise and exploit effectively in order to develop optimised orthopaedic components. As a pre-clinical assessment tool, the ability to predict in silico the mechanical response of UHMWPE components is valuable for the orthopaedic implant industry. Nevertheless, the computational simulation of the mechanical and wear characteristics of UHMWPE implants by means of finite element (FE) analyses requires sound constitutive models capable of simulating their viscoelastic properties.

To date, FE analyses of TKR systems have often been used to predict contact stresses, contact areas (CAs) and stress distribution within the UHMWPE insert during static or quasi-static loading conditions. A possible approach to evaluate the stresses within the tibial insert in TKRs has been to estimate where the location of the contact between the femoral component and tibial insert is likely to occur at a specific instant of the gait cycle. By considering this approach, the contact may be assumed to occur at the deepest contact point of the tibial proximal surface, or estimating where the contact may take place in a well-functioning knee [7, 8].

A variety of studies that used FE analyses based on explicit solving schemes (thereafter referred as 'explicit FE') [9–16] and multibody dynamics [17–19] simulations were conducted to gain insights into the mechanical behaviour (i.e., wear resistance) of new

UHMWPE formulations, as well as to compare kinematics of knee simulators such as the Stanmore (force-controlled system; Instron Corp., Canton, MA) [20] and AMTI (displacement-controlled system; Advance Mechanical Technology Inc., Watertown, MA) [21].

A number of explicit FE models [9, 10] have used an interactive, adaptive remeshing technique based on phenomenological-based laws such as the classical Archard's law for abrasive wear [22], where wear is assumed to be linearly dependent on contact pressure (CP) and sliding distance, to qualitatively estimate UHMWPE wear occurring at the articulating surface.

These in silico studies did not consider the effect of creep at the early stage of deformation unlike the computational study of Bevill et al. [23] that demonstrated that failing to account for creep in the constitutive formulation led to underestimation of the wear rates.

More recently, Willing et al. [11] developed a TKR FE model capable of predicting strain hardening, damage, wear and creep of the UHMWPE tibial insert using an adaptive remeshing approach controlled by the evolution of damage at the contact surface. Their quasi-static FE model was validated using two sets of testing waveforms based on ISO standards from a displacement-controlled AMTI knee simulator. The quasi-static FE model predicted the measured forces in the AMTI knee simulator and the amount of wear damage with reasonable accuracy, although neglecting inertial effects. The major limitation in such a phenomenological approach is that it takes into account parameters that depend on the contact only (i.e., CP, sliding distance) hence neglecting all subsurface meso/micro deformation mechanisms.

Another approach for explicit FE analyses was to predict both kinematic and contact mechanics of TKR during dynamic loading conditions [13, 14, 24]. These FE analyses looked at the magnitude and distribution of stresses within the UHMWPE insert that is of particular interest because excessive stress may lead to delamination and wear [20].

D'Lima et al. [15] calculated contact stresses using a FE analysis driven by measured in vivo tibial forces and kinematic data during activities of daily living after TKR. The tibial insert was modelled as an elasto-plastic material, and the predicted contact measures were compared to those found experimentally. The input motions in the AMTI knee simulator consisted of vertical loads (1.5–3.6 kN), and flexion-extension rotation (0–120°). From their results, lunge activity generated the highest contact stresses when compared to walking and stair climbing.

Recently, Lanovaz et al. [25] reported a dynamic FE model of an AMTI knee wear simulator, where the tibial insert was modelled as an elasto-plastic material; the simulated contacted areas were compared to experimental data collected in vitro using a Fuji pressure-sensitive film. Moreover they followed a 'force feedback' approach: a FE model based on a standard force-based cyclic wear-testing protocol (ISO 14243-1) was created, then the resulting displacement waveforms were extracted from the model and used as control inputs in a physical wear test experiment. They demonstrated that a force-driven FE model can provide the displacement waveforms for a displacement-driven mechanical testing system.

To date, with few exceptions [11, 12, 26], phenomenological computational models of wear of UHMWPE have neglected creep deformations that are believed to play an important role in driving sub-surface damage, particularly in the early days of a new implantation. Moreover, FE approaches to simulate wear of UHMWPE components use calculated CP and sliding distances as inputs to calculate the resulting wear volume. These (naturally time-dependent) response variables are intrinsically linked to the extent and deformation of the CAs in response to external loads applied to the joint. Inelastic deformation mechanisms such as creep and plasticity, which do occur in vivo, are therefore critical because of their role in damage initiation and wear.

The aim of the present study is to characterise the respective role of elastic, creep and plastic deformations of a UHMWPE insert under compression on the CA characteristics and associated kinematic and kinetic quantities such as deformation and CP.

This is achieved by developing an experimentally based FE model replicating the mechanical characteristics of a AMTI Knee Simulator (Advance Mechanical Technology Inc.). This study is part of an on-going project aiming to develop a mechanistic constitutive wear model.

10.2 Materials and Methods

10.2.1 *Experimental Setup*

An AMTI knee simulator (Advance Mechanical Technology Inc.) was used to create a simplified articulation using a size 3 right CoCrMo Sigma® femoral component (DePuy Orthopaedics, Warsaw, IN) with a modified Sigma® fixed bearing polyethylene insert secured in a CoCrMo Sigma modular tray. The proximal (articulating) surface of the insert was machined flat to provide a uniform 8 mm thickness that reduced confounding factors that may obfuscate measurements of CA and penetration depth (PD) [27]. A constant vertical load (0.25, 0.5, 0.9, 1.8, 3.6 kN) was applied with flexion (0° to 30°) for 1000 seconds with a 10-second ramp-up and ramp-down. The flexion was applied at 1 Hz about the distal sagittal femoral radius with no internal-external (I–E) rotation, anterior–posterior (A–P) translation or A–P tilt. Varus–valgus (V–V) motion was constrained while medial–lateral (M–L) motion was left unconstrained. To help identify the CA, the proximal surface of the insert was coated with blue ink prior to each loading condition. The experiment was conducted at room temperature and the surface was lubricated with water. It is important to note that the femoral component design has a constant radius within the applied flexion range, so the contact geometry should stay constant.

The resulting deformation was determined by quantifying the PD and CA after each loading condition. The PD was defined as the maximum deviation from an idealised best fit of the proximal surface and was measured using a contact profilometer (Zeiss Surfcom 5000, Maple Grove, MN) and a three-dimensional laser scanner (Metron G2-24, Snoqualmie, WA) with post-processing performed in Geomagic Qualify (Research Triangle Park, NC). The CA was assessed through photographs showing the removal of toolmakers ink from the insert as well as images of the deformed inserts generated from the laser scanner. These images of CA were analysed using ImageJ (NIH, Bethesda, MD). Recovery of the PD was also measured 1 and 2 days after the 3.6 kN loading cycle (using the laser-scanner and profilometer).

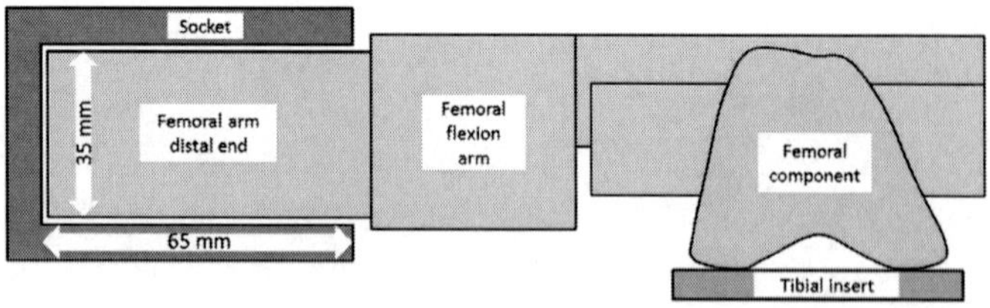

Figure 10.1 Schematic highlighting the geometrical details of the AMTI knee simulator arm placed within its socket together with the placement of the implant components (frontal view and not to scale).

Although the loading was intended to be symmetrical (i.e., 50/50 medial/lateral load split), this was not achieved due to pliancy in the flexion arm, resulting in asymmetry between the two condyles. Consequently, care was taken to model this pliancy in silico. The arm extends into the socket by about 65 mm ($2^1/_2$") and has a diameter of about 35 mm ($1^3/_8$") (Fig. 10.1).

The arm is actually undersized relative to the socket, hence a small amount of motion is possible and this can include translation and/or rotation, see Fig. 10.2.

The motions can take place along the I–S and V–V directions, or in the A–P and I–E directions (Fig. 10.2). Its not possible to get neither the maximum rotation at the same time as the maximum translation nor the maximum translation in two directions at the same time (e.g., I–S and A–P directions) since the socket is circular in cross-section. One can work out the difference in sizes between the arm and the socket using basic trigonometry and measurements from the rig. The maximum possible V–V rotation is up to 0.2°, and

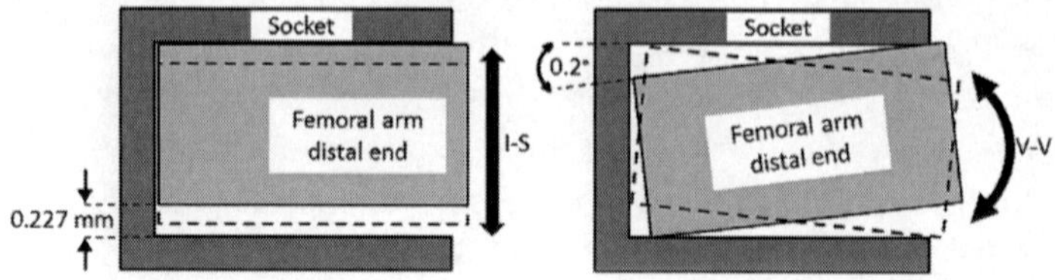

Figure 10.2 Schematic highlighting the (exaggerated) fitting tolerance of the femoral arm within its socket using the same frontal view as Fig. 10.1. The type and magnitude of possible motions are indicated as inferior-superior (I–S) translation and V–V.

was measured via manual manipulation of the rig. This introduces a degree of measurement uncertainty and there is probably a station-to-station variability associated with this measurement.

Because the arm extends inside of the socket by about 65 mm, the difference between the arm and the socket diameters can be calculated as A–B $= 65\text{ mm} \times \sin(0.2^\circ) = 65\text{ mm} \times 0.0035 = 0.227\text{ mm}$, see Figs. 10.1 and 10.2. The possible translation motion is therefore 0.227 mm, which can be seen as about $\pm$ 0.11 mm relative to a fixed central reference point in the A–P and I–S directions. The motion in the M–L direction is different; it is governed by the pliancy of the mechanism locking the arm into the socket, and has been experimentally measured to have a range of 0.5 mm, that is, $\pm$ 0.25 mm relative to a neutral fixed reference point.

10.3 Computational Modelling

A FE model simulating an AMTI knee simulator articulating a PFC® Sigma® femoral component (size 3) on a flat UHMWPE tibial insert (size 3) under a series of monotonically increasing vertical loads was developed in Abaqus/CAE 6.11 (Dassault Systèmes Simulia Corp., Providence, RI) according to the experiment described in the previous section [27] (Fig. 10.3).

From this generic FE model, three different FE models, each with a different material model for UHMWPE, were generated:

- **E–FE model** (Elastic–FE): isotropic linear elastic properties (Young's modulus of 575.5 MPa and Poisson's ratio of 0.46 [28])
- **EP–FE model** (ElasticPlastic–FE): isotropic elastoplastic properties with kinematic hardening [29]
- **EV–FE model** (ElasticViscoelastic–FE): isotropic viscoelastic properties featuring creep measured at 15 MPa compressive load [30]

In the FE simulation, the mechanical response of the femoral component (CoCrMo) was assumed to be linear elastic with a 220 GPa Young's Modulus and Poisson's ratio equal to 0.33. Similar constitutive assumptions were used for both the socket and

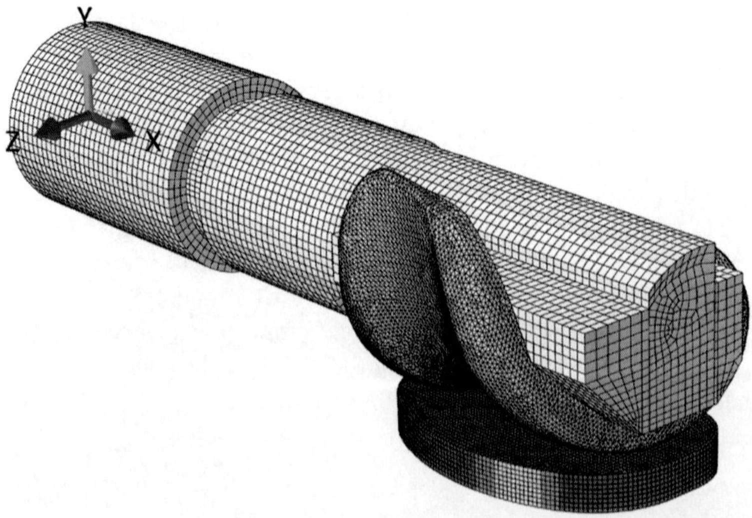

Figure 10.3 Perspective view of the finite element model of the AMTI knee simulator setup. Both tibial and femoral components were of size 3. The tibial insert was machined flat 8 mm thick.

cantilever arm (steel) and a 200 GPa Young's modulus and a 0.3 Poisson's ratio were used.

The forces driving the simulated AMTI system consisted of constant vertical loads (0.25, 0.45, 0.9, 1.8 and 3.6 kN) modulated by a time-dependent amplitude curve (Fig. 10.4) and applied sequentially.

There were no input motions such as I–E rotation, A–P/V–V/M–L motions and A–P tilt. Unlike for the physical experiment, the sinusoidal flexion about the distal sagittal femoral radius was not considered in the FE analyses. It was assumed that the flexion cycle operated at a constant deformation of the UHMWPE surface, and consequently, this meant that the CA was not affected by the sinusoidal flexion.

The pliancy effect of the cantilevered arm due to imperfect conformity between the femoral flexion arm and the socket was computationally replicated through direct geometrical constraints induced by recreating the identical geometrical tolerances of the mechanical parts as measured in the physical setup.

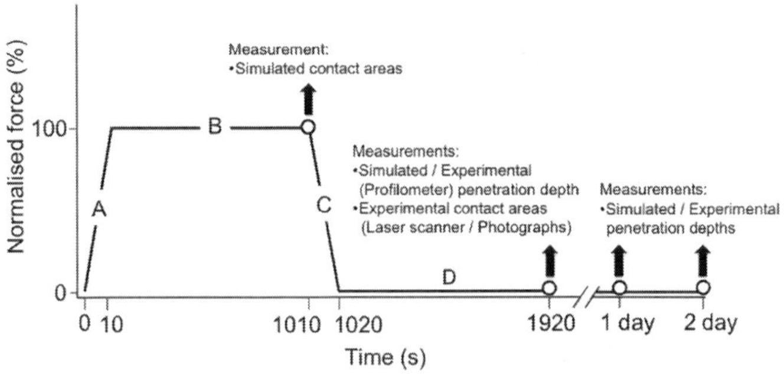

Figure 10.4 Force amplitude profile for each of the five loads considered (0.25, 0.45, 0.9, 1.8 and 3.6 kN). The loading cycle is split into four stages (a: ramping-up of the load, b: holding of the load, c: ramping-down of the load, d: deformation recovery). The final loading cycle (3.6 kN) features two additional stages corresponding to extended recovery periods (1920 seconds to 24 hours and 24 to 48 hours).

The nodes of the external surface of the socket were fixed in translation while the vertical load was applied at the bottom surface of the tibial insert assumed to be a rigid surface. Each experimental loading cycle in the FE models was simulated using four distinct analysis steps (Fig. 10.4) (option *STEP in Abaqus):

- **Step A:** ramp loading over 10 seconds,
- **Step B:** load holding for 1000 seconds,
- **Step C:** ramp unloading over 10 seconds,
- **Step D:** deformation recovery over 900 seconds.

The total cycle length per loading cases is therefore 1920 seconds. This four-step cycle was applied sequentially for the five load cases: 0.25, 0.5, 0.9, 1.8 and 3.6 kN. This meant that a full FE analysis consisted of 5 $\times$ 4 steps with the addition of two steps simulating deformation recovery between 9600 seconds (the end of the five cycles) and 1 day and between 1 and 2 days (Fig. 10.4).

Tie constraints were created between the femoral component and the AMTI cantilever arm to simulate the experimental constraint created through cementation of these parts. A horizontal spring with a stiffness of 10 N (Fig. 10.1a) was connected to the distal end of the beam to the inside of the socket in order to simulate the mechanism

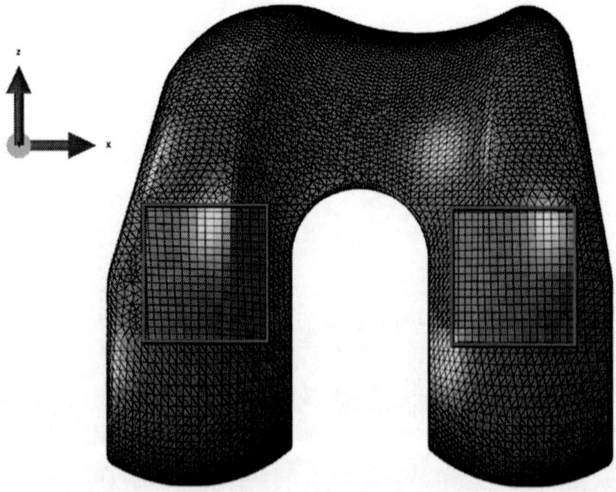

Figure 10.5 Inferior–superior view of the femoral condyle hybrid mesh highlighting the boundary between the sub-regions (tetrahedral and hexahedral elements).

locking the arm into the socket and to prevent horizontal sliding of the beam. Gravity loads were not accounted for.

Abaqus/standard surface-to-surface contact algorithm [31] was used to enforce the contact between the socket and flexion arm, and between the femoral and tibial components. Static coefficients of friction for lubricated and greasy surfaces of 0.08 (steel-on-steel) [32] and 0.04 (CoCrMo-on-UHMWPE) [14] were used. The whole mechanical system had a total of 226,949 elements. Both socket and AMTI flexion arm were meshed with fully integrated linear hexahedral (C3D8) elements, with global mesh size of 2 mm. The femoral component geometry was partitioned so that a hybrid tetrahedral–hexahedral mesh could be built (Fig. 10.5). The rationale for this was to use linear hexahedral elements for the surfaces of the femoral condyles contacting the tibial insert, also meshed with fully-integrated eight-noded hexahedrons (C3D8). Because of the complexity of the femoral condyle geometry it was not possible to mesh it automatically using exclusively hexahedral elements. The remaining part of the femoral condyle was meshed using linear tetrahedral (C3D4) elements. A mesh sensitivity study to assess the accuracy of results was performed on the femoral and tibial components using three different global mesh sizes: 0.6, 1 and 1.5 mm. The best compromise between accuracy and analysis time was found to be for a global mesh size of 1 mm. The femoral and tibial components used the same global mesh size in all simulations.

CAs and PDs were measured at the lateral condyle because it had the larger loads and PDs in both experiments and FE simulations.

10.4 Results

10.4.1 *Penetration Depth*

The PD for each of the five vertical loads (peak values at the end of the loading phase) and after 1 and 2 days of deformation recovery are depicted in Fig. 10.6 for the laser scanning and profilometry measurements as well as the three series of FE analyses (E–FE, EP–FE and EV–FE models, which will also subsequently be referred to as, respectively, the 'elastic model,' 'elastoplastic model' and 'viscoelastic model').

The variation of PD with increasing load follows an almost linear trend for both experimental measurements that are very similar (Fig. 10.6). At 1-day deformation recovery, the PDs are 0.293 and 0.320, respectively, for the laser scanning and profilometry measurements. At 2 days recovery, these values are 0.326 and 0.306.

As expected, and by construction, the E–FE model fully recovers the deformation in each loading cycle after 900 seconds as well as after 1 and 2 days of recovery.

The FE models under-predict PDs compared to those measured experimentally. The variation of PD for the viscoelastic model exhibits a characteristic creep-like curve (plateauing exponential) with increasing load whilst that curve is also non-linear for the elastoplastic model (Fig. 10.6a). At loads higher than 0.5 kN, the PD for the EV–FE model increases more slowly than that of the EP–FE model. After 1 and 2 days of recovery, the deformation predicted by the EP–FE is the closest to that of the experimental data. The E–FE and EV–FE models, on the other hand, fully recover their deformation (Fig. 10.6). The overall relative difference between the EV–FE model and the laser scanner data range between 72% and

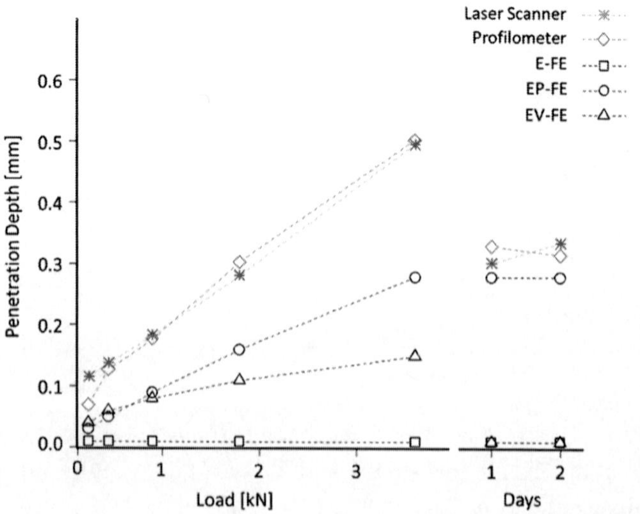

Figure 10.6 Maximum penetration depths within the UHMWPE tibial insert for both simulations and experiments and for the five different loading scenarios (0.25 to 3.6 kN). Penetration depths at 1 and 2 days following the last loading cycle (3.6 kN) are also plotted.

59%, while the relative difference between the EV–FE model and the profilometer data range between 72% and 50%. For the EP–FE model, these differences lie, respectively, in the ranges 8–81% and 12–66%. At 1 and 2 days deformation recovery, the residual deformation in the EP–FE model is 0.27 mm, which is very close to the 0.326 and 0.306 mm values obtained experimentally (Table 10.1).

10.4.2 *Contact Area*

CA measurements are compared for the FE models, laser scanning and photographic measurements on Fig. 10.7. Laser scanning acquisitions of the CA were only performed for the 1.8 and 3.6 kN loads. The photogrammetric (ink staining) method only provides an indication of the peak experimental CA, whereas the laser-scanning method provides additional measurements of the subsequent recovery phase. CAs are generally higher for the viscoelastic model than for the plastic and the elastic models. As expected, the CA increases with the load. The CAs for the inelastic (EV–FE and

Table 10.1 Relative difference (%) in the PD between FE simulations (E–FE, EP–FE and EV–FE) and experiments (laser scanner and profilometer)

	Relative difference in PD (%)			
Load (kN)	EV–FE vs. laser scanner	EP–FE vs. laser scanner	EV–FE vs. profilometer	EP–FE vs. profilometer
0.25	−72	−81	−50	−66
0.45	−61	−69	−57	−66
0.9	−59	−53	−58	−52
1.8	−63	−45	−66	−49
3.6	−71	−45	−72	−45
1st day recovery		−8		−16
2nd recovery		−17		−12

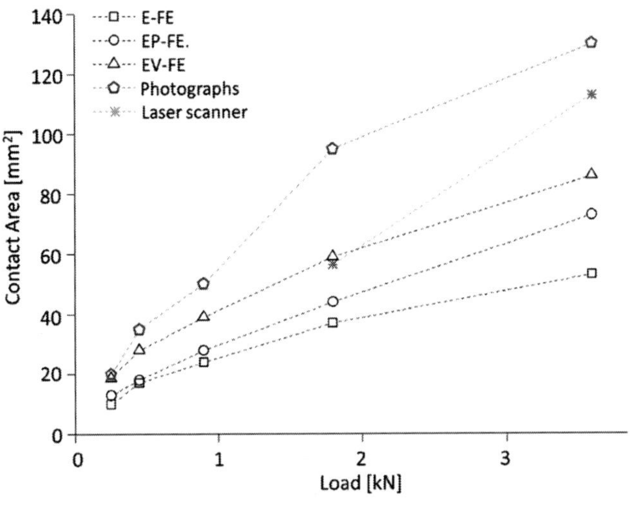

Figure 10.7 Maximum contact areas on the UHMWPE tibial insert (reported for the lateral condyle) for both simulations (E–FE, EP–FE and EV–FE) and experiments (photographs of contact scars and profilometry measurements).

EP–FE) and elastic models monotically deviate from each other as the load increases, which highlights the progressive dominant role of inelastic deformation mechanisms. The numerical models consistently under-predict the CAs measured experimentally except

Table 10.2 Simulated total CAs for the three FE models and for each loading cycle and relative distribution of load (%) between the medial and lateral condyles of the tibial insert

	Simulated total CA (mm^2)			Relative distribution of load (%) between tibial condyles					
Load (kN)	E–FE	EP–FE	EV–FE	Lateral cond. E–FE	Medial cond. E–FE	Lateral cond. EP–FE	Medial cond. EP–FE	Lateral cond. EV–FE	Medial cond. EV–FE
0.25	10	13	21	100	0	100	0	90	10
0.45	17	19	37	100	0	95	5	76	24
0.9	29	35	58	83	17	80	20	67	33
1.8	49	62	90	76	24	71	29	66	34
3.6	76	108	138	70	30	68	32	62	38

for the viscoelastic model which matches the CA obtained via photographic measurements at 0.25 kN and mostly that obtained by laser scanning at 1.8 kN. In both experimental acquisitions and simulations, CAs are found to be different between the medial and lateral condyles. This imbalanced load distribution between them is caused by non-negligible geometrical tolerances magnified by the pliancy effect of the arm (Figs. 10.2 and 10.9).

Simulated total CAs with the relative distribution of the load between the lateral and medial condyles for every load cycle are listed in Table 10.2. Total CAs were used for comparison because the majority of TKR studies found in the literature report the total area in contact (i.e., both condyles). In all FE models, most or the totality of the load is bourne by the lateral condyle, particularly at small loads (Table 10.2).

Unlike the elastic and plastic models, in the EV–FE model contact was always distributed on both the medial and lateral condyles. The overall CA difference between the EV–FE model and photographs ranges between 5% and 38%, while the difference between the EP–FE model and photographs varies between 35% and 54% (Table 10.3).

10.4.3 *Contact Pressure*

In the experiments, CPs were not measured but they are standard output of any contact FE analysis. As highlighted in Section 10.1, CPs

Table 10.3 Relative difference (%) in CA between FE simulations (E–FE, EP–FE and EV–FE) and experiments (photographs and laser scanner) for each loading cycle

	Relative difference in CA (%)			
Load (kN)	EV–FE vs. photographs	EP–FE vs. photographs	EV–FE vs. laser scanner	EP–FE vs. laser scanner
0.25	−5	−35	—	—
0.45	−20	−49	—	—
0.9	−22	−44	—	—
1.8	−38	−54	5	−22
3.6	−34	−44	−24	−35

are critical in computational wear algorithms based on Archard's law. They are therefore reported for the three FE models and for each load interval on Fig. 10.8.

CP values were output for all the surface nodes part of the potential contact zone on the tibial insert (pre-defined before each FE analysis and identical for the three FE models). This means that, nodes not in contact with the femoral condyles have a null CP.

In stark contrast with other published studies, and instead of reporting only peak CPs that can be misleading due to numerical artefacts, minimum, maximum, first, second and third quartiles as well as the mean values are reported. The mean value has the advantage of not being influenced by the extreme ones. The mean CP (averaged over the nodes in contact) increases with the load magnitude in all FE simulations.

The mean and median CPs are very close to each other indicating a relatively homogeneous spread of values, which was also corroborated by homogeneous distributions of CP on colour plots (not presented here). Mean CPs for the E–FE model ranges from 18.73 MPa (0.25 kN load) to 34.19 MPa (3.6 kN load) (Fig. 10.8a). For the EP–FE and EV–FE models, the ranges are, respectively, 12.46–26.68 MPa (Fig. 10.8b) and 7.92–20.43 MPa (Fig. 10.8c).

Overall for the elastic model, CP values are more spread around the median than those of the viscoelastic and plastic models suggesting larger gradients of deformation.

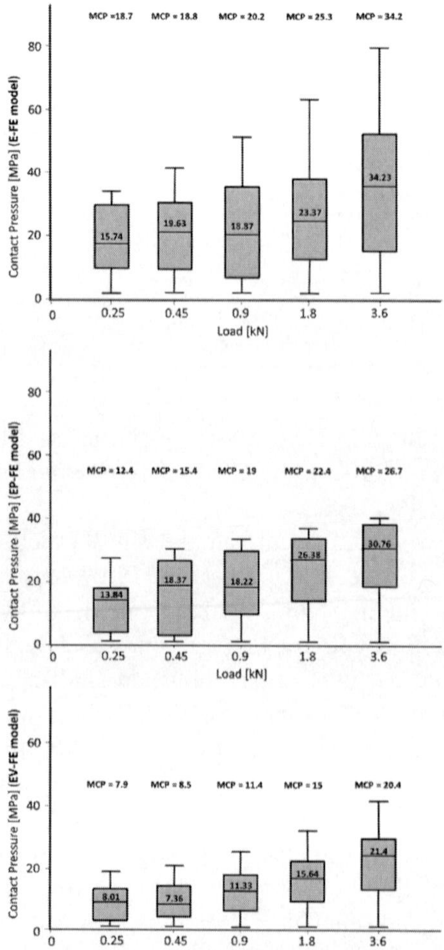

Figure 10.8 Box plots of the calculated contact pressure as a function of the compressive load applied (0.25 kN to 3.6 kN) for the three finite element models considered (a) elastic UHMWPE, (b) elastoplastic UHMWPE and (c) viscoelastic UHMWPE. Minimum, maximum, first, second and third quartile of the contact pressure are reported together with the mean contact pressure (MCP).

10.5 Discussion

10.5.1 *Penetration Depth*

Results of the FE analyses indicate that permanent plastic deformations (EP–FE model) and residual creep deformations (EV–FE model) are observed at all load levels. This qualitatively corroborates the laser scanner and profilometry measurements that also highlight the existence of inelastic residual deformations. At low loads (0.25 kN and 0.45 kN) residual deformations for the viscoelastic model are higher than those for the elastoplastic model. However, it should be noted that the constitutive parameters for the creep formulation were based on a creep experiment at a 15 MPa compressive stress level. This means that the creep response of the EV–FE model is probably overestimated for loads that induce a CP lower than 15 MPa [30]. Conversely, at stress levels lower than 15 MPa would mean underestimation. From 0.9 kN to 3.6 kN loads, the residual deformations of the elastoplastic model are higher than those experienced by the viscoelastic model. The residual deformations of the EP–FE model are the closest to those measured experimentally and follow a similar mostly linear evolution. After 1 and 2 days recovery, the plastic deformations are very close to the experimental ones (Fig. 10.6) indicating that, by deduction, the viscous effects in the UHMWPE insert under compression have a negligible influence after 1 and 2 days. This is also confirmed by the full deformation recovery (zero residual PD) obtained for the EV–FE model after 1 day. From these results, it is evident that viscous and plastic effects are key in the mechanical response of a UHWMPE tibial insert mounted on a AMTI knee simulator and subjected to the loads described in Section 10.2.

Significant viscous effects are still present after 900 seconds recovery and this has an important significance for in vivo clinical conditions where joints are subject to high frequency loads with a time period two order of magnitude smaller than 900 seconds. Depending on the magnitude of compressive loads, up to 0.14 mm penetration (at 3.6 kN) can be observed at the surface of the UHMWPE insert after 900 seconds recovery. That means that this figure can be significantly higher in the immediate time vicinity of when the maximum load is applied. These surface deformations have the potential to be detrimental to the kinematics of the TKR but this remains to be proved. Any deviation from the optimal kinematics may lead to high stress (particularly for cross-shear motions), which could exceed the plastic yield point of UHMWPE. The potential consequence of that is an acceleration of the wear process particularly if UMWPE debris are released in the joint.

In an FE study looking at the influence of creep deformation on the contact mechanics of TKR, Rullkoetter and Gabriel [33] calculated a PD of 0.44 mm before unloading with an applied compressive load of 3 kN. They used a quasi-static FE simulation to replicate a tibio-femoral contact above 30° knee flexion. The UHMWPE tibial insert was modelled as a linear viscoelastic solid similar to the current study but with different constitutive parameters. In the EV–FE model, at the end of the load holding phase, a 0.7 mm PD was observed. This value has a comparable magnitude to that found by Rullkoetter and Gabriel [33] in their computational analysis.

10.5.2 *Contact Area*

It is known that the contact patches on the tibial insert changes according to the degree of knee flexion [34]. It is therefore worth pointing out that the present study considered the TKR system for a zero degree flexion configuration (which corresponds to a standing position).

The EV–FE simulations showed the highest CA among all FE simulations (Fig. 10.7), and, as a logical consequence, the lowest CP (Fig. 10.8).

Using a FE model, Rullkoetter and Gabriel [33] simulated a compressive creep experiment on a TKR with a 3 kN load. They measured a total CA of 140 mm^2, which was very close to 138 mm^2 at 3.6 kN found in our EV–FE model (summing together the medial and lateral condyles). Combining physical experiments and FE analyses, Liau et al. [35] studied the effects on the contact mechanics of inserting a Fuji film into an artificial tibio-femoral articulation (TKR). In their simulation the contact was set at 0° flexion and the UHMWPE tibial insert was modelled as linear elastic.

They found a CA of 200 mm^2 at 3 kN, while a CA of 76 mm^2 for the 3.6 kN load case was measured in our elastic E–FE model, which was about 2.5 times smaller. This significant difference in CA could be attributed to a number of factors including implant size, mechanical properties, implant conformity; we used a flat tibial insert (low conformity) while Liau et al. [35] used an implant with higher conformity. The FEs composing the mesh of the UHMWPE insert in Liau et al.'s study [35] appears to have a sub-optimal aspect ratio which could have led to erroneous (not converged) solutions. Since the version 5.8 of Abaqus/Standard used by Liau and co-authors [35] significant advances have been made in the accuracy of contact algorithms. Frictionless contact was also used in their FE model but this effect is believed to be negligible in explaining the difference in CA.

In the present study, the observed difference in CA distribution between condyles (Table 10.2 and Fig. 10.9) is likely due to the pliancy effect on the femoral arm of the AMTI simulator. The effect was observed in both the physical and computational experiments. The FE results demonstrating this physical observation

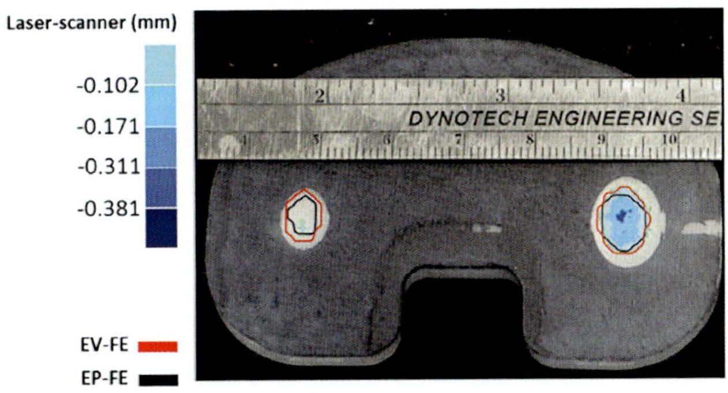

Figure 10.9 Photograph of contact scars (white regions) on the UHMWPE tibial insert after application of the 1.8 kN load cycle. The scars are produced by friction-induced removal of the blue tooling ink. Colour contours (red: viscoelastic model; black: elastoplastic model) indicate contact area boundaries predicted by the finite element analyses while the filled blue area represents the surface plot of the contact area obtained by laser scanning. The colour intensity of the area corresponds to the maximum penetration depth measured using laser scanner (mm).

were obtained by replicating as accurately as possible the exact dimensions and geometrical tolerances of the AMTI setup and not by tweaking (for best fit) FE parameters controlling the stiffness of the whole assembly such as mechanical properties of the individual components or by creating artificial constraints. This certainly provides confidence in the general methodology followed in this study.

As a general rule, it can prove challenging to compare apparently similar computational studies as they generally differ in terms of geometries, boundary and loading conditions, material properties, numerical solving schemes (implicit versus explicit), choice of element formulations, density of the mesh or contact algorithms. However, our CA results are of the same order of magnitude of what has been found by Rullkoetter and Gabriel [33] and Liau et al. [35].

10.5.3 *Contact Pressure*

In the experiment conducted by Metcalfe et al. [27], no pressure measurement system (e.g. Tekscan) was used so simulated CP obtained in the current work in both condyles could only be compared to data available in the literature.

The fact that the inter-quartile range—which represent half of the data spread around the median—of the CP is much smaller for the viscoelastic model (Fig. 10.8c) indicates that the CP values are more homogeneous than those computed for the elastic and plastic model. Given that the mean and median CP values of the viscoelastic model are the smallest of the three FE models one can infer that viscoelastic deformations offer a stress reduction mechanism that could be beneficial on one hand. On another hand, this is accompanied—or rather induced—by larger UHMWPE deformations that could be detrimental to the kinematics of the joint when it is no longer loaded in simple axial compression.

Halloran et al. [24] simulated a TKR using an elastoplastic constitutive law to model the mechanical response of the tibial insert. They measured a CP of 5 MPa at 0% gait cycle for an applied load of 0.25 kN, and a value of 10 MPa at 25% gait cycle for an applied load of 0.4 kN. The EP–FE model predicted a mean CP of 12.46 MPa at

0.25 kN and 15.45 MPa at 0.45 kN, which is comparable to Halloran et al.

In their study simulating the compressive creep behaviour of a tibial insert, Rullkoetter and Gabriel [33] found a maximum CP of 28 MPa for an applied load of 3 kN, which is comparable to the mean CP of 20.43 MPa found in the EV–FE model.

Liau et al. [35] found a CP of 28 MPa for an applied load of 3 kN, which value was comparable to 34 MPa at 3.6 kN found in our F–FE model. The differences in CP between Liau et al.'s study [35] and this one are not so dramatic as those found for CA. This would suggest that the main source of discrepancy for CA is probably related to differences in size of the implant components considered.

In the physical experiment [27], the tibial surface articulating against the femoral component have probably experienced shear forces in the A–P direction due to the femoral F–E rotation rather than pure I–S compression. These F–E rotations would be likely to be induced by kinematic geometric constraints arising from bringing in contact two surfaces featuring different curvatures (possibly misaligned). In an in vivo setting, the shear forces generated at the tibial insert surface through the rotation of the femoral component could lead to cyclic fatigue damage and consequently to a more severe degree of deformation of the tibial insert than that predicted by the FE simulations, particularly if considering inelastic effects such as creep or plastic flow.

Explicit and implicit FE methods have been widely used in the last two decades to simulate TKR kinematics and kinetics with special emphasis on predicting contact mechanics. UHMWPE contact stresses and deformations are key quantities to evaluate the potential for wear and/or damage occurring in TKRs. Indeed, they are the main input variables that drive algorithmic phenomenological wear models based on the Archard's law of abrasive wear [22]. Although these types of model have been very successful at replicating or even predicting wear of orthopedic implants [9–11, 26, 36, 37], they suffer from their very empirically based nature, which makes abstraction of the underlying physics of wear. Consequently, these phenomenological models of wear are restricted to specific type of loading conditions and tend to generalise poorly. It is the authors' opinion that the future of wear prediction of orthopedic implants lies in the development of more mechanistic approaches that will capture the underlying physics of the material (e.g., temporary and permanent viscous realignment of UHMWPE polymer chains with slip of crystalline planes) where wear becomes part of the constitutive formulation of the material instead of being a system property that depends, for example, on the direction and magnitude of sliding and on an associated empirical factor.

This hybrid experimental–computational study has demonstrated that, under physiological compressive loads, creep and plastic deformations must be accounted for to predict more faithfully the inelastic deformations and CPs experienced by the UHMWPE tibial component of a TKR. Failure to predict accurately CPs limit the validity of most current computational wear models that centrally rely on this quantity and the sliding distance to compute wear volume. The study is part of a long-term research program aiming to develop a physics-based constitutive model of UHMWPE featuring wear and valid for arbitrary kinematics.

10.6 Conclusion

This experimentally based computational study demonstrated the need to account for the creep and plastic properties of UHMWPE in order to capture the residual deformation of a tibial insert under compressive loads. This work has laid down the foundations for a more advanced mechanistic constitutive model under development that will capture the coupled time-dependent anisotropic inelastic mechanical response of UHMWPE from small to finite deformations.

Acknowledgements

The first author was financially supported by DePuy Orthopaedics, a Johnson & Johnson company. Georges Limbert is partly funded through an Engineering and Physical Sciences Research Council (EPSRC) grant [Grant EP/F034296/1]. This financial support is gratefully acknowledged. The authors would like to thank Giorgos Ragkousis and Orestis Katsamenis at the University of Southampton for their assistance in editing the CAD geometries used in the finite element simulations as well as Jennifer Tikka at DePuy for her assistance with the laser scanner analysis.

References

1. J. Charnley, Surgery of the Hip-joint, *Br. Med. J.*, 1(5176) (1960), 821–826.
2. J. Charnley, Total prosthetic replacement of the hip, *Reconstr. Surg. Traumatol.*, 11 (1969), 9.
3. O.K. Muratoglu, C.R. Bragdon, D.O. O'Connor, M. Jasty, W.H. Harris, A novel method of cross-linking ultra-high-molecular-weight polyethylene to improve wear, reduce oxidation, and retain mechanical properties: Recipient of the 1999 HAP Paul Award, *J. Arthroplasty*, 16(2) (2001), 149–160.
4. L.A. Pruitt, Deformation, yielding, fracture and fatigue behavior of conventional and highly cross-linked ultra high molecular weight polyethylene, *Biomaterials*, 26(8) (2005), 905–915.
5. D. Casey, J. Cottrell, E. DiCarlo, R. Windsor, T. Wright, PFC knee replacement: Osteolytic failures from extreme polyethylene degradation, *Clin. Orthop.*, 464 (2007), 157.
6. M. Sobieraj, C. Rimnac, Ultra high molecular weight polyethylene: Mechanics, morphology, and clinical behavior, *J. Mech. Behav. Biomed. Mater.*, 2(5) (2009), 433–443.
7. D.L. Bartel, V. Bicknell, T. Wright, The effect of conformity, thickness, and material on stresses in ultra-high molecular weight components for total joint replacement, *J. Bone Joint Surg. (Am)*, 68(7) (1986), 1041.
8. J. Mottershead, P. Edwards, M. Whelan, R. English, Finite element analysis of a total knee replacement by using Gauss point contact constraints, *Proc. Inst. Mech. Eng., Part H: J. Eng. Med.*, 210(1) (1996), 51–63.
9. L.A. Knight, S. Pal, J.C. Coleman, F. Bronson, H. Haider, D.L. Levine, et al. Comparison of long-term numerical and experimental total knee replacement wear during simulated gait loading, *J. Biomech.*, 40(7) (2007), 1550–1558.

10. S. Pal, H. Haider, P.J. Laz, L.A. Knight, P.J. Rullkoetter, Probabilistic computational modeling of total knee replacement wear, *Wear*, 264(7) (2008), 701–707.
11. R. Willing, I.Y. Kim, A holistic numerical model to predict strain hardening and damage of UHMWPE under multiple total knee replacement kinematics and experimental validation, *J. Biomech.*, 42(15) (2009), 2520–2527.
12. A. Abdelgaied, F. Liu, C. Brockett, L. Jennings, J. Fisher, Z. Jin, Computational wear prediction of artificial knee joints based on a new wear law and formulation, *J. Biomech.*, 44(6) (2011), 1108–16.
13. M. Soncini, L. Vandini, A. Redaelli, Finite element analysis of a knee joint replacement during a gait cycle, *J. Appl. Biomater. Biomech.*, 2(1) (2004), 45–54.
14. A. Godest, M. Beaugonin, E. Haug, M. Taylor, P. Gregson, Simulation of a knee joint replacement during a gait cycle using explicit finite element analysis, *J. Biomech.*, 35(2) (2002), 267–275.
15. D.D. D'Lima, N. Steklov, B.J. Fregly, S.A. Banks, C.W. Colwell Jr., In vivo contact stresses during activities of daily living after knee arthroplasty, *J. Orthop. Res.*, 26(12) (2008), 1549–1555.
16. S. O'Brien, Y. Luo, C. Wu, M. Petrak, E. Bohm, J.M. Brandt, Computational development of a polyethylene wear model for the articular and backside surfaces in modular total knee replacements, *Tribol. Int.*, 59 (2013) 284–291.
17. B.J. Fregly, Y. Bei, M.E. Sylvester, Experimental evaluation of an elastic foundation model to predict contact pressures in knee replacements, *J. Biomech.*, 36(11) (2003), 1659–1668.
18. B.J. Fregly, W.G. Sawyer, M.K. Harman, S.A. Banks, Computational wear prediction of a total knee replacement from in vivo kinematics, *J. Biomech.*, 38(2) (2005), 305–314.
19. Y. Bei, B.J. Fregly, Multibody dynamic simulation of knee contact mechanics, *Med. Eng. Phys.*, 26(9) (2004), 777–789.
20. P.S. Walker, G.W. Blunn, D.R. Broome, J. Perry, A. Watkins, S. Sathasivam, et al. A knee simulating machine for performance evaluation of total knee replacements, *J. Biomech.*, 30(1) (1997), 83–89.
21. T. Johnson, M. Laurent, J. Yao, L. Gilbertson, The effect of displacement control input parameters on tibiofemoral prosthetic knee wear, *Wear*, 250(1–12) (2001), 222–226.
22. J. Archard, Contact and rubbing of flat surfaces, *J. Appl. Phys.*, 24 (1953), 981.

23. S.L. Bevill, G.R. Bevill, J.R. Penmetsa, A.J. Petrella, P.J. Rullkoetter, Finite element simulation of early creep and wear in total hip arthroplasty, *J. Biomech.*, 38(12) (2005), 2365–2374.
24. J.P. Halloran, A.J. Petrella, P.J. Rullkoetter, Explicit finite element modeling of total knee replacement mechanics, *J. Biomech.*, 38(2) (2005), 323–331.
25. J. Lanovaz, R. Ellis, Dynamic simulation of a displacement-controlled total knee replacement wear tester, *Proc. Inst. Mech. Eng.: Part H, J. Eng. Med.*, 222(5) (2008), 669–681.
26. D. Zhao, H. Sakoda, W.G. Sawyer, S.A. Banks, B.J. Fregly, Predicting knee replacement damage in a simulator machine using a computational model with a consistent wear factor, *J. Biomech. Eng.–Trans. ASME*, 130(1) (2008), 011004–011010.
27. A.D. Metcalfe, J. Tikka, M.A. Strickland, M.R. Dressler, Quantifying deformation and recovery of polyethylene in total knee replacement, *Trans. Soc. Biomater.*, 803 (2011), ISBN: 9781617825606.
28. J.J. Liau, C.K. Cheng, C.H. Huang, W.H. Lo, The effect of malalignment on stresses in polyethylene component of total knee prostheses-a finite element analysis, *Clin. Biomech.*, 17(2) (2002), 140–146.
29. T. Said, S.L. Gomaa, Testing and modeling the nonlinear behavior of UHMWPE used in orthopaedic implants, 0104 (2010), ASB.
30. E.A. Reeves, D.C. Barton, D.P. FitzPatrick, J. Fisher, A two-dimensional model of cyclic strain accumulation in ultra-high molecular weight polyethylene knee replacements, *Proc. Inst. Mech. Eng.: Part H, J. Eng. Med.*, 212(3) (1998), 189–198.
31. ABAQUS, U. M. 6.12.1 documentation. Dassault Systemes Simulia Corporation., Providence, RI, USA (2012).
32. V.I. Johannes, M.A. Green, C.A. Brockley, The role of the rate of application of the tangential force in determining the static friction coefficient, *Wear*, 24(3) (1973), 381–385.
33. P. Rullkoeter, S. Gabriel, Viscoelastic behavior of UHMWPE TKR components, *46th Annual Meeting of the Orthopaedic Research Society*, 0562 (2000).
34. B.C. Carr, T. Goswami, Knee implants—Review of models and biomechanics, *Mater. Design*, 30(2) (2009), 398–413.
35. J.J. Liau, C.K. Cheng, C.H. Huang, W.H. Lo, Effect of Fuji pressure sensitive film on actual contact characteristics of artificial tibiofemoral joint, *Clin. Biomech.*, 17(9) (2002), 698–704.

36. M. Strickland, M. Dressler, T. Render, M. Browne, M. Taylor, Targeted computational probabilistic corroboration of experimental knee wear simulator: the importance of accounting for variability, *Med. Eng. Phys.*, 33(3) (2010), 295–301.
37. D. Zhao, W.G. Sawyer, B.J. Fregly, Computational wear prediction of UHMWPE in knee replacements, *ASTM, Special Technical Publication*, 1472 (2006), 45–50.

Index